terra australis 36

Terra Australis reports the results of archaeological and related research within the south and east of Asia, though mainly Australia, New Guinea and island Melanesia — lands that remained terra australis incognita to generations of prehistorians. Its subject is the settlement of the diverse environments in this isolated quarter of the globe by peoples who have maintained their discrete and traditional ways of life into the recent recorded or remembered past and at times into the observable present.

List of volumes in Terra Australis

terra australis 36

Transcending the Culture–Nature Divide in Cultural Heritage

Views from the Asia-Pacific Region

Edited by Sally Brockwell, Sue O'Connor & Denis Byrne

Australian
National
University

E PRESS

ANU
E PRESS

© 2013 ANU E Press

Published by ANU E Press
The Australian National University
Canberra ACT 0200 Australia
Email: anuepress@anu.edu.au
Web: http://epress.anu.edu.au

National Library of Australia Cataloguing-in-Publication entry

Title:	Transcending the culture - nature divide in cultural heritage : views from the Asia-Pacific region / edited by Sally Brockwell, Sue O'Connor & Denis Byrne.
ISBN:	9781922144041 (pbk.) 9781922144058 (ebook)
Notes:	Includes bibliographical references.
Series:	Terra Australis ; 36.
Subjects:	Cultural property--Protection--Pacific Area.. Historic preservation--Pacific Area. Pacific Area--Antiquities
Other Authors/Contributors:	Brockwell, Sally O'Connor, Sue. Byrne, Denis.

Dewey Number: 363.69091823

Series Editor: Sue O'Connor

Cover image: Sue O'Connor

Back cover map: Hollandia Nova. Thevenot 1663 by courtesy of the National Library of Australia.
Reprinted with permission of the National Library of Australia.

Contents

Sally Brockwell is a Research Fellow in the Department of Archaeology and Natural History, College of Asia and the Pacific at The Australia National University, Canberra. She has worked extensively in northern Australia and East Timor focusing on hunter-gatherer adaptations in the mid to late Holocene. Currently she is researching earth mound sites and environmental change in northern Australia.

Steve Brown is a Cultural Heritage Researcher with the New South Wales government, in Sydney, Australia and a PhD candidate at the University of Sydney. His research interests include the intangible values of landscape (particularly around attachment, belonging and place); the heritage of ephemeral and 'ordinary' physical traces of history across landscapes; applied approaches to managing heritage values of bio-cultural landscapes; and the heritage of landscapes with the imprint of Indigenous and colonial settler interaction. Steve is the author of *Cultural Landscapes: A Practical Guide for Park Management* (2010).

David Bulbeck is a Senior Research Associate at the Department of Archaeology and Natural History in the College of Asia and the Pacific at The Australian National University, Canberra. His primary research interests are the Holocene archaeology and biological anthropology of southwest Sulawesi and the Malay Peninsula. His recent sole authored publications include: 'Biological and cultural evolution in the population and culture history of Malaya's anatomically modern inhabitants' (2011); 'Uneven development in southwest Sulawesi, Indonesia during the Early Metal Phase' (2010); and 'An archaeological perspective on the diversification of the languages of the South Sulawesi stock' (2008).

Denis Byrne leads the research program in cultural heritage at the Office of Environment and Heritage (New South Wales) in Sydney, Australia. He is also Adjunct Professor at the Transforming Cultures Centre, University of Technology, Sydney. His interests include the materiality of popular religion, the everyday engagement of people in Asia and Australia with their material past, and fictocritical archaeological writing, the latter resulting in his 2007 book, *Surface Collection*.

Boonyarit Chaisuwan graduated in Archaeology from the Silpakorn University, Bangkok, Thailand. Currently, he is with the 15th Regional Office of Fine Arts, Phuket. Most recently he has authored the chapter 'Early contacts between India and the Andaman Coast in Thailand from the Second Century BCE to Eleventh Century CE' and his 2009 book *Thung Tuk: A Settlement Linking Together the Maritime Silk Route*.

Harley Coyne is a Traditional Owner and project officer with the Department of Indigenous Affairs (Southern Region, Western Australia) and an active community member with a variety of programs throughout southern Western Australia.

Tim Denham is a Research Fellow in Archaeology at La Trobe University, Melbourne, Australia. His research focuses on the emergence and transformation of agriculture in the highlands of Papua New Guinea. He was the lead author and organiser of Papua New Guinea's successful nomination of the Kuk Swamp site to UNESCO's World Heritage List (2008).

Vernice Gillies is chair of the Albany Heritage Reference Group Aboriginal Corporation (AHRGAC) in Western Australia and is recognised widely as a community leader and facilitator. Vern has held a number of important and diverse positions within health, education, heritage, natural resource management and community programs throughout southern Western Australia.

David Guilfoyle is a Project Archaeologist at Northern Land Use Research Inc., Alaska, USA and Managing Director of Applied Archaeology International, Albany, Australia. He is a Research Associate with the Western Australian Museum and the Centre of Excellence in Natural Resource Management, University of Western Australia, Perth. David is also a coordinator of two community heritage foundations, the Gabbie Kylie Foundation and the Dowark Foundations, hosted under the auspices of the National Trust of Australia (Western Australia). He recently coordinated and delivered an award-winning, community natural resource management project throughout southern Western Australia and continues to focus his research interests on the integration of traditional knowledge, landscape archaeology, and the interplay of cultural and natural resource management

Anna Karlström finished her PhD in archaeology at Uppsala University in 2009 and is currently a Postdoctoral Research Fellow at the Aboriginal and Torres Strait Islander Studies Unit, University of Queensland, Brisbane, Australia. Her research interests include heritage studies, contemporary perspectives on archaeology, the past and its materialities, and Southeast Asian and Australian indigeneity. Her PhD thesis resulted in the book *Preserving Impermanence: the creation of heritage in Vientiane, Laos* (2009).

Chaowalit Khowkhiew is a Lecturer in prehistoric archaeology in the Faculty of Archaeology, Silpakorn University, Bangkok, Thailand. His research interests include Pleistocene archaeology in Peninsula Thailand.

Suengki Kwak is a graduate student in the Department of Anthropology at the University of Washington, Seattle, USA. His research interests include the transition from hunter-gatherers to farmers in East and Southeast Asia, especially South Korea and archaeological applications of organic geochemistry.

Ian Lilley has worked in Australasian and Indo-Pacific archaeology and heritage for over 30 years. He is an ICOMOS World Heritage Assessor and Secretary-General of ICOMOS/ICAHM. His interests include indigenising archaeology and heritage management, migration and trade, social identity, ethics, and the role of archaeology and heritage in contemporary society.

Nick McClean is a researcher and freelance radio producer currently based in the School of Culture, History and Language at The Australian National University, Canberra. As part of his doctoral thesis he is undertaking oral history and cultural heritage research in a number of communities in Australia, India, Indonesia and East Timor. His key areas of interest are cross-cultural and comparative analyses of conservation, community conservation initiatives, and the co-management of protected areas with indigenous communities. He has also produced feature length radio documentaries for the Australian Broadcasting Commission in collaboration with Indigenous groups in Australia and East Timor.

Andrew McWilliam is a Senior Fellow and Head of the Department of Anthropology in the College of Asia and the Pacific at The Australian National University, Canberra. He has published widely on the ethnography of Timor, and has continuing research interests in East Timor, Eastern Indonesia and northern Australia. Recent work has focused on issues of governance, community economies and ritual exchange, resource tenures and plantation histories. He is co-editor with E.G. Traube of the volume, *Land and Life in Timor-Leste: Ethnographic Essays* (2011).

Ben Marwick is an Assistant Professor in the Department of Anthropology, University of Washington, Seattle, USA. His interests include hunter-gatherer archaeology, technology and human ecology in mainland Southeast Asia and Australia. His co-edited volumes include *Keeping your Edge: Recent Approaches to the Organisation of Stone Artefact Technology* (2011) and *New Directions in Archaeological Science* (Fairbairn *et al.* 2009). He is co-editor of the *Bulletin of the Indo-Pacific Prehistory Association*.

Myles Mitchell is a research and consulting archaeologist with Applied Archaeology International and a PhD candidate at the National Centre for Indigenous Studies, The Australian National University, Canberra. Maintaining a focus on contemporary culture, identity and values within custodian communities, Myles is committed to working towards an archaeology that is socially and culturally relevant to custodian communities as well as being solutions-oriented for cultural heritage management.

Cat Morgan is currently working on her Master of Arts thesis at the University of Leicester, Leicester, United Kingdom, with her research based on the Lake Pleasant View heritage complex in southern Western Australia. She is assisting with the development of an interactive mapping program with the Esperance Traditional Owners as well as several other projects with Applied Archaeology International.

Sue O'Connor is Professor in the Department of Archaeology and Natural History in the College of Asia and the Pacific at The Australian National University, Canberra. Her research focuses on the evidence for migration and colonisation in the Indo-Pacific region, as well as on theoretical issues surrounding early human migration such as maritime capacity, technological innovation and symbolling. She has undertaken numerous research projects in Australia, Indonesia and East Timor. Her recent co-edited volumes include *Rethinking Cultural Resource Management in Southeast Asia* (Miksic *et al.* 2011) and *Islands of Inquiry: Colonisation, Seafaring and the Archaeology of Maritime Landscapes* (Clark *et al.* 2008).

Sandra Pannell has a doctorate in anthropology from the University of Adelaide, Australia and she currently works as a consultant anthropologist in the field of Native Title and cultural heritage research, as well as being an Adjunct Senior Research Fellow with the School of Arts and Social Sciences at James Cook University in Cairns, Australia. She has held lecturing and research positions at the University of Adelaide, James Cook University, at the Centre for Resource and Environmental Studies at The Australian National University, and at the Rainforest Cooperative Research Centre in Cairns. She has undertaken anthropological research in Indonesia, Timor-Leste, and throughout Aboriginal Australia. Sandra Pannell is the author of two books, one on World Heritage (2006) and the other on an Indigenous environmental history of North Queensland (2006). She is also the editor of a book on violence, society and the state in Indonesia (2003) and the co-editor of two publications – one on resource management in eastern Indonesia (1998) and the other on Indigenous planning in northern Australia (Larsen and Pannell 2006). She is the author of a number of articles on kinship, intellectual property rights, native title, resource management and marine tenure.

Rasmi Shoocongdej is an Associate Professor in the Faculty of Archaeology at Silpakorn University, Bangkok, Thailand. She specialises in late Pleistocene to post-Pleistocene tropical foragers and works along the western border of Thailand including Mae Hong Son and Kanchanaburi provinces. She has also published on nationalism and archaeology, looting and public education. She holds several professional positions, including senior representative for the World Archaeological Congress, advisory board member for *World Archaeology*, *Asian Perspectives*, *Bulletin of the Indo-Pacific Prehistory Association*, and *Archaeologies* (World Archaeological Congress).

Anita Smith is Research Associate in Archaeology at La Trobe University, Melbourne, Australia with 20 years research experience in the Pacific Islands and Australia. Since 2002 she has worked with Pacific Island communities and Governments and as consultant to UNESCO in the World Heritage Pacific 2009 Program. In 2008-2011 she was a member of the Australian delegation to the World Heritage Committee. Anita is co-author of the ICOMOS thematic study *Cultural Landscapes of the Pacific Islands* (2007) and editor of *World Heritage in a Sea of Islands: the Pacific 2009 Program* (2012).

Daud Aris Tanudirjo (PhD 2001 Australian National University) is an archaeologist who is currently Vice Dean of Research, Community Service and Cooperation Affairs in the Faculty of Cultural Sciences at Gadjah Mada University, Jogjakarta, Indonesia. His research focuses on prehistoric archaeology and cultural heritage management in the Indonesian archipelago.

Cholawit Thongcharoenchaikit is a researcher in vertebrate palaeontology at the Natural History Section of the Thailand National Science Museum, Bangkok. His interests include human-environment relations in peninsular Thailand during the late Holocene.

Cate Turk is undertaking PhD research at the Institute for Geography, University of Erlangen-Nürnberg in Erlangen, Germany, where she is a Lecturer in the Cultural Geography and Orient research group. She was awarded a Masters of Philosophy in cultural geography, Geography Department, University of Edinburgh, Edinburgh, Scotland in 2001. From 2003-2006 Cate was a Policy Officer in the Australian Department for Environment and Heritage where she worked at Kakadu National Park, Northern Territory, Australia and as an Australia Youth Ambassador for Development in the Department of Culture and Heritage, National Government of Fiji Islands.

References

Brown, S. 2010. *Cultural Landscapes: A Practical Guide for Park Management*. Department of Environment, Climate Change and Water, NSW Government, Sydney.

Bulbeck, D. 2011. Biological and cultural evolution in the population and culture history of Malaya's anatomically modern inhabitants. In: Enfield, N. (ed.), *Dynamics of human diversity.* Pacific Linguistics 627. Australian National University, Canberra.

Bulbeck, D. 2010. Uneven development in southwest Sulawesi, Indonesia during the Early Metal Phase. In: Bellina, B., Bacus, E. Pryce, T. and Wisseman Christie, J. (eds.), *Fifty years of archaeology in Southeast Asia: Essays in honour of Ian Glover*, pp. 26-39. River Books, Bangkok.

Bulbeck, D. 2008. An archaeological perspective on the diversification of the languages of the South Sulawesi stock. In: Simanjutak, T. (ed.), *Austronesians in Sulawesi*, pp. 185-212. Center for Prehistoric and Austronesian Studies, Jakarta.

Byrne, D. 2007. *Surface collection: Archaeological travels in Southeast Asia*. Altamira Press, Lanham, Maryland, USA.

Chaisuwan, B. 2011. Early contacts between India and the Andaman Coast in Thailand from the Second Century BCE to Eleventh Century CE. In: Manguin, P-Y, Mani, A. and Wade, G. (eds), *Early interactions between South and Southeast Asia: Reflections on cross-cultural exchange,* pp. 83-112. Institute of South East Asian Studies, Singapore.

Chaisuwan, B. and Naiyawat, R. 2009. *Thung Tuk: A settlement linking together the maritime silk route.* Trio Creation, Songkhla.

Clark, G.R., Leach, F. and O'Connor, S. (eds) 2008. Islands of inquiry: Colonisation, seafaring and the archaeology of maritime landscapes. Terra Australis 29. ANU E Press, Canberra.

Fairbairn, A., O'Connor, S. and Marwick, B. (eds) 2009. *New Directions in Archaeological Science*. Terra Australis 28. ANU EPress, Canberra.

Karlström, A. 2009. *Preserving Impermanence: The creation of heritage in Vientiane, Laos.* Department of Archaeology and Ancient History, Uppsala University, Uppsala, Sweden.

McWilliam, A. and Traube, E.G. 2011. *Land and life in Timor-Leste: Ethnographic essays*. ANU EPress, Canberra.

Marwick, B. and MacKay, A. 2011. *Keeping your edge: Recent approaches to the organisation of stone artefact technology*. BAR S2273. Archaeopress, Oxford.

Miksic, J., Goh, G. and O'Connor, S. (eds) 2011. *Rethinking cultural resource management in Southeast Asia; Preservation, development, and neglect*, Anthem Press, London.

Pannell, S. 2006. *Reconciling Nature and Culture in a Global Context? Lessons from the World Heritage List*. Cooperative Research Centre for Tropical Rainforest Ecology and Management. Rainforest CRC, Cairns, Australia.

Pannell, S. 2006. (with contributions from Ngadjon-Jii Traditional Owners) *Yamani Country: A spatial history of the Atherton Tableland, North Queensland*. Research Report No. 43. Cooperative Research Centre for Tropical Rainforest Ecology and Management. Rainforest CRC, Cairns, Australia.

Pannell, S. (ed.) 2003. *A state of emergency: Violence, society and the state in Eastern Indonesia. CDU Press, Darwin*.

Pannell, S. and von Benda-Beckmann, F. (eds) 1998. *Old World places, New World problems: Exploring issues of resource management in Eastern Indonesia*. Australian National University Centre for Resource and Environmental Studies, Canberra.

Larsen, L. and Pannell, S. 2006. *Developing the wet tropics Aboriginal cultural and natural resource management plan: Workshop proceedings*. Report No. 45. Cooperative Research Centre for Tropical Rainforest Ecology and Management. Rainforest CRC, Cairns, Australia.

Smith, A. and Jones, K. 2007. *Cultural landscapes of the Pacific Islands: ICOMOS thematic study*. UNESCO World Heritage Centre, Paris.

Smith, A. (ed.) 2012. *World Heritage in a sea of islands: The Pacific 2009 Program*. UNESCO, Paris.

Introduction
Engaging culture and nature

Denis Byrne, Office of Environment and Heritage, NSW, Australia

and University of Technology, Sydney, Australia

Sally Brockwell, The Australian National University, Canberra, Australia

Sue O'Connor, The Australian National University, Canberra, Australia

This volume began life as a session at the 2010 Australian Archaeological Conference on the cultural heritage of protected areas in the Asia-Pacific region. Our particular concern was with the proposition that the discourse of nature conservation was predisposed to a vision of protected areas (in the form of national parks and other 'nature' reserves) as pristine nature. According to such a vision, protected areas represent wildernesses that, having escaped the ravages of human exploitation, had now to be preserved as the last reservoirs of biodiversity on a planet threatened with ecological disaster. To what extent, we asked, did such a mindset eclipse the history and heritage of protected areas as human habitats, not to mention effacing the contemporary presence in them of living human cultures?

These questions engage the much larger issue of the culture-nature divide as an ontological marker of Western modernity. This divide should more properly be described as a 'duality' or a 'dyad', terms that have been used to describe a habit of thought which developed in the West from the seventeenth century but with older antecedents. This habit of thought also produced the distinctive mind/body split which is particularly identified with the Cartesian worldview. In recent years there has been increasing interest in the real world ramifications of the culture-nature duality from those working in fields ranging from sociology (for example, Latour 1993) and anthropology (Ingold 2000; Haraway 2006), to geography (Whatmore 2006; Head 2007). In Australia it has occupied the attention of Debbie Rose (1996, 2011) and others in the Ecological Humanities group.[1] There has been a growing consciousness of the extent to which culture-nature dualism is foundational to Western modernity and thus seminal to the West's encounter with the non-Western world.

In modern Western thought, society (or culture) has been understood to be not just radically separate from nature but situated in a hierarchical relationship of dominance to it. This ontological model facilitated the vastly accelerated exploitation and despoilation of natural 'resources' (nature being considered a resource for humanity) which accompanied the Industrial Revolution, the development of capitalism and the European colonial-imperial project. Closer to home in terms of the present volume, culture-nature dualism has been found to be a major impediment to breaking out of the silos represented by nature conservation and heritage conservation. This, of course, is the view from the position of the experts. The view from the ground includes that of Indigenous people who find that in the struggle to maintain their land there are willing allies to be found among environmental and heritage conservation organisations but that these alliances mean having to internalise the culture-nature dichotomy, at least for the purposes of dealing with these experts, accessing their funding, and participating in their conservation programs. In other words such alliances entail an internalisation of an alien ontology.

---

1   The Ecological Humanities: http://www.ecologicalhumanities.org/ [accessed: May 2012].

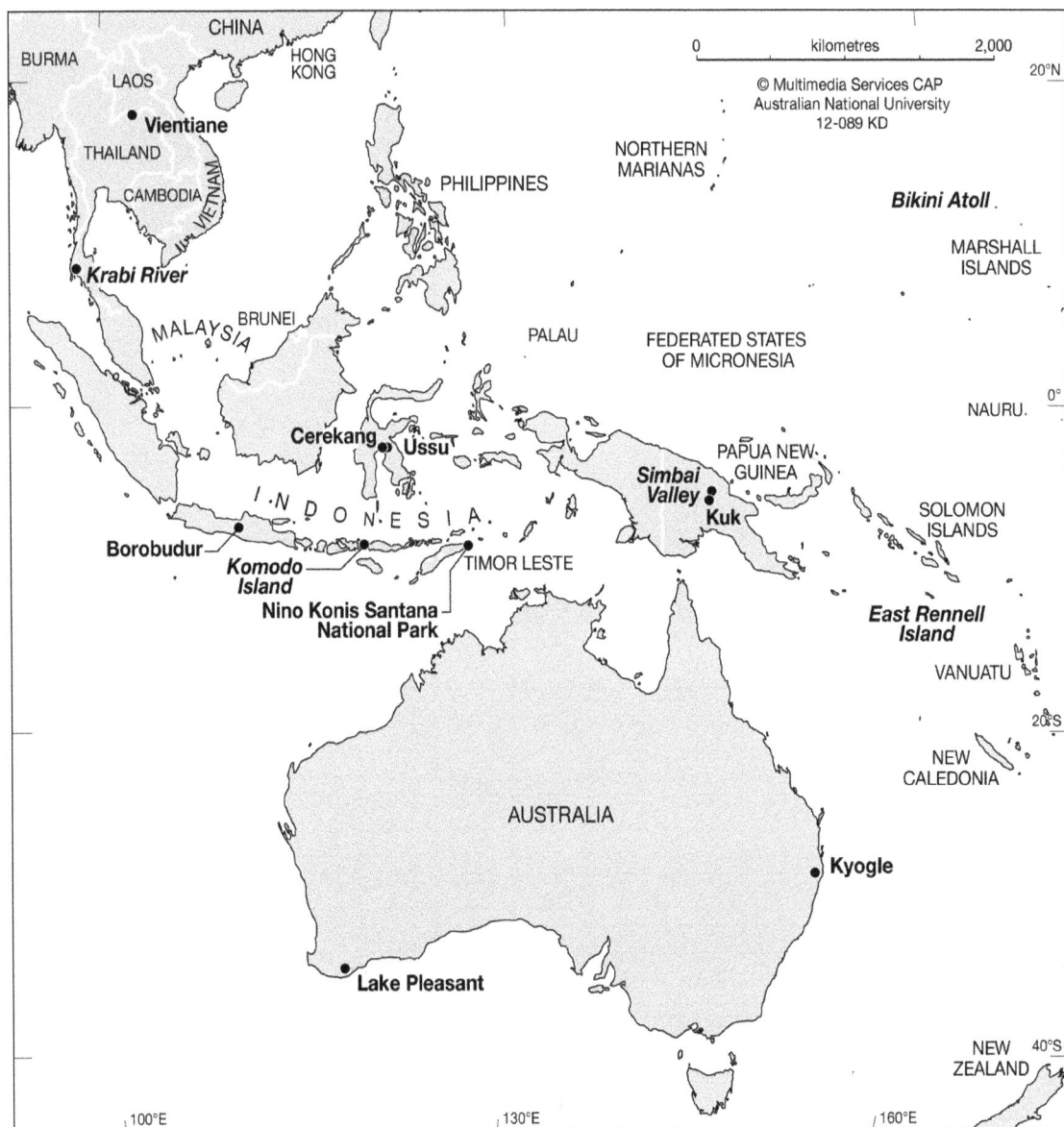

Map of Study Areas

Source: CartoGIS, College of Asia and the Pacific, ANU

A people-less nature

Protected areas, including national parks, World Heritage areas, and other categories of conservation reserve, have become the theatres in which many archaeologists and heritage practitioners, including a number of the contributors to this volume, find themselves working either in a research capacity or as professional advisors. The culture-nature split is manifest in several ways in the protected area sphere. National parks, which predominate in the world's protected area system, for a long time embodied the Fortress Conservation ideology (Brockington 2002) in which protected areas were 'planned and managed against the impact of people (except for visitors), and especially to exclude local people' (Phillips 2003:12). The Fortress Conservation model originated with Yellowstone and Yosemite National Parks in the USA (Dowie 2011:1–22), the forest reserves of colonial India (Beinart and Hughes 2007:111–129) and the game reserves of colonial Africa. The fortress model fell out of favour in the 1970s when conservationists had to concede that in tropical-zone protected areas (a majority of which are inhabited by humans) exclusion policies were simply not working since in practical terms it was impossible to police the activities of local inhabitants. A raft of new strategies was developed for negotiating and collaborating with local groups for conservation outcomes. While elements of the nature conservation movement conceded the enculturated character of protected areas, the wilderness myth remained powerfully attractive to other elements. In recent years attention has been directed to the role of certain global conservation NGOs in depicting protected areas as people-less spaces (Brockington and Igoe 2006). Far from discredited, the wilderness trope remains alive and well in the rhetoric of global conservation NGOs (Brockington and Igoe 2006:445; Ross *et al.* 2011).

One of the discursive habits that stem from culture-nature dualism is encapsulated in the concept of human impact, which as Lesley Head (2007:837) points out, positions humans as acting on nature from the outside. While Head credits the human impact concept as having done useful work in the mid-twentieth century in countering environmental determinism (Head 2007:838), she sees it as doing 'different work' in the early twenty-first century:

> It paradoxically reinforces the view of humans as external to the natural system, and encourages explanatory focus on simple correlations in time and/or space rather than on mechanisms of connection. It is neither conceptually nor empirically strong enough for the complex networks of humans and non-humans now evident, in prehistoric as well as in contemporary timeframes. (Head 2007: 838–39)

Another, more recent, dualism that could be argued useful work while at the same time having debilitating effects is that of tangible-intangible heritage which, according to Michael Herzfeld, is a separation 'that perpetuates a fundamentally Cartesian and colonial model' (in Byrne 2011:148).

One of the many ramifications of the culture-nature duality in Australia is played out in the field of Native Title where most determinations exclude rights and interests of an economic or commercial nature (Weir 2012:7). In the frame of Native Title, it is the suspension (or reversal) of the duality which has had debilitating effects. This goes deep into the history of Europe's relations with non-Western others, particularly those others deemed to be living close to nature who were 'exempted' from the culture-nature divide and, instead, were 'collapsed into nature as part of the flora and fauna' (Weir 2012:7). Aboriginal Native Title claimants frequently find they are constrained to participate in a European value frame in which traditionality and authenticity are defined in terms of a static symbiosis with nature. Indeed the legal requirement to demonstrate connection pre- and post-sovereignty perpetuates an ideal of immutable sameness in Indigenous cultural engagement with landscape which differs little from early European views of Aboriginal cultural stasis and stagnation captured by the well-known soubriquet of an 'unchanging people in an unchanging environment' (Pulleine 1929:310).

A corollary of the naturalisation of Aboriginal people is that while nature conservationists and national park managers are mostly happy to conserve and promote Aboriginal archaeological heritage sites in parks they have been much less enthusiastic about European heritage sites. Aboriginal heritage sites are seen as continuous with wilderness whereas European sites are frequently seen as incompatible with it. It was, in fact, not uncommon in the 1960s and 70s in places like New South Wales for old huts, homesteads, stockyards, and other traces of white habitation to be removed in order to 'restore' park landscapes to a state of nature, or, in our terms, to fabricate the fiction of pure nature.[2]

---

2   Office of Environment and Heritage NSW (2009). *Challenges in the landscape: Memories of conserving historic heritage in the NSW park system*, 1967-2000. http://www.environment.nsw.gov.au/nswcultureheritage/NPWSCorporateHistory.htm

Nature-less cultural landscapes

The notion of people-less landscapes has its counterpart in those visions of landscape which construe them as being purely a cultural product or construct. The issue here is perhaps not so much one of failing to concede the 'natural' dimension or the 'natural' elements of a cultural landscape as it is one of casting nature in a passive light. It is to do with a tendency to see nature as merely a setting or stage upon which human projects are enacted.

The post-structuralist turn in the humanities and social sciences helped reveal the social contingency of our place in the world; it revealed the extent to which the lived world of humans is socially constructed. And yet an extreme social constructionist position that would envisage humans never being able to engage directly with nature, but only with their social constructions of nature, is a perspective which denies nature an independent reality and an agency of its own. It also implies that we humans do not experience nature as embodied beings whose senses respond to stimuli given off by the organisms and the inorganic entities in nature. Clearly the ecologies that we conceive of as making up our 'natural' environment, our setting, in actuality incorporate us. Tim Ingold (2000:5) goes so far as to insist that what we choose to call 'social relations' are actually 'but a sub-set of ecological relations' (see also Tilley 1994:23).

This latter view of the world is one that is proving to be enormously productive for environmental anthropologists in the Asia-Pacific region (e.g. Zerner 2003a; Brosius *et al.* 2005). The ethnographic work of Anna Tsing (1993) and Marina Roseman (2003), for example, describes richly dialectical relations between human and non-human actors in the forests of Southeast Asia, as does Charles Zerner's (2003b) in the Indonesian seascape. In the field of archaeology, Ian McNiven (2008) and McNiven and Feldman (2003) show how the archaeological record of Torres Strait reflects the negotiated relationships between hunters and the animals they hunt. Archaeologically, mounds composed of ritually treated dugong skulls inform us of this relationship and its magical dimension. With the exception of rock art studies, archaeologists have traditionally given little consideration to 'the ontological status of animals' and the spiritual connections between people and animals that such sites encode (McNiven and Feldman 2003:189). O'Connor *et al.* (2011 and this volume) and Pannell and O'Connor (in press) discuss the mutuality and embeddedness of ecological and spiritual relationships with nature for Fataluku people living in East Timor's first National Park. They describe the physical manifestations of this ritual engagement in the form of built structures marking the presence of non-human beings or spirits, and they argue the need for recognition of this mutuality in heritage governance and management.

The habit of binary thinking makes it difficult to grasp and assimilate the notion of ecological relations and the dialectical entanglement of humans and non-humans in nature. The cultural landscape concept which derives from the field of geography and has been adopted with enthusiasm by many in the heritage field (e.g. Longstreth 2008; Taylor 2012) has been subject to the criticism that instead of transcending the culture-nature binary it simply builds up the cultural side of the equation, giving a misleading emphasis to a human agency construed as separate from, rather than networked with, non-human agency. Val Plumwood (cited in Head 2007:839) critiqued the cultural landscape concept in similar terms, seeing it as obscuring non-human agency.

Lesley Head (2007, 2010) has developed a critical commentary on the cultural landscape concept, which is valuable for its penetrating insight into the genealogy of the concept in geography, archaeology, and heritage but also for its appreciation of the valuable work the concept has done, particularly in the sphere of Indigenous rights. For example, she points to the way it has been able to 'put people back in' to landscapes that had been constructed by settler societies, such as Australia's, as wilderness (Head 2010:430). Also, though strongly advocating a networked, dialectical understanding of human embeddedness in nature, she confronts the historical reality that humans have had a hugely disproportionate role in shaping landscapes. She observes that: 'Given the human role in recent changes in earth surface processes, particularly climate change, and the need for human action to reverse the situation, there is surely a case for recognition of some strong human agency?' (Head 2010:439). She also makes the observation that 'the cultural landscape concept appears to be a tool [landscape] managers find useful' (Head 2010:433). This is highly relevant in the context of this volume's key concern with protected areas as a theatre for transcending the culture-nature divide and it takes us to the issue of dialogue between the professions of archaeology and heritage and those of nature conservation and protected area management.

The need for mediating discourse

The reality is that virtually all protected areas in the Asia-Pacific region are managed by departments of government whose brief is nature conservation (e.g. environmental protection agencies) or natural resource management (e.g. forestry departments). Overwhelmingly, protected area managers come to the job with a background in environmental management or the biological sciences or with training in natural resource management (NRM). On-ground management regimes are geared towards biodiversity management and nature conservation outcomes and while these regimes are not necessarily averse to also managing protected areas for their 'cultural values' the challenge for us is to make culture 'legible' to them (McWilliam, this volume).

It is not enough to simply get cultural heritage sites and contemporary human land use patterns 'on the map' to complement existing mapping of soils, vegetation, and non-human species distribution. The problem with such an 'add-on' approach is that the relative ease of getting 'culture' onto the map in this way is matched by the ease with which it can subsequently fall off. Unless culture is represented as integral to the ecology of an area it tends to lose out: it tends to be seen as separable. For instance, it may become subject to separate management plans and these may be deferred or set aside when funding is short. As inequitable as it may seem, the reality is that protected area management is currently overwhelmingly in the hands of nature conservationists and natural resource managers, meaning the onus is on us to come up with a discourse of culture which is not just intelligible to our counterparts in nature conservation and NRM but which articulates and networks with the language and technologies of 'natural values' management. Such a mediating discourse might be described as mapping culture *into* landscape rather than onto it. This does not have to mean subsuming culture within the existing discursive frameworks of nature conservation which, as Smith and Turk (this volume) point out, often do engage with local communities in ways that seek to draw them in as collaborators on projects that are grounded in Western value systems. 'Community-based' rhetoric could, in such cases, be said to be a cover for what are actually attempts to extend Western value systems into local settings.

We would argue that the kind of critical analysis undertaken by Smith and Turk (this volume) exemplifies the deepening understanding of the history and politics of conservation which is now evident in the fields of heritage and archaeology. This understanding and political sophistication has often been gained through extensive fieldwork in which anthropologists, archaeologists and heritage experts have witnessed what happens when 'global conservation' interests interface with Indigenous community interests. But it is also built upon the work of environmental historians (e.g. Grove 1995; Beinart and Hughes 2007), environmental anthropologists (e.g. Brockington 2002; Tsing 2005), and indeed archaeologists (e.g. Meskell 2012), who have researched the history of Western nature conservation, the culture of its institutions and networks, and the dynamics of its reach into the non-Western world. One might say we are in a better position than ever before to transcend the culture-nature divide in protected areas, and indeed elsewhere, because we know more about where this dualism comes from and how it has been operationalised.

Many conservation biologists and ecologists have made an effort to interface with local communities in a more equitable way (see, for instance, the IUCN journal, *Policy Matters*). The 'values' approach to protected area management (Lockwood 2006) is often manifest in a somewhat mechanical documenting of heritage and contemporary cultural values that are then managed alongside and separate from 'natural values'. But the values approach does appear to have the potential to evolve into a framework that integrates the 'cultural' and 'natural' into a holistic ecology. In relation to heritage sites, instead of simply recording archaeological sites and managing 'around' them, as now generally happens, a values approach could be evolved in a way that enables them be seen as integral to the ecological history of an area. Sites of contemporary cultural significance should be seen as imbricated in and networked into the area's 'natural values', which in turn are imbricated in the cultural world of the human inhabitants or neighbours of the protected area. Though this kind of integrative approach has so far proven elusive (Ross *et al.* 2011) it remains an ideal worth striving for.

Overview of the chapters

The chapters in this volume have been grouped under the three following themes:

1. World Heritage: Ian Lilley on nature and culture in World Heritage management; Anita Smith and Cate Turk on customary systems of management and World Heritage in the Pacific Islands; Steve Brown on the poetics and politics of the World Heritage listing of Bikini Atoll; Daud Tanudirjo on management and community tensions at the Borobudur; and Sandra Pannell on nature and culture in the context of the World Heritage listed Komodo Island in Indonesia.

2. Community archaeology: Nick McClean on Githabul community approaches to mapping culture in northern New South Wales; David Guilfoyle *et al.* on embedding community control in cultural heritage management using a case study from south western Western Australia; Tim Denham on cultural heritage management in Papua New Guinea; Ben Marwick *et al.* on community engagement during archaeological excavations in Thailand's Krabi River Valley; and Anna Karlström on local heritage and the problem with conservation in Laos.

3. Performative landscapes and conservation: Denis Byrne with a critique of conservation biology's view of popular religion; David Bulbeck on history, ecology and pre-Islamic relations with sacred places in South Sulawesi; Andrew McWilliam on cultural heritage performative modalities in relation to the new National Park in East Timor; and Sue O'Connor *et al.* on the dynamics of culture and nature in a protected East Timorese landscape.

1. World Heritage

In the four decades since its introduction, the World Heritage Convention has been immensely influential. It can be argued that one aspect of its influence is that many heritage sites and areas, particularly in the non-Western world, have only come into what might be called the 'global conservation estate' because of the perceived potential of the World Heritage List to garner international profile (and tourism revenue) for countries, many of which have few opportunities to attract that kind of attention. As Raj Isar (2011:42) has recently noted, UNESCO 'possesses its own forms of symbolic capital' and this is able to be deployed by states for their own ends.

In practical terms, the World Heritage Convention and the World Heritage List sit squarely in the middle of the issue of the culture-nature divide. In its original 1972 form, the Convention reified the conceptual divide in the form of two separate lists for cultural and natural properties, only modifying this position in 1992 by introducing the 'cultural landscapes' category. Ian Lilley (this volume) notes that even in the case of 'mixed properties', the cultural and natural aspects of areas proposed for listing are still assessed completely separately. One might say that a discourse that mediates culture and nature is nowhere more needed that in the World Heritage field. At the outset of her chapter, Sandra Pannell (this volume) makes the sobering observation that 'the globalisation of the nature-culture distinction … often encourages the very threats and dangers the Convention seeks to ameliorate through listing'.

Pannell, in her Komodo case study, shows how World Heritage listing has seen the interests of people resident in or living adjacent to the Komodo National Park de-authenticated in various ways. The dynamic-adaptive qualities of local culture have meant cultural practices have changed over time and people's patterns of movement and regional interactions have been fluid. These very qualities, however, as Pannell shows, have been pathologised by park managers and conservation advisors in a manner that represents contemporary locals as inauthentic and allows internationally-defined biodiversity values to trump local cultural values. UNESCO's language of 'cultural property' represents cultural heritage as a thing and has the effect of freezing local practices in a way that posits them as losing authenticity if they change (Byrne 2009; De Cesari 2010). Pannell shows us in painful detail how this has played out on the ground in Komodo.

While Pannell's case study depicts a resident population having its land redefined in a manner that discursively displaces it, Steve Brown, in his chapter on the listing of Bikini Atoll, relates the case of a population that was physically removed from its home-habitat as a prelude to a historical event that is now memorialised by World Heritage listing. The event in question was, of course, the nuclear test explosions carried out on the atoll in 1946 by the US military. Brown shows how the evacuation of the atoll prior to the test has now made the terrain available to be reinscribed with a history which is not that of its prior human inhabitants. The narrative of the island's use for nuclear testing has effaced its prior human history.

In a further example of displacement, Daud Tanudirjo relates how the local Muslim population occupying the landscape around the (circa) ninth-century AD Buddhist stupa, Borobudur, located in Central Java, has over the last decade developed innovative approaches to counter its marginalisation by the bureaucrats and commercial interests charged with managing the World Heritage site. The resettlement of locals from five villages close to the monument, the prevention of on-site hawking of souvenirs, and the curtailment of local religious practices that had integrated the monument into the local Islamic world all effectively excised the monument from its contemporary cultural setting. Local activists, collaborating with archaeologists, have now developed cultural tourism ventures which encourage visitors to engage with local communities in the larger landscape in which the Borobudur is set. In a fascinating and highly innovative move, this initiative explicitly incorporates the radiating symbolism of the Mandala (the Borobudur itself was constructed as a three-dimensional Mandala) to reconnect the monument to its surrounding social landscape.

Smith and Turk, drawing on their experience of the East Rennell Island World Heritage area, make the critical point that the 'customary systems' of local inhabitants constitute a conservation management system in themselves and in their own terms. When places like Rennell Island are brought into the fold of 'global conservation', outside conservation experts work to reframe these customary systems in terms of Western conservation practice and scientific discourse. An impression is created that local systems are being raised to the level of conservation systems. Smith and Turk argue the case for a reversal of impetus in which global conservation would adapt itself to local Indigenous customary systems, recognising and accepting that conservation will very likely be conceived by these systems in terms different to those of the Western conservation movement.

2. Community archaeology

Annie Clarke (2002) discusses the emergence of community-based approaches to archaeology in Australia since the 1970s, citing a history which seems likely to resonate with that of a number of other countries. She points out that '...the development of research strategies designed to meet Indigenous concerns about the practice of archaeology can be seen to have two interlinked aims: first, to work towards achieving informed consent to practice and second, to establish meaningful processes of involvement and interaction between archaeological practitioners and Indigenous people' (Clarke 2002:250). The chapters in this volume that fall within the community archaeology theme very much address the latter point. Although separated widely by geography, these experiences share several common threads including the need to engage local communities in heritage protection, recognise the integration of cultural and natural heritage values, and employ appropriate methodologies in community engagement.

There are two chapters discussing Australian examples. In south west Western Australia, David Guilfoyle and his co-authors present a case study that demonstrates the importance of local communities taking control for effective cultural heritage management. They describe the processes involved with managing heritage associated with the Yoolbeerup wetlands system through which the community became the effective leader of a project that integrated cultural and natural heritage management strategies through various agencies. The success of this project empowered the community to become involved in other similar projects. Guilfoyle *et al.* argue that this case study demonstrates that ownership and control are fundamental to the successful outcomes of cultural and natural heritage management and that the methodology that was developed through this project can be applied effectively elsewhere. In a case study from northern NSW, Nick McClean discusses the uses of cultural mapping as an effective technique for recording cultural heritage. He explains how a local Indigenous community, the Githabul, developed two cultural mapping approaches to record local aspects of heritage. McClean emphasises the importance of 'being on country' as a key element of maintaining cultural identity and recording intangible aspects of cultural heritage.

Beyond Australia, examples of community archaeology in action are presented from Papua New Guinea, Thailand and Laos. Tim Denham discusses two different types of cultural resource management project with which he has been involved in PNG. He describes a community heritage project among the Kalam of the Simbai Valley in Madang Province and issues associated with the World Heritage nomination of the Kuk Early Agricultural Site. He also discusses the increasing commercialisation of cultural heritage management. Denham builds his case from the specific to the general to illustrate a broad range of experiences within the cultural heritage sector in PNG. Anna Karlström notes that in the past, nature conservation discourse has informed cultural heritage discourse and she proposes that the direction of influence can be reversed. She contrasts the traditional Western heritage discourse based on the conservationist ideal of the preservation of 'original fabric' with her experience in Laos where local communities often value intangible aspects of heritage which tend to blur natural and cultural qualities. Like Guilfoyle *et al.*, Karlström argues for the importance of local community involvement for successful heritage protection outcomes and suggests a number of methodolgies to make this work. Marwick and co-authors present the tailored strategies they employed to engage with different sections of the local community during an archaeological excavation in the Krabi River Valley in southern Thailand. They provide background by reviewing current Thai cultural heritage practices, observing that the Thai government tends to favour a tourist-oriented approach promoting monumental heritage. However Marwick *et al.* highlight how effectively cultural heritage protection can be achieved via community archaeology. They discuss their methodologies for promoting local community engagement in the Krabi River Valley, suggesting that these could have application in similar situations elsewhere.

3. Performative landscapes and conservation

Natural historians and nature conservationists map and classify landscapes according to their biodiversity, vegetation, soils, and other physical attributes. The final four chapters of this volume are concerned, in different ways, with the sacred, historical, and mythological attributes of landscapes and the cultural performativity that keeps these attributes known and alive. Reporting on research in south Sulawesi, David Bulbeck describes areas of land more or less closed to human access and economic activity due to the quality of sacredness entailed in their ancestral histories and the presence of spirit beings, graves, and sacred objects. Sacredness, as an aspect, has the effect of protecting the natural values of these areas and even protecting graves and archaeological deposits from looting. Similar exclusion zones – for example, the sacred groves of India (Kothari 2003), the *feng shui* woodlands of China (Zhuang and Corlett 1997), or the *lulic* forests of East Timor (McWilliam 2003) – have been shown to have considerable conservation value for biodiversity and for this reason in recent years they have attracted a great deal of interest from conservationists.

The growing interest of sections of the conservation biology field in what are referred to as 'natural sacred sites' is the focus of Denis Byrne's chapter. He argues that while this interest gives support to local groups in the maintenance of their religious systems against outside pressure it is also an interest that tends to domesticate certain elements of local religious belief and practice. Conservation scientists tend to focus on the 'rational' elements of popular religion in places like Thailand; they appear to be uncomfortable with the magical and supernatural dimension of religion. Byrne suggests this represents a projection onto non-Western popular religion of the secular-rationalist 'disenchanted' worldview that has been predominant in the West since the Protestant Reformation and the Enlightenment.

It is undeniably the case that when the habitats of Indigenous groups attract the interest of outside conservation interests, local inhabitants often find their lifestyle constrained and even find their ownership rights voided. In their respective chapters, Andrew McWilliam and Sue O'Connor *et al.* describe the situation in a new National Park at the eastern end of East Timor where the state's valuation of the Park in terms of its biodiversity and its association with the resistance struggle has tended to edge out the interests of the Park's 15,000 inhabitants. While for the state, and often also for Western heritage practitioners, the past is memorialised in static terms, for the local inhabitants the past is a living quality of the contemporary landscape. O'Connor *et al.* for instance describe how *téi* (spirit beings) must be regularly 'fed' with offerings in order for people to gain their protection. The merging of the historical and sacred realms is illustrated by the local belief that a particular *téi* was instrumental in the defeat of the Indonesian military forces. Both Bulbeck and McWilliam make the point that history is understood in these cultures in non-linear terms: historical figures and events are spiritually and poetically embodied in the contemporary landscape rather than confined to the past as points in a linear narrative (or to layers of a stratigraphic sequence, to use an archaeological image).

Acknowledgements

We would like to thank the College of Asia and the Pacific, The Australian National University, for providing funding for our session at the 2010 Australian Archaeological Conference. The production and indexing of this volume is supported by the Research School of Asia and the Pacific at The Australian National University.

References

Beinart, W. and Hughes, L. 2007. *Environment and Empire*. Oxford University Press, Oxford.

Brockington, D. 2002. *Fortress Conservation: The preservation of the Mkomazi Game Reserve, Tanzania*. James Currey, Oxford.

Brockington, D. and Igoe, J. 2006. Eviction for conservation: A global overview. *Conservation and Society* 4(3): 424–470.

Brosius, P.J., Tsing, A.L. and Zerner, C. (eds). 2005. *Communities and conservation*. AltaMira Press, Walnut Creek.

Byrne, D. 2009. A critique of unfeeling heritage. In: Smith, L. and Akagawa, N. (eds), *Intangible Heritage*, pp. 229–252. Routledge, London.

Byrne, D. 2011. Archaeological heritage and cultural intimacy: An interview with Michael Herzfeld. *Journal of Social Archaeology* 11(2):144–157.

Clarke, A. 2002. The ideal and the real: Cultural and personal transformation of archaeological research on Groote Eylandt, northern Australia. *World Archaeology* 34(2):249–264.

De Cesari, C. 2010. World Heritage and mosaic universalism. *Journal of Social Archaeology* 10(3): 299–324.

Dowie, M. 2011. *Conservation refugees: The hundred-year conflict between global conservation and native peoples*. MIT Press, Cambridge.

Grove, R. 1995. *Green imperialism: Colonial expansion, tropical island edens and the origins of environmentalism*. Cambridge University Press, Cambridge.

Haraway, D. 2006. *When species meet (posthumanities)*. University of Minnesota Press, Urbana.

Head, L. 2007. Cultural ecology: the problematic human and the terms of engagement. *Progress in Human Geography* 31(6):837–846.

Head, L. 2010. Cultural landscapes. In: Hocks, D. and Beaudry, M.C. (eds). *The Oxford Handbook of Material Culture Studies*, pp. 427–539. Oxford University Press, Oxford.

Ingold, T. 2000. *The perception of the environment: Essays in livelihood, dwelling and skill*. Routledge, London.

Isar, Y.R. 2011. UNESCO and heritage: global doctrine, global practice. In: Anheier, H.H. (eds), *Heritage, memory and identity*, pp. 39–52. Sage, London.

Kothari, A. 2003. Protected areas and social justice: the view from South Asia. *The George Wright Forum* 20(1): 4–17.

Latour, B. 1993. *We have never been modern* (Translated into English by C. Porter). Harvard University Press, Cambridge.

Lockwood, M. 2006. Values and benefits. In: Lockwood, M., Worboys, G. and Kothari, A. (eds). *Managing protected areas: a global guide*, pp.101–115. Earthscan, London.

Longstreth, R. (ed). 2008. *Cultural landscapes: Balancing nature and heritage in preservation practice*. University of Minnesota Press, Minneapolis.

McNiven, I. 2008. Sentient sea: seascapes and spiritscapes. In: B. David and J. Thomas (eds), *Handbook of landscape archaeology*, pp. 149–157. Left Coast Press, Walnut Creek.

McNiven, I. and Feldman, R. 2003. Ritual orchestration of seascapes: hunting magic and dugong bone mounds in Torres Strait, NE Australia. *Cambridge Archaeological Journal* 13(2):169–194.

McWilliam, A. 2003. New beginnings in East Timor forest management. *Journal of Southeast Asian Studies* 34(2): 307–327.

Meskell, L. 2012. *The nature of heritage: The new South Africa*. Wiley-Blackwell, Chichester.

O'Connor, S., Pannell, S. and Brockwell, S. 2011. Whose culture and heritage for whom? The limits of national public good protected area models in East Timor. In: Miksic, J., Goh, G. and O'Connor, S. (eds), *Rethinking cultural resource management in Southeast Asia: Preservation, development and neglect*, pp. 39–66. Anthem Press, London.

Pannell, S. and O'Connor, S. In press. Where the wild things are: an exploration of sacrality, danger and violence in confined spaces. In: Moyes, H. (ed.), *Sacred darkness: A global perspective on the ritual use of caves*. University of Colorado Press, Boulder.

Phillips, A. 2003. Turning ideas on their head: the new paradigm for protected areas. *The George Wright Forum* 20(2): 8–32.

Pulleine, R.H. 1929. The Tasmanians and their stone-culture. In: Lord, C.E. (ed.), *Report of the nineteenth meeting of the Australasian Association for the Advancement of Science (Australia and New Zealand)*, pp. 294–314. Government Printer, Hobart.

Rose, D. 1996. *Nourishing terrains: Australian Aboriginal views of landscape and wilderness*. Australian Heritage Commission, Canberra.

Rose, D. 2011. *Wild dog dreaming: Love and extinction*. University of Virginia Press, Charlottesville.

Roseman, M. 2003. Singers of the landscape: Song, history, and property rights in the Malaysian rainforest. In: Zerner, C. (ed.), *Culture and the question of rights*, pp. 111–141. Duke University Press, Durham.

Ross, A., Sherman, K.P., Snodgrass, J.G., Delcore, H. and Sherman, R. 2011. *Indigenous peoples and the collaborative stewardship of nature*. Left Coast Press, Walnut Creek.

Taylor, K. (ed.). 2012. *Managing cultural landscapes*. Routledge, London.

Tilley, C. 1994. *A phenomenology of landscape: Places, paths and monuments*. Berg, Oxford.

Tsing, A. 1993. *In the realm of the diamond queen*. Princeton University Press, Princeton.

Tsing, A. 2005. *Friction: An ethnography of global connection*. Princeton University Press, Princeton.

Weir, J. 2012. Country, Native Title and ecology. In: Weir, J. (ed.), *Country, Native Title and ecology*, pp. 1–19. ANU Epress, Canberra.

Whatmore, S. 2006. Materialist returns: practising cultural geography in and for a more-than-human world. *Cultural Geographies* 13: 600–609.

Zerner, C. (ed.) 2003a. *Culture and the question of rights*. Duke University Press, Durham.

Zerner, C. 2003b. Sounding the Makassar Strait: the poetics and politics of an Indonesian marine environment. In: Zerner, C. (ed.), *Culture and the question of rights*, pp. 56–108. Duke University Press, Durham.

Zhuang, X.Y. and Corlett, R.T. 1997. Forest and forest succession in Hong Kong, China. *Journal of Tropical Ecology* 14: 857–866.

1

Nature and culture in World Heritage management: A view from the Asia-Pacific (or, never waste a good crisis!)

Ian Lilley, University of Queensland, Brisbane, Australia

The gulf between natural and cultural World Heritage management in the Asia-Pacific region – and indeed right around the world – remains wide. This situation obtains from the top to the bottom of the World Heritage (WH) system and persists despite the now well-worn arguments against it and despite continual if still somewhat fitful efforts to find a remedy. Taking a broad view of what we call 'management' to encompass everything upstream, during and downstream of a nomination, this chapter[1] discusses the background to the issue, the current positions on the matter of UNESCO, ICOMOS and IUCN,[2] and what might be done at both the institutional level and on the ground to help find a concrete solution in the near future.

To disclose my interests in the matter, I am Secretary-General of the ICOMOS International Scientific Committee on Archaeological Heritage Management (ICAHM) and an ICOMOS World Heritage Assessor. In this latter capacity I do desktop and field assessments of properties nominated for World Heritage listing and have worked closely with IUCN in these contexts. I am also Secretary-General of the Indo-Pacific Prehistory Association (IPPA) and Convenor of the International Heritage Group (IHG), both mentioned towards the end of the chapter.

What's the issue?

In a nutshell, the impetus to draw natural and cultural heritage management closer together flows from related and broadly parallel decisions in UNESCO, ICOMOS and IUCN to expand definitions and categories of WH places and protected natural areas to include what UNESCO and ICOMOS call 'cultural landscapes' and what IUCN calls 'Category V [i.e. 5]' and 'Category VI protected areas'. The similarities and differences between these two major kinds of heritage place are discussed in more detail below. Here it is enough to note that cultural landscapes and protected landscapes are both centrally concerned with relationships between people and nature but that only the former are a recognised class of World Heritage.

---

1 An early draft of this paper is published in the Proceedings of the 1st International Conference on Best Practices in World Heritage: Archaeology. Menorca, Spain, 9-13 April 2012.

2 ICOMOS – International Committee on Monuments and Sites, IUCN – International Union for the Conservation of Nature; together with ICCROM, the International Centre for the Study of the Preservation and Restoration of Cultural Property, these groups form the statutory Advisory Bodies to UNESCO on World Heritage.

UNESCO got into the business of cultural landscapes as part of a long-term effort to expand the World Heritage List in terms of both its topical and its geographical diversity (see Jokilehto 2005 for background on 'Filling the Gaps'). To put it crudely, the idea was to have fewer European cathedrals, and more things such as natural places, archaeological sites and cultural landscapes in all regions as well as more of any kind of natural, cultural and especially 'mixed' cultural and natural properties in non-European locations. As UNESCO's assessment of *The State of World Heritage in the Asia-Pacific Region 2003* (Feng 2003:15) puts it:

> the bias towards monumental architecture as well as the preponderance of cultural over natural properties, has been repeatedly scrutinised by the World Heritage Committee and Advisory Bodies. However, the World Heritage List of properties is far from fully representing the rich ethno-cultural and biogeographical diversity of the Asia-Pacific region.

As a corollary of broadening the List, non-Western approaches to heritage and its management were to be encouraged and accommodated by the World Heritage nomination process. Amongst other things, this means encouraging and accommodating perspectives on heritage and its management which do not separate nature and culture but rather treat them holistically as indivisibly inter-related aspects of the world in which people live. Mixed nominations and cultural landscapes (the distinction remains somewhat blurry) are seen as keys to advancing this agenda. UNESCO's *The State of World Heritage in the Asia-Pacific Region 2003* (Feng 2003:18) draws attention to the fact that:

> the Asia-Pacific region is at the origin of the development of the concept of cultural landscapes on the World Heritage List. The first three cultural landscapes inscribed on the List, Tongariro National Park in New Zealand, Uluru Kata Tjuta National Park in Australia, and the Banaue Rice Terraces in the Philippines are all located in Asia and the Pacific. The recognition of the Maori spiritual attachment and veneration of the sacred mountain peaks at Tongariro[3] represented a turning point for the [World Heritage] Convention in further emphasising the importance of interaction between people and their environment. The introduction of the category of associative cultural landscape has encouraged the submission of mixed nominations throughout the world, as well as stimulating Pacific Island Countries to see the applicability of the World Heritage Convention in their countries, where customary land ownership and Indigenous knowledge form the basis for heritage protection.

This all sounds commendably postcolonial and emancipatory, and has resulted in globally groundbreaking work such as Smith and Jones's (2007) assessment of Pacific cultural landscapes, but what does it mean for people at the coalface around the region? Not a lot, from my observation, though not because of any lack of effort or goodwill. More sites are certainly being nominated as 'mixed properties', at least in the Pacific, which ostensibly means their cultural and natural dimensions are considered together. In reality, though, at least in my recent experience, the natural and cultural aspects of such nominations are treated completely separately, even when the technical assessors of the two 'sides' of the nominations are in-country together and get on well, personally and professionally. In addition, the mixed properties with which I have been involved were very obviously originally 'natural' nominations with culture added on later, in a nod to the new more encompassing imperatives.

I am not saying that those concerned with the nominations were not genuine in their interest in cultural matters, and I do not want to diminish their efforts to be inclusive. Far from it. It must be said, though, that their enthusiasm generally outstripped their specialist subject knowledge and management capacity regarding cultural heritage. It was also clear that the 'cultural sides' of these

---

3   Tongariro National Park in New Zealand was first listed in 1990 as a natural property but in 1993 was re-inscribed as the first-ever World Heritage cultural landscape in recognition of its 'associative' cultural values.

mixed nominations had been put together after and more hastily than the 'natural sides', and that the proposed management of the properties remained almost entirely in the hands of natural heritage managers. Those natural heritage managers were also better organised, better funded and *far* better supported by external civil society organisations than the cultural heritage managers.

In countries with little income to begin with, this relative poverty of resources for cultural matters greatly exacerbates and is, recursively, further worsened by the split between nature and culture in heritage management. The nations we are encouraging to nominate cultural landscapes and mixed sites so that local perspectives on World Heritage are recognised and valued often really can't afford to nominate and then manage such properties, even with external assistance. This means that when offered a choice by a process divided between nature and culture, they quite pragmatically tend to put most of their eggs in the better-built basket, namely natural heritage management, to help ensure access to the supposed benefits of World Heritage recognition that we tout. Thus despite the official inclusive rhetoric about fostering non-Western perspectives which do not separate nature and culture, this means that people can see their holistic community-based perspectives actively disregarded – or only given a token nod – in favour of quintessentially Western approaches which very definitely separate humans from their environments. The effort to expand the List and the ways in which listed properties are managed is thus suborned and the List doomed to just replicate itself along Western lines, albeit in exotic locations.

This is not just a problem with sites coming on-stream now but which were 'in process' for a long time and were thus nominated before discussion of nature vs culture had progressed very far. The dominance of natural heritage extends to many 'mixed' properties currently awaiting nomination and assessment on Asia-Pacific World Heritage Tentative Lists. A quick scan of such lists shows that many if not most 'mixed' tentative listings focus almost entirely on the natural dimensions of the property. Thus, for instance, the tentative listing of the Huon Peninsula Terraces in Papua New Guinea (PNG), one of the earliest known sites of human settlement in the Asia-Pacific, devotes just two words to the matter out a 514-word text. The remainder concerns the evolution and biodiversity of the physical landscape. One has to be sincerely thankful such sites have been tentatively listed on any basis, but it is hard not to think that we are going 'back to the future' rather than making any genuine headway.

To wrap up this litany of concern about barriers to cooperation across the nature-culture divide, there is no formal capacity for managers of properties nominated as either 'just' cultural – such as Kuk in PNG – or 'just' natural – such as East Rennell in the Solomons – to have input from the 'other side', despite the rhetoric about holistic approaches to management which are sensitised to local perceptions.

I have stressed that the foregoing observations are made on the basis of my own recent experience with World Heritage management, broadly defined, but none of these concerns is new. Nor are such complaints made only by people such as me from the 'ICOMOS side'. IUCN advocates of protected landscapes make precisely the same sorts of comments. Phillips (2005:20), for instance, points out:

> that landscape has usually been seen as a second class member of the environmental club. 'Lacking a coherent philosophy, thin on quantification and without a strong, unified disciplinary core, it has often been viewed as a 'soft' topic, to be swept aside in the rush to develop and exploit the environment, a trend that is justified by that trite commentary: 'jobs before beauty" (Phillips and Clarke 2004). Compared to the wilderness movement in North America, and its equivalents in Australia and other countries, the idea of taking an interest in lived-in, working landscapes was slow to emerge, and confined to relatively few countries for many years. In this it contrasts with the demands of wildlife conservation or pollution control. The protection, management and planning of landscape has generally been a less powerful movement, and has taken longer to emerge as a political force.

The contrast is particularly evident at the international level.

He goes on (2005:26) to explain:

> The World Heritage Convention…combines two ideas: cultural heritage and natural heritage, and in operating the convention two separate streams of activity have developed…Over the years, the sharp separation and differentiation of these two approaches has been found less and less helpful in understanding the world's heritage and its needs for protection and management…the separation of the cultural and natural world – of people from nature – makes little sense. Indeed it makes it more difficult to achieve sustainable solutions to complex problems in the real world in which people and their environment interact in many ways.

The discourse(s)

The two approaches referred to by Phillips are exemplified by the differing discourses of UNESCO and ICOMOS on the one hand and IUCN on the other. I will sketch the gist of each rather than attempt to chart every twist and turn of the sometimes ill-defined and confusing decision-making processes.

Cultural landscapes

UNESCO recognised 'cultural landscapes' as a category of World Heritage in 1992, specifically to overcome obvious conceptual and practical difficulties with the cultural/natural dichotomy recognised in the original World Heritage Convention and with the 'intellectually flaccid idea of the 'mixed site''(Fowler 2003:17) that subsequently emerged to bridge the nature-culture divide. Fowler (2003:18) notes that the notion is of nineteenth century origin but was brought to prominence by the Berkeley geographer Carl Sauer in the early decades of the twentieth century. For Sauer (1925, cited in Mitchell *et al.* 2009:15):

> The cultural landscape is fashioned out of the natural landscape by a culture group. Culture is the agent, the natural area is the medium, the cultural landscape is the result.

Archaeologists mainly know Sauer for his work on the origins of agriculture, and Fowler (2003:18) goes on to point out that conservationists – including heritage archaeologists – only adopted Sauer's idea in the 1990s. Fowler contends that the 'cultural landscapes' designation "remains … an uncommon term for an opaque concept". Fowler wrote those words almost a decade ago now. To advance the state of play, the World Heritage Centre published Mitchell, Rössler and Tricaud's edited volume *World Heritage Cultural Landscapes. A Handbook for Conservation and Management* in 2009. Yet the term and its practical application remain a work in progress, not least because of the differing perspectives brought to the discussion by the non-Western societies such as those in the Asia-Pacific, which continue to be strongly encouraged to participate in World Heritage affairs (cf. Smith and Jones 2007). Somewhat surprisingly in this context, the definitions of cultural landscapes in the Operational Guidelines for the Implementation of the World Heritage Convention have not changed since 1992 despite several revisions to the Guidelines in that time. This wording specifies three types of cultural landscape (UNESCO 2011:88):

1. The clearly-defined landscape (such as gardens and parklands);

2. The organically-evolved landscape, in which culture and nature have co-evolved (to use an IUCN term). There are two sub-types, the relict or fossil landscape in which the material results of a past co-evolution are still visible, and a continuing landscape, where the co-evolution carries on; and,

3. The associative cultural landscape, where a natural landscape is invested with largely or entirely intangible cultural values.

To be inscribed on the World Heritage List, any place nominated under any of these categories must exhibit Outstanding Universal Value, using the same six criteria as other categories of eligible cultural properties.

Protected areas

Confusingly, although IUCN advises UNESCO on natural aspects of World Heritage, and thus like ICOMOS is enjoined to use the cultural landscapes designation in relation to World Heritage matters, the organisation has its own nomenclature which is not particularly consistent with UNESCO's conceptualisations and terminology.

IUCN works with the notion of 'protected areas', of which there are six kinds with differing management goals. Categories I-IV are purely natural heritage areas with decreasing levels of restriction on human activity within their boundaries (e.g. Category I is called a 'Strict Nature Reserve/Wilderness Area'). Categories V and VI, on the other hand, recognise the place of humans in the world. The two are sometimes combined under the rubric of "protected landscapes", though that term technically applies only to Category V, Protected Landscape/Seascape. In conception, Category V is closest to a cultural landscape. Indeed, the main IUCN publication on the matter states that '*protected landscapes are cultural landscapes* that have co-evolved with the human societies inhabiting them' (Brown *et al.* 2005:3, my emphasis). Category VI, 'Managed Resource Protected Area', is designed to manage sustainable harvesting of natural resources. It thus also encompasses the protection of human interaction with the environment, but on a somewhat different basis.

Although IUCN's protected landscapes are similar to UNESCO and ICOMOS's cultural landscapes, they are not the same. As Brown *et al.* (2005:9–10) write:

> there are important distinctions between the two designations, in particular related to how they are selected. In designation of Category V Protected Landscapes, the natural environment, biodiversity conservation, and ecosystem integrity have been the primary emphases. In contrast, the emphasis in World Heritage Cultural Landscape designation has been on human history, continuity of cultural traditions, and social values and aspirations (Mitchell and Buggey, 2001). As Adrian Phillips further notes in his chapter [in the same 2005 volume], "outstanding universal value" is a fundamental criterion in recognising a World Heritage Cultural Landscape, while the emphasis in Category V Protected Landscapes is on sites of national, or sub-national significance.

Phillips (2005:27) also points out that the IUCN Category V lacks UNESCO/ICOMOS's 'designed landscape' type.

ICOMOS's Rössler (2005:46) regards cultural landscapes 'as a role model paralleling the development of the IUCN Category V Protected Landscape/Seascape. In cultural landscapes specifically, the local communities are acknowledged with the (co)responsibility in managing the sites'. IUCN's Brown *et al.* (2005:10) are at pains to declare, however, that 'central to the protected landscape approach, though not expressed in any formal designation, are the array of strategies that Indigenous and local communities have been using for millennia to protect land and natural and cultural resources important to them.'

Biocultural diversity

Given what seems to be the 'furious agreement' among UNESCO, ICOMOS and IUCN regarding what we might lump as 'protected cultural landscapes', it is hard to understand why such a gulf remains between the 'natural' and 'cultural' camps when it comes to managing what everyone concedes is effectively the same thing. Remain it does, though, and in an attempt to bridge it, some within IUCN have begun promoting the concept of 'biocultural diversity' (BCD). Although it dates back to at least the 1980s, BCD in its current form is primarily the initiative of Canadian linguist Luisa Maffi (2005), of the NGO Terralingua, generously supported by the Christensen Fund (e.g. Persic and Martin 2007). Maffi (2010:74) writes that the:

> Central tenets of this field are that the diversity of life is diversity in both nature and culture and that the two diversities are co-evolved and interdependent.

> Recognition of this link has significant implications for conservation practice and for the policies that regulate access to and use, management, and protection of biodiversity and natural resources. Conservation discourse and policies are moving away from certain preservationist and exclusionary approaches toward ones that increasingly promote the full and effective participation of Indigenous peoples and local communities. A biocultural approach provides an integrative framework that further strengthens and motivates this shift. A new focus is becoming apparent in the statements of principle and, in some instances, the programmes of work of various international organizations that make specific reference to the importance of cultural diversity and traditional knowledge in relation to biodiversity.

Although almost entirely an IUCN project – the Christensen-funded UNESCO volume on *Links between biological and cultural diversity* (Persic and Martin 2007) makes no mention of ICOMOS – a team including an Australian cultural heritage practitioner with an executive role in ICOMOS has begun engaging with the concept with promising results (e.g. Hill *et al.* 2011). It seems, though, that IUCN itself is not as keen on BCD as its own rhetoric suggests. Maffi (2010:75) laments that the IUCN Council did not make provision in its 2009-2012 program to implement the organisation's own resolution on 'integrating culture and cultural diversity into IUCN's policy and programme'. Nor did the Council see the issue 'as a priority for additional fundraising'.

In IUCN's in-house journal *Policy Matters,* Maffi (2010:77) frankly admits that:

> existing intellectual and institutional frameworks pose barriers to greater and more concrete progress towards adopting a biocultural perspective in policy and practice. Some conservation organizations are not yet wholly sensitive to people-centered conservation. This presents obstacles to the implementation of an integrative approach that incorporates an understanding of cultural dynamics, as well as the full and equal participation of Indigenous peoples and local communities in conservation decisions that affect them. As yet, there is even greater reluctance to accept the idea of Indigenous peoples and local communities as stewards of the biodiversity and ecosystems of their territories. In turn, cultural institutions have tended to remain rather insular, instead of seeking meaningful connections and collaborations with conservation organizations. Funding limitations preventing more integrative work are also reason for organizations in the respective realms of nature and culture not to "stray" into the others' institutional territory. Finally, policy-makers (especially at the national level) tend not to act on an issue unless there is a groundswell of support for it.

Where to from here?

Plainly there is a need to act if change is to occur and we are to effect a meaningful shift in the kinds of properties successfully nominated to the World Heritage List, especially on the basis of holistic, local community-based approaches to their management. The need for change is particularly acute in the Asia-Pacific. Although this region is at the forefront of developments in the nomination and management of mixed sites and cultural landscapes, the current situation on the ground indicates that there is still a long way to go if local concerns and aspirations concerning integrated approaches to natural and cultural heritage are to be met in a manner than gels satisfactorily with global norms. ICOMOS and IUCN need to cooperate more effectively on the matter while also getting their own houses in order so that their various structural elements (and political/ideological factions) are mutually supportive, and supportive in action as well as in theory. As noted earlier, IUCN has made nearly all of the running on this matter to date, so although ICOMOS people periodically appear in IUCN publications discussing the issues, ICOMOS itself has not engaged with IUCN in any consistent and productive way to advance matters. This means that the two organisations continue to operate in 'two separate spheres of activity', as Phillips puts it. In turn, this means that World Heritage managers on the ground, as well as the sites they try to protect and the communities they endeavour to engage, continue to suffer the consequences, including communal division and the loss of knowledge associated with properties as well as the loss of physical fabric in such places. This is exactly the opposite of what more expansive views of World Heritage are intended to produce. In my personal experience it is already seeing people vocally reject World Heritage as yet another neo-colonial imposition that causes them even more disadvantage than they already suffer in a globalised world.

Time and resources are too short to permit this situation to go on. Something new is needed if we believe World Heritage is of any value. I do not underestimate the difficulties in overcoming the problems so clearly identified by Maffi, and I certainly do not believe we can solve the matter overnight. By the same token, one should 'never waste a good crisis'. Nothing will be achieved unless concrete steps are taken to maintain the momentum that Maffi and others have built up. Work has to proceed simultaneously on a number of fronts, to 'join up' all the players to move things forward effectively. In this connection, the ICOMOS International Committee on Archaeological Heritage Management (ICAHM) is working with the International Heritage Group (IHG), a recently formed NGO, and with the Indo-Pacific Prehistory Association (IPPA) to achieve locally appropriate results on the ground in the Asia-Pacific and elsewhere.

Bridging the nature-culture divide is a key IHG objective, and so to get some early 'runs on the board' IHG is organising a 'knowledge café' on the matter at the 2012 IUCN Congress in Korea, entitled *Cultural Heritage Management Capacity and Enhanced Biocultural Resilience in High-Value Landscapes*. The event – a form of specialised workshop – is co-sponsored with ICOMOS-ICAHM and IUCN and features the participation of Tim Badman, Head of IUCN's World Heritage Programme, two colleagues cited earlier – Jessica Brown, Chair of the IUCN Protected Landscapes Specialist Group, and Terralingua's Luisa Maffi – as well as other senior members of IUCN with high profiles in this area. The objective is to come up with a shortlist of achievable proposals for concrete action that will enhance capabilities in the integrated management of natural and cultural World Heritage in places where heritage managers identify themselves as in need of such assistance.

Another pivotally important aspect of the larger practical effort will be structural change in the form of a return to actually doing what the 1998 'Berlin Agreement' between ICOMOS and IUCN specifies should be done to integrate evaluation of World Heritage cultural landscapes. Amongst other things, the accord 'agreed to co-ordinate working practice towards producing a common evaluation report, agreement over recommendations and harmonisation of presentation' (Fowler 2003:16). To my knowledge such goals have never been realised. If they were they would provide local managers with a much more coherent start than they get now with new nominations of cultural landscapes and mixed properties. My recent experience also suggests that it would be timely to extend such agreement to cover any assistance to local heritage managers to ensure harmonised approaches to World Heritage nominations and the management of natural and cultural dimensions of sites in their care, and to the co-ordination of the World Heritage Centre's monitoring of listed sites. It would also help immeasurably if the process explicitly required integrated local community input on natural and cultural dimensions of the nominations to ensure its inclusion on more than a token or *ad hoc* basis.

Communication is another crucial area of action at the coalface. Issues of mutual translation of IUCN and ICOMOS ideas and terms will loom large in such efforts to integrate approaches to natural and cultural heritage management, as will translation both literal and metaphorical of 'World Heritage-speak' for local communities involved with nominations and, if successful, the management of listed properties. As made clear in the foregoing discussion, it is natural heritage managers who have made most of the advances so far, conceptually as well as on the ground, where at least in the Asia-Pacific they are usually better organised and resourced than cultural heritage managers owing to the well-known 'cuddly panda effect' in conservation. The fact that they may often have to do much of the 'heavy lifting' in conserving the cultural as well as natural aspects of a listed place does not mean however that their normative approaches – including biocultural diversity – should inevitably hold sway. It is important that ICOMOS, ICAHM and civil-society organisations such as the International Heritage Group press to have specialist cultural heritage approaches integrated into any management plans at both the conceptual and technical levels.

This can only be done by developing a common terminology. BCD does not seem to fit the bill in this connection. It has not attracted much attention from cultural heritage specialists despite the hopes of its proponents (e.g. Harmon 2007 in addition to Massi and Hill *et al.* cited earlier) and despite calls from UNESCO for "an anthropological approach to the definition of cultural heritage and people's relationship with the environment" (Mitchell *et al.* 2009:25). To quote Strathern (2006:192) quoting Galison (1996:14), the job thus remains to 'work out an intermediate language, a pidgin, that serves a local, mediating capacity'. As MacEachern (2010:350) points out in a related context, this is because past failures to advance satisfactorily in such circumstances are not a matter:

> of bad faith on the part of one group of people or another...[but rather result from] the difficulties of translation, of groups of people who in many cases wished to work productively together, but who found themselves frequently at odds or misdirected because of a failure to appreciate the presumptions and the constraints on other actors in what was supposed to be a shared endeavour.

Harmonisation and coordination between natural and cultural World Heritage management, broadly defined to capture everything upstream, during and downstream of a nomination, are essential if we are to avoid such 'difficulties of translation', which lead to such undesirable consequences as the minimal attention to cultural matters in tentative listings, the belated addition of cultural considerations to 'mixed' nominations and the unquestioned dominance of 'natural' approaches to the management of listed properties. As a start, matters of translation and harmonisation should be included in the capacity-building options in UNESCO's emerging plans for 'creative responses' to 'upstream processes to nominations', as discussed by a meeting of global experts in April 2010 in the lead-up to the 40th anniversary of the World Heritage Convention in 2012. The meeting was held in Thailand and co-sponsored by Australia and Japan and was another example of the Asia-Pacific taking the lead in World Heritage matters. Those of us working in this region should take advantage of the momentum that meeting created. If this opportunity is missed, the gulf between nature and culture will continue to undermine other progress made in our part of the world and globally in preparing the Convention for its next 40 years.

Integrated capacity-building is the key, but only if it is approached in ways that promote and support rather than undermine and suppress local interest and initiative. The moves canvassed above are one step in that direction.

Acknowledgements

Sincere thanks to the universities of Queensland and Oxford – and to Chris Gosden in particular – for backing the International Heritage Group. Through Oxford, the Leverhulme Trust provided a generous Visiting Professorial Fellowship for me to work on IHG's creation during a sabbatical in 2011. The University of Queensland will provide a base for the IHG network. Thanks too to all my friends and colleagues in IHG, ICAHM, ICOMOS, IUCN and IPPA (etc., etc.) for their unflagging interest and support.

Two anonymous referees made invaluable suggestions to improve this chapter, but I am of course entirely responsible for its final form.

References

Brown, J., Mitchell, N. and. Beresford., M. 2005. Protected landscapes: a conservation approach that links nature, culture and community. In: Brown, J., Mitchell, N. and Beresford, M. (eds), *The protected landscape approach linking nature, culture and community*, pp. 3–18. IUCN, Gland.

Feng, J. (coordinator). 2003. *The state of World Heritage in the Asia-Pacific Region 2003*. World Heritage papers 12. UNESCO, Paris.

Fowler, P. 2003. *World Heritage cultural landscapes 1992-2002*. World Heritage papers 6. UNESCO, Paris.

Harmon, D. 2007. A bridge over the chasm: Finding ways to achieve integrated natural and cultural heritage conservation. *International Journal of Heritage Studies* 13(4–5):380–392.

Hill, R., Cullen-Unsworth, L., Talbot, L. and McIntyre-Tamwoy, S. 2011. Empowering Indigenous peoples' biocultural diversity through World Heritage cultural landscapes: a case study from the Australian humid tropical forests. *International Journal of Heritage Studies* 17(6):571–591.

Jokilehto, J (compiler). 2005. *The World Heritage List. Filling the gaps – an action plan for the future*. ICOMOS, Paris.

Maffi, L. 2010. Policy for biocultural diversity: Where are we now? *Policy Matters* 17:74–77.

Maffi, L. 2005. Linguistic, cultural, and biological diversity. *Annual Review of Anthropology* 29:599–617.

MacEachern, S. 2010. Seeing like an oil company's CHM programme. Exxon and archaeology on the Chad export project. *Journal of Social Archaeology* 10(3):347–366.

Mitchell, N., Rössler, M. and Tricaud, P.M. (eds) 2009. *World Heritage Cultural Landscapes. A Handbook for Conservation and Management*. World Heritage Papers 26. UNESCO, Paris.

Persic, A. and Martin, G. (eds). 2008. *Links between biological and cultural diversity*. UNESCO, Paris.

Phillips, A. 2005. Landscape as a meeting ground: Category V protected landscapes/seascapes and World Heritage cultural landscapes. In: Brown, J., Mitchell, N. and Beresford, M. (eds), *The protected landscape approach linking nature, culture and community*, pp. 19–35. IUCN, Gland.

Phillips, A. and Clarke, R. 2004. Our landscape from a wider perspective. In: Bishop, K. and Phillips, A. *Countryside planning*, pp. 49–67. Earthscan, London.

Rössler, M. 2005. World Heritage cultural landscapes: a global perspective. In: Brown, J., Mitchell, N. and Beresford, M. (eds), *The protected landscape approach linking nature, culture and community*, pp. 36–46. IUCN, Gland.

Smith, A. and Jones, K. 2007. *Cultural landscapes of the Pacific Islands*. ICOMOS, Paris.

Strathern, M. 2006. A community of critics? Thoughts on new knowledge. *Journal of the Royal Anthropological Institute* (N.S.):191–209.

UNESCO. 2011. *Operational guidelines for the implementation of the World Heritage Convention* http://whc.unesco.org/archive/opguide11-en.pdf [accessed:1/2/11].

2

Customary systems of management and World Heritage in the Pacific Islands

Anita Smith, La Trobe University, Melbourne, Australia

Cate Turk, Universität Erlangen-Nürnberg, Erlangen, Germany

'My observation … of the often bewildering scene of contested conservation initiatives for the Marovo Lagoon has enabled me to watch how major international environmentalist organizations have risen and then fallen as their simplistic concept of community proved soundly incompatible with the Marovo people's time honoured ways of organising themselves.' (Hviding 2006:83)

Introduction

In recent decades local communities have been increasingly engaged in protecting heritage sites through the development of new models of conservation practice, such as co-management, joint management and community management. Our interest in heritage conservation in the Pacific leads us to examine a related form of management arrangement, customary management. In this paper we examine how systems of customary land tenure prevalent throughout the region require a method for establishing and managing heritage fundamentally different to that employed in the Yellowstone model of state-managed protected areas, or the co-management of Indigenous landscapes such as that of Uluru-Kata Tjuta in Australia. We suggest a fine distinction between customary systems *of* management and customary systems used *in* management; and consider the implications customary land tenure has for governance in heritage management; the definition of heritage values; and the need for sustainable livelihoods.

Our discussion is centred around customary management and the uptake of the UNESCO World Heritage Convention in the Pacific region. Although the Pacific Islands continue to be the least represented geo-cultural region on the World Heritage List, the island communities and governments have made a substantial contribution to the World Heritage process, in expanding the understanding of heritage and through their insistence on the recognition of community and culture as central to all World Heritage initiatives (Smith in press). By examining how this international instrument intersects with local practices that protect places of significance, we highlight the struggles between different ways of conceiving conservation, tensions between local and global conservation agendas as well as the common ground where interests coincide.

World Heritage and Pacific Island states

Since the mid 1990s the World Heritage Committee has actively sought to increase the representation of the Pacific Islands on the World Heritage List (Smith in press). In a landmark decision the Committee inscribed East Rennell Island, in the Solomon Islands, on the World Heritage List in 1998. East Rennell being the first site inscribed in an independent Pacific Island nation and the first site anywhere to be inscribed on natural values under customary ownership and protection (UNESCO 1999; see also Smith 2011). Not until 2008 were there further successful nominations in the Pacific Island states, being the cultural properties of Chief Roi Mata's Domain in Vanuatu and Kuk Early Agricultural Site in Papua New Guinea, followed in 2010 by the inscription of Bikini Atoll in the Republic of the Marshall Islands as a cultural property and the Phoenix Islands Protected Area in Kiribati, a natural property.

The World Heritage system poses a number of challenges for the Pacific Island states. In these developing nations, human, financial and technical resources are very limited, and heritage conservation is a low priority for governments. Most importantly in the context of this paper, the rights of customary land owners are enshrined in the constitutions of many Pacific Islands states, creating tensions in the development of national legislation to protect World Heritage properties (in compliance with the state party's obligations under the World Heritage Convention Representatives of Pacific Island states have argued that the World Heritage Committee must recognise not only these challenges but also the unique character of the region's heritage that is reflected in the inseparable relationship between Pacific Island people and their environments) (Smith and Jones 2007; te Heuheu *et al.* 2012). A statement to the Committee in 2007, known as the 'Pacific Appeal', identifies key elements that underpin this heritage including the region having amongst the highest proportions of people living within traditional governance systems and land and sea remaining under traditional management of any region of the world:

> Protection of our heritage must be based on respect for and understanding and maintenance of the traditional cultural practices, Indigenous knowledge and systems of land and sea tenure in the Pacific. (UNESCO 2007a)

These recommendations have been taken forward in various ways, affecting not only the Pacific but how World Heritage conservation is practiced more generally. In particular, community participation in all stages of implementation of the Convention was championed by Aotearoa/New Zealand during their term as Chair of the World Heritage Committee in 2007 in which they represented interests of the Pacific Island region. Under the leadership of Tumu Te Heuheu, paramount chief of Ngati Tuwharetoa, Aotearoa/New Zealand successfully proposed the addition of a fifth 'C' for Community to the four strategic objectives of the World Heritage Committee – Credibility, Conservation, Capacity-building and Communication – that frame planning and funding of World Heritage programs (UNESCO 2007b).

Customary systems in management

The elevation of 'community' as a strategic objective of World Heritage Committee stands alongside efforts to incorporate community, and cultural aspects more broadly, in other international conservation programs. Several programs focussed on conservation of natural heritage have worked to incorporate cultural aspects: as values to be managed (e.g. sacred aspects of 'natural' sites); as ways of managing (drawing upon traditional ecological knowledge); and as ways of engaging community (e.g. community conserved areas). There is an increasing bank of resources that describe best practice management in this regard including: *Indigenous and Traditional Peoples and Protected Areas - Principles, Guidelines and Case studies* (Beltrán *et al.* 2000); *Indigenous and Local Communities and Protected Areas: Towards Equity and Enhanced Conservation* (Borrini-Feyerabend *et al.* 2004) or *Indigenous Peoples and Community Conserved Areas(ICCAs) - Guidelines towards appropriate recognition and support* (IUCN 2008); *Guidelines for the protection of Sacred Natural Sites – Guidelines for Protected area Managers* (Wild and

McLeod 2008); and *Integrated Conservation-Development Project methodologies* (ICDPs) (see West and Brockington 2006; and Clark, Bolt and Campbell 2008, for recent discussions of these). Changing institutional practices relate not only to state-managed protected areas, but affect the way NGOs go about their in-country activities (see Hails 2007 for a discussion of community based conservation within World Wildlife Fund). These approaches are evolving as projects run their course and attract critical attention (see West and Brockington 2006 for a review of relevant critiques). A register of Indigenous and Community Conserved Areas currently being established by the World Conservation Monitoring Centre (see Corrigan and Granziera 2010) should enable a greater appreciation of the types and range of these initiatives.

This body of conservation literature offers useful tools for heritage managers including ways of doing conservation that are sensitive to community aspirations. These guidelines are not written to be politically correct, they draw on real cases to provide ways of working between different world-views, with the overall aim of improving conservation practices. It is worth emphasising however the hybrid nature of these initiatives. The fuzziness of the language sometimes obscures the outside involvement in these local projects: 'community'-based conservation projects are often the result of an internationally developed methodology when initiated through non-government and government aid programmes. For example, Govan *et al.* (2006:63) speak of locally managed marine areas in contrast to 'global approaches', thus localising what has been an initiative driven from outside.

Following his observation of conservation projects in Melanesia, Simon Foale notes:

> ... conservation of the scientific, cultural, and other heritage values of coral reefs in these countries is and will continue to be inextricably bound up with aid programs and NGO projects, and therefore becomes the responsibility of a broad range of actors, requiring the synthesis of a broad range of intellectual disciplines. (Foale 2008:32)

It is the multiple interests bundled together that lead to what we identify these approaches as a 'customary systems *in* management' approach. Although these interests coalesce around a local community, very often the governance arrangements and the articulations of management are those fashioned by outside interests and expectations. Where locals have control and are comfortable operating within a Western management framework, the use of customary systems of decision-making and environmental knowledge in management can be a powerful combination. Lisa Palmer (2006) highlights an example of this in her discussion of natural resource management, where under 'nation to nation' agreements in Canada members of the Cree nation exercise the right to determine how timber is harvested on their lands.

Since 1994 (for cultural sites) and 1999 (for natural sites, following inscription of East Rennell Island)[1] the World Heritage Committee has afforded customary management of World Heritage properties the same status as management by government institutions:

> All properties inscribed on the World Heritage List must have adequate long-term legislative, regulatory, institutional and/or traditional protection and management to ensure their safeguarding ... States Parties should demonstrate adequate protection at the national, regional, municipal, and/or traditional level for the nominated property. (UNESCO 2011 Paragraph 97)

---

1   Previous versions of the Operational Guidelines had required that sites nominated to the World Heritage List must have 'adequate legal protection and management mechanisms (cultural sites)' or 'adequate long-term legislative, regulatory or institutional protection (natural sites)'. Very early versions of the guidelines had requirements for legal protection but not for management. This was amended to 'adequate legal and/or traditional protection and management' in 1994 for cultural sites and in 1999 for natural sites.

When a site is nominated to the World Heritage List, the State Party is required to provide 'a clear explanation of the way this protection operates to protect the property' (UNESCO 2009, Paragraph 97). The Guidelines specify how various aspects of a site should be managed, but deliberately do not provide a concrete definition for traditional management, stating rather:

> Management systems may vary according to different cultural perspectives ... They may incorporate traditional practices, existing urban or regional planning instruments, and other planning control mechanisms, both formal and informal. (UNESCO 2011, Paragraph 110)

To date, only a handful of sites have been inscribed under customary management systems. For the most part these are systems of participatory management that include one or more local communities, such as examples of joint and co-management, in Uluru-Kata Tjuta and Kakadu National Parks in Australia, Traditional Owners are formally integrated within a park management system making decisions about conservation of their lands. In these cases a national park (with all the associated apparatus of government) was established prior to land being handed back to traditional custodians. Tongariro National Park, in Aotearoa/New Zealand provides a different case study, where *tangata whenua* ('the people of the land') made the case for establishing a national park from the outset. Protection of the sacred mountains of Tongariro was negotiated in the 1870s during colonisation, when the then Paramount chief 'gifted' his *tapu* lands to be *tapu* ('sacred and protected') under Queen Victoria. As one of the first *national* parks, following not long after Yellowstone, management is through national government administration, and the paramount chief (at present Sir Tumu te Heuheu) sits on the park board.

These examples highlight an important distinction between 'customary systems used *in* management' and 'customary systems of management'. Customary systems with*in* a Western conservation management framework are continually negotiated but framed by outside agendas and concepts of conservation practice. 'Customary systems of management' mean living in landscapes, conserving important places and negotiating uses of resources through local customary practices. Integrating local customary systems within Western models of conservation may be an effective approach in many parts of the world, and especially for places where traditional owners are no longer resident and/or dependent on access to land and resources for their livelihoods. In the Pacific Islands, on the other hand, continuing traditional land tenure and resource use means conservation is framed by the local cultural practices in decision making and ways of understanding and being in the landscape. Across the Pacific region customary owners cannot be invited to participate in conservation of their traditional lands, rather they mediate any activities that take place on their land. Discussing natural heritage conservation in the Solomon Islands, in particular, anthropologist Eduard Hviding stresses:

> ... biodiversity conservation initiatives in most cases are bound to focus on species and environments that are already culturally and economically significant for somebody else: those who live in, subsist on and, in many cases claim property rights to the 'species' and 'areas', which for them are more likely to be 'resources' and 'territories'. (Hviding 2006:72)

Towards customary systems of management in the Pacific

We return here to World Heritage and the example of the East Rennell World Heritage site in the Solomon Islands. East Rennell is a small and remote upraised coral island in the south west of the Solomon Islands that was inscribed in the World Heritage List for its natural values in 1998 under the then Criterion N (ii), now known as Criterion ix. The formal justification for inscription states:

East Rennell, as a stepping stone in the migration and evolution of species in the western Pacific, is an important site for the science of island biogeography. Combined with the strong climate effects of frequent cyclones, East Rennell is a true natural laboratory for scientific study. (http://whc.unesco.org/en/list/854)

Although internationally recognised only for these natural values, the customary owners of the property, approximately 800 people, live within the World Heritage area in four villages along the edge of Lake Tegano, a freshwater lake that occupies over half of the World Heritage area. For them, East Rennell is their cultural landscape, one that their Polynesian ancestors settled from somewhere to the east in the distant past. Subsistence agriculture, fishing and hunting underpin the economy and the community relies on forest products for most construction materials. The traditional land tenure system divides the land into tribal areas (*Kakaiangá*) each under the authority of one of the island's chiefs (*Hakahua*). The lake is regarded as common property.

The World Heritage nomination for East Rennell briefly mentioned the history and culture of the island however neither the cultural values of the landscape nor the traditional system of authority and land tenure were described. The International Union for the Conservation of Nature (IUCN), the advisory body to the World Heritage Committee for properties of natural values, advised that the nomination 'breaks new ground in terms of nominating a natural site that is under customary land ownership, that has no formalised legal basis and for which the object is sustainable resource use' (IUCN 1998). On inscribing the property, the Committee agreed that customary protection and management of the natural values of the property should be supported and noted that the rights of customary owners and customary law are acknowledged in the Constitution of the Solomon Islands and the Solomon Islands *Customs Recognition Act (1995)*. However the Committee on the advice of IUCN reiterated the need for a locally-developed management plan that documented the customary system of protection and a national World Heritage Protection Bill to be prepared by the Solomon Islands government.

Civil unrest in the Solomon Islands resulted in almost no communication between the customary owners on East Rennell and the outside world from 2000–2005. In 2005 a delegation from UNESCO visited the island and found that although no progress had been made in development of a management plan, the intrinsic conservation values of the site were not unduly threatened (Tabbasum and Dingwall 2005). IUCN noted at the time of inscription, customary ownership 'can be more conducive to conservation than if the land was under control of a distant government office' (IUCN 1998:81).

A management plan for East Rennell was finally completed in 2007 utilising the 'Resource Management Objectives and Guidelines for East Rennell' prepared at the time of nomination and with support from the UNESCO World Heritage Fund (Wein 2007). The management plan was prepared through a participatory process with the community and recognises the central role of traditional owners in conservation of the biodiversity of East Rennell. However, it takes what we identify as a customary system *in* management approach, integrating customary owners within management framed by the Western scientific values about 'nature'. The customary systems that have protected the values of the property are not documented or utilised in the management plan although it does suggest that 'documentation of past traditional management practices for Rennell is needed to provide an appropriate blend of traditional practices and contemporary community-based management practices' (Wein 2007:16).

On East Rennell, like many other places in the Pacific, the conservation initiative including the development of the World Heritage nomination has been driven by non-local interests. In satisfying the requirements of the World Heritage Committee and IUCN, the values of the property have been articulated according to Western scientific discourse with little investigation of how these values may intersect with local understandings of landscape and resources. Under the 2007 management plan, the local management authority is the East Rennell World Heritage Trust Board (now reformed as the Lake Tegano World Heritage Site Association or LTWHSA), a representative community organisation established in a previous form during the development of the nomination in the late 1990s. In accordance with the management plan, the LTWHSA with the assistance of international volunteers has implemented a community engagement process and education programs (Gabrys in press), raising community awareness of World Heritage and conservation of natural resources. But many in the community continue to be unaware of the existence of the management plan and those that are, are unclear about what it may mean in their daily lives. In discussions with the community of East Rennell in 2010, Smith (2011) found a widespread misconception that Rennellese culture was included in the values of the World Heritage inscription – the community does not distinguish nature from culture – and an urgent need for all documentation relating to the World Heritage values and their management to be translated into the Rennellese language. Further although LTWHSA is an inclusive and representative body in a Western democratic sense, the relationship of LTWHSA to the traditional authority, the East Rennell Council of Chiefs is unclear. This appears fundamental to future management of the property as it is the Council of Chiefs who are the decision making authority in relation to land and resource disputes.

Although the inscription of East Rennell on the World Heritage List was a landmark in international recognition of customary land tenure, at the time there were no established precedents or processes within World Heritage for articulating or assessing customary management systems for the protection of 'natural' values. Customary access to, use and management of resources is embedded within wider cultural practices of the Rennellese that have conserved the 'natural' values of the island. In East Rennell the local community, with the support of LTWHSA and the Council of Chiefs, is now considering a project to record their traditional knowledge to provide a framework in which World Heritage values of the site are articulated and managed through cultural practices of the community. This will be a 'catch-up' process initiated and directed by the community. The livelihoods of the community are under pressure, food security is tenuous and the increasing need and desire to participate in the cash economy is placing pressure on the island's resources. The World Heritage status of the island has not and will not alter this situation but the local community consider that greater recognition of their unique culture in the World Heritage system will protect not only the values for which their island is considered to be World Heritage but the reasons why those values exist.

Customary management within the holistic cultural framework of Pacific communities makes sense, yet it continues to be at odds with the way institutional structures approach heritage conservation, within national governments; in international non-governmental organisations; and through international conventions. Organisations such as the South Pacific Regional Environment Program and the Secretariat of the Pacific Community have responsibility for natural or cultural heritage conservation respectively at the regional level. Most Pacific Island governments have separate ministries or agencies responsible for natural and cultural heritage and the funding programs of many international non-government organisations reinforce these divisions.

This is also the case in the World Heritage system despite the original logic behind the World Heritage Convention, bringing together nature conservationists and those concerned with the protection of cultural property and monuments. To some extent this has been addressed through the inclusion of 'cultural landscape' as a category of site in the World Heritage system in 1993. Cultural landscapes are defined as properties that reflect 'the combined works of man [sic] and nature' and explicitly recognise the cultural values that have created and given meaning to landscapes. Not surprisingly, a number of Pacific Islands states have included cultural landscapes in their tentative lists of potential World Heritage sites. Various places in the Pacific undoubtedly have outstanding 'natural' values but there are no 'natural' landscapes in the Pacific Islands and as the recent 'Maupiti Declaration' (2009) affirms, for many Pacific Islanders, the ocean is a cultural seascape.[2]

As mentioned previously, since the inscription of East Rennell, four further properties in independent Pacific Island states have been successfully nominated to the World Heritage List. Chief Roi Mata's Domain, Vanuatu and Kuk Early Agricultural site Papua New Guinea were inscribed as cultural landscapes in 2008 and managed according to what we identify as customary systems of management. Bikini Atoll in the Marshall Islands, also a cultural landscape, and Phoenix Islands Protected Area, Kiribati, a primarily marine property, were inscribed in 2010 with management plans taking customary systems *in* management approaches. Notwithstanding that these four properties have very different management issues for example the tiny country of Kiribati is attempting to manage over 40 million ha of the Pacific Ocean while in Vanuatu a major threat to the values of Chief Roi Mata's Domain is the leasing of adjacent land for development, in each there is negotiation between customary land (or sea) owners and non-customary legal mechanisms for protection is ongoing. There has not yet been sufficient time to assess the implications of their different management approaches for long-term protection of the sites.

In developing the management plan for Chief Roi Mata's Domain, Meredith Wilson writes that the act of engaging in conservation management planning can be an empowering one:

> Throughout this process the project team has supported the community in developing a Plan of Management that is based upon the reinforcement of existing customary systems of marine and land tenure. The engagement of local communities in the development of appropriate management regimes ... is an approach that ensures that the local community remains in control of the management of its own resources. (Wilson 2006:7)

'Remaining in control' is key here. If the establishment of a formal heritage property/ protected area requires new management entities (and it does not follow that this is necessarily the case) these entities should be coherent with existing local governance structures. Decision making within the Lelepa region of Vanuatu in which Chief Roi Mata's Domain (CRMD) is located is vested in the power of the chiefs. CRMD is managed as a joint venture of the land owning groups, the Chiefs and landowners having approved the formation of the World Heritage and Tourism Committee to implement the management plan. The World Heritage and Tourism Committee is composed of representatives from the villages and external advisors including the Vanuatu Cultural Centre (Wilson 2006; Wilson *et al.* in press). Similarly, the management plan for the Kuk Early Agricultural site specifies the use of traditional clans and relationship to land to appoint people responsible for management. Under the plan, heritage officers are drawn from, and proposed by, the three main clans that comprise the Kawelka customary owners. Each officer will be responsible for overseeing day-to-day activities and for ensuring compliance with the management plan within a discrete area of their clan's land (Muke *et al.* 2007). At the heart of the CRMD management plan is customary protection through the system of *nafsan natoon* ('local lore') that refers to socially prescribed behaviour in relation to certain people, places and things (Wilson 2006:22). The management plan attempts to document this system. The Kuk early agricultural site is of international significance as a site of early agriculture, inscribed on

---

2   The full text of the Ocean Declaration is available from: http://www.temanaotemoana.org/UserFiles/File/Maupiti%20Ocean%20 Declaration-Final.pdf

the World Heritage List as an organically evolved cultural landscape that demonstrates changing land use practices through time. As Muke *et al.* (2007:332) point out, management of the site through continuation of traditional agriculture and its transformations is considered to enhance the significance of the site.

A challenge, particularly for conservationists (and for the World Heritage Committee), is examining how traditional management structures do the equivalent work of a Western conservation programme. Community approaches to conservation may seem inconsistent or incompatible when viewed through the lens of an international conservation regime. Assessment is coloured by preconceived expectations of best-practice in management. This includes requirements for governance, such as committees, reports and plans, which stand as representative of proper management. For example, when assessing a nomination the World Heritage Committee takes the presence of an explicit heritage administrative structure and planning documents as signifying that the site is being 'managed'. These may not be superfluous to a coherent customary management system, but a site being well managed under customary tenure may not have the need for such heritage management structures and tools (cf. Govan *et al.* 2009).

National governments too require bureaucratic flexibility in this regard. To fulfil reporting requirements under the World Heritage Convention, Pacific governments are required to explain to the World Heritage Committee how sites are being protected and provide evidence of this, while at the same time, acknowledging a devolved local customary management system. National legislation for the protection of natural and cultural heritage in the region is at best limited and *ad hoc*. Given that the rights of customary owners are enshrined in the Constitutions of many Pacific Island States, the usefulness of heritage legislation developed outside the region as a model for the Pacific states is limited.

The approach adopted for the Kuk Early Agricultural Site provides an example of how customary law may support and may itself be supported through legislation. Under the Constitution of Papua New Guinea land is automatically owned by local communities who traditionally resided in an area. Although at Kuk the land has been formally owned by the Government under a lease arrangement since 1968, through the management plan the government acknowledges the rights of the Kawelka customary owners and other groups to occupy and use the land in traditional ways (Muke *et al.* 2007:330). The framework for legal protection for the World Heritage property is a combination of national legislation including the *National Cultural Property (Preservation) Act 1965 (PNG)*, the Papua New Guinea *Conservation Areas Act* (1978) in combination with an Organic Law, enacted by the local Provincial Government, provides a means of reinforcing customary management in legal terms:

> The Organic Law empowers local communities to generate binding laws to protect their own cultural and natural resources (Section 43). Laws are generated by local communities, approved by the local government council…endorsed by the provincial government…and subsequently made into law by the NEC. (Government of Papua New Guinea 2007:63–4)

This Organic Law (not yet in force but see Denham in press) provides a structure for representing local practices in terms expected in conservation management planning. It is also a means of legally binding and promoting ongoing traditional uses within the local community (who, we should be clear to emphasise, have varied and conflicting attitudes to ownership and use of the site. See Strathern and Stewart 1998; Denham in press). In the Solomon Islands there is still no national heritage legislation to protect World Heritage sites. However, legislation in some ways similar to an Organic Law is being drafted for East Rennell, in the form of a Provincial Government Resource Management Ordinance. This will provide for protection of natural resources through prohibition on activities having adverse environmental effects identified in a Resource Order requested by customary owners and enforced by the Provincial Government. Under the draft Ordinance, a customary land owning group may make its own policy statements and plans regarding the use of resources.

The Provincial Government Ordinance may serve to reinforce the central role of customary owners in management, but the system of customary land tenure on East Rennell (essential to the Ordinance) and the relationship of cultural understandings of the landscape to this legal framework remain to be documented. Bringing differing world views together is not straightforward. Paige West's work in Papua New Guinea describes the confusion between locals and conservationists participating in a Wildlife Conservation project where notions of 'species' and 'biodiversity' have no equivalent in local understandings of the environment (West and Brockington 2006). However, Simon Foale's work in Melanesia led him to reflect that rather than contribute to the potential loss of local ecological knowledge through assessing and articulating heritage values in non-local terms, the act of defining heritage values in local terms may be a useful means of recording and (re)discovering local environmental knowledge (Foale 2008).

Given this, a more appropriate approach to protection through legislation with its genesis in the region may be that of the *Model Law for the Protection of Traditional Knowledge and Expressions of Culture* developed by the Secretariat of the Pacific Community (2006) as a framework for governments considering national legislation for the protection of Indigenous culture. Although the emphasis is on intangible and movable heritage, the Model Law has been designed to recognize and strengthen the continuing traditional land tenure in the region and the provisions supporting customary and traditional governance are relevant to our discussion of landscapes. These include:

> Encourage the use of customary laws and systems and traditional governance and decision making systems as far as possible, and recognise that communities will always be entitled to rely exclusively or in addition upon their own customary and traditional forms of protection … ; and,

> Recognise that the continued uses, exchange, transmission and development of Traditional Knowledge and Expressions of Culture within the customary context by the relevant traditional community, as determined by customary laws and practices, should not be restricted or interfered with. (SPC 2004:13)

Discussion

The realities of living in an area and conserving it at the same time require negotiation, whatever the system of management. Local social and economic circumstances dictate the priorities of a community and in turn inform the processes and outcomes of conservation. In the case of East Rennell the remoteness of the island and difficulty of access has limited development, fostered the continuation of traditional gardening and use of local resources and, as a consequence, protected the island's biodiversity. On the other hand, the remoteness of the island limits communication and makes access to education, employment opportunities and the cash economy extremely difficult and, in the face of increasing food insecurity on the island, is a potential trigger for unsustainable resource use such as logging and mining.

The designation of World Heritage means little in itself if the customary owners are not satisfied. Local priorities, not least of which is sustainable livelihoods, direct community engagement with non-local heritage conservation projects and regimes. Customary systems of decision making and management may assist communities to negotiate pressures such as those faced by East Rennell. However, any romanticised notions of traditional ecological practices held by outsiders need to be tempered with concern for livelihoods and the realisation that conservation, as it is understood by non-locals, might not necessarily be sympathetic with local expectations of heritage conservation. Furthermore, aspects of customary management may conflict with outsider interests especially in relation to social equity. Opportunities for the participation of all community members may be lacking in customary management approaches especially given that in the Pacific traditional authority may be inherited and older men are commonly privileged in decision making.

More broadly, provincial and national government and regional organisation (under resourced as they may be) need to ensure that the legislative and institutional structures they create promote and integrate customary owners and traditional practices. To achieve this, there is a real need for better coordination and integration of cultural and natural heritage protection and conservation in the region. A practical beginning is to work on methodologies for conserving cultural and natural values in an integrated system of heritage management that would complement the ecological landscape assessment work of conservationists with cultural mapping techniques and community definition of values and aspirations. Members of the community of East Rennell have expressed their wish to develop a project to record their cultural values including land tenure, environmental knowledge, traditional resource use, crafts, songs and dance to provide an umbrella framework in which their 'natural' values would be managed (Smith 2011).

To date the World Heritage sites in the Pacific have not been inscribed for sufficient time for longer-term issues in the protection and management regimes of the properties to arise. It is perhaps too early even to examine whether the structures defined on paper actually represent the way decisions are made in practice. Nevertheless, engagement has been productive in the sense that it has helped to broaden practices under the World Heritage Convention. Over time this may lead to different approaches within the Pacific through new ways of structuring the relationship between customary land managers and the state, and new ways of forming alliances in the conservation of sites of local and global heritage value. Considering Community as the 5th 'C' strategic objective of the World Heritage Committee is more than promoting community consultation or including community issues in management frameworks, it is also about who is driving the conservation project, who initiated it, and whether it meets local needs and aspirations. In defining mechanisms for customary management we seek a meaningful and practical articulation of community.

References

Beltrán, J. 2000. *Indigenous and traditional peoples and protected Areas: Principles, guidelines and case studies*. IUCN World Commission on Protected Areas Best Practice Protected Area Guidelines Series No.4.

Borrini-Feyerabend, G., Kothari, A. and Oviedo, G. 2004. *Indigenous and local communities and protected areas: Towards equity and enhanced conservation*. IUCN, Gland.

Clark, S., Bolt, K. and Campbell, A. 2008. Protected areas: an effective tool to reduce emissions from deforestation and forest degradation in developing countries? World Conservation Monitoring Centre: UNEP Working paper, 16 May 2008 Revision.

Corrigan, C. and Granziera, A. 2010. *A handbook for the Indigenous and community conserved areas registry*. World Conservation Monitoring Centre – United Nations Environment Program. Version 1.

Denham, T. In press. Building institutional and community capacity for World Heritage in Papua New Guinea: The Kuk Early Agricultural Site and Beyond. In: Smith, A. (ed.), *World Heritage in a sea of islands*. World Heritage papers. UNESCO, Paris.

Foale, S. 2008. Conserving Melanesia's coral reef heritage in the face of climate change. *Historic Environment* 21(1):29–36.

Gabrys, K. In press. Community and governance in the World Heritage property of East Rennell. In: Smith, A. (ed.), *World Heritage in a sea of islands*. World Heritage papers. UNESCO, Paris.

Govan, H., Tawake, A., Tabunakawai, K. Jenkins, A., Lasgorceix, A., Techera, E., Tafea, H., Kinch, J., Feehely, J., Ifopo, P., Hills, R., Alefaio, S., Meo, S., Troniak, S., Malimali, S., George, S., Tauaefa, T., Obed, T. 2009. *Community conserved areas: A review of status and needs in Melanesia and Polynesia*. ICCA regional review for CENESTA /TILCEPA /TGER /IUCN/ GEF-SGP.

Govan, H., Tawake, A. and Tabunakawai, K. 2006. Community-based marine resource management in the South Pacific. *Parks* 16(1):63–67.

Hails, C. 2007. The evolution of approaches to conserving the World's Natural Heritage: The experiences of WWF. *International Journal of Heritage Studies* 13(4):365–379.

Hviding, E. 2006. Knowing and managing biodiversity in the Pacific Islands: Challenges of environmentalism in Marovo Lagoon. *International Social Studies Journal* 187.

IUCN. 1998. Documentation on World Heritage Properties (Natural). http://whc.unesco.org/en/sessions/22COM/documents/

IUCN. 2008. Indigenous Peoples' and Community Conserved Areas (ICCAs) –Guidelines towards appropriate recognition and support.

Muke, J., Denham, T. and Genorupa, V. 2007. Nominating and managing a World Heritage Site in the highlands of Papua New Guinea. *World Archaeology* 39(3):324–338.

Palmer, L. 2006. 'Nature', place and the recognition of Indigenous Polities. *Australian Geographer* 37(1):33–43.

Papua New Guinean Government. 2007. *The Kuk early agricultural site: A cultural landscape*. Nomination for consideration as World Heritage Site. Ministry of Environment and Conservation, Port Moresby.

Secretariat of the Pacific Community. 2006. *Guidelines for developing national legislation for the protection of traditional knowledge and expressions of culture based on the Pacific Model Law 2002*. Noumea.

Smith A. In press. The Pacific 2009 World Heritage Program. In: Smith, A. (ed.), *World Heritage in a sea of islands. World Heritage papers*. UNESCO, Paris.

Smith A. 2011. East Rennell World Heritage Site: misunderstandings, inconsistencies and opportunities in the implementation of the World Heritage Convention in the Pacific Islands. *International Journal of Heritage Studies* 17(6):592–607.

Smith, A. and Jones, K. 2007. *Cultural landscapes of the Pacific Islands*. ICOMOS, Paris.

Strathern, A.J. and Stewart, P.J. 1998. *Kuk Heritage: Issues and debates in Papua New Guinea*. PNG National Museum, Port Moresby.

Tabbasum A. and Dingwall, P. 2005. Report on the mission to East Rennell World Heritage Property and Marovo Lagoon, Solomon Islands. 30 March–10 April, 2006 Unpublished report, UNESCO World Heritage Centre, Paris.

te Heuheu, T. M., Kawharu, M. and Tuiheiva, R. 2012. World Heritage and Indigeneity. *World Heritage* 62:44–50.

UNESCO. n.d. World Heritage Centre Activities – Pacific 2009 Programme. http://whc.unesco.org/en/pacific2009

UNESCO. 1999. Report of the 22nd session of the World Heritage Committee, Tokyo. WHC-98/CONF.203/18. UNESCO, Paris.

UNESCO. 2007a. *Appeal to the World Heritage Committee from Pacific Island state parties.* http://whc. unesco.org/en/sessions/31COM/documents/ (Document: whc07-31com-11ce[1] Annex 1).

UNESCO. 2007b. *Proposal for a 'Fifth C' to be added to the strategic objectives.* http://whc.unesco.org/en/ sessions/31COM/documents/ (Document: whc07-31com-13be[1].

UNESCO. 1972. *Convention concerning the protection of the world's cultural and natural heritage.* Adopted by the UNESCO general conference at its seventeenth session.

UNESCO. 2009 *Operational guidelines for the implementation of the World Heritage Convention.* World Heritage Centre.

Vanuatu, Republic of. 2006. *Chief Roi Mata's Domain.* Nomination by the Republic of Vanuatu for inscription on the World Heritage List.

Wein, L. 2007. *East Rennell World Heritage Site Management Plan.* East Rennell World Heritage Trust Board, Solomon Islands.

West, P. and Brockington. D. 2006. Some unexpected consequences of protected areas: an anthropological perspective. *Conservation Biology* 20(3):609–616.

Wild, R. and McLeod, C. 2008. *Sacred natural sites – Guidelines for protected area managers.* Task Force on the Cultural and Spiritual Values of Protected areas in collaboration with UNESCO's Man and the Biosphere Programme. Best Practice Protected Area Guidelines Series No.16.

Wilson, M. 2006. *Plan of Management for Chief Roi Mata's Domain (CRMD).* Written on behalf of the World Heritage and Tourism Committee (WHTC), Lelepa Region.

Wilson, M., Ballard, C., Matanik, R. and Warry, T. In press. Community as the First C: Conservation and development through tourism at Chief Roi Mata's Domain, Vanuatu. In: Smith, A. (ed.), *World Heritage in a sea of islands.* World Heritage papers. UNESCO, Paris.

3

Poetics and politics: Bikini Atoll and World Heritage Listing

Steve Brown, Office of Heritage and Environment, NSW, Australia

'No literary work of imagination could create a more monstrous evil.' (Zinn 2010:29)

"Social Poetics': an approach that opens up the tensions between official models of national culture and the lived experience of ordinary citizens.' (Herzfeld 1997)

Bombs and bikinis

It is strange the way that 'bikini' simultaneously references nuclear test site and itsy-bitsy bathing costume. Yet the two are intimately connected – *le bikini* was revealed at a public pool in Paris on 5 July 1946, four days after the globally publicised Able Atomic Test explosion was unleashed at Bikini Atoll, Republic of the Marshall Islands (Figure 1). The atoll's name was appropriated for *le bikini*, a garment promoted as a vision of the *vestiges* (meaning traces, remnants or relics in French) of clothing that would remain after experiencing an atomic explosion (Cameron 1970 in Davis 2005a:615–616). Ironically, the bathing costume would, during the 1950s, come to be associated with untouched and romanticised sun-drenched tropical islands.

The original *Bikini* translates from Marshallese as 'the lands of many coconuts' and refers to the huge groves of trees visible on the horizon to ancient mariners as they approached the atoll islands (Jack Niedenthal pers. comm., 2011) (Figure 2). The reality of humanly-willed catastrophe of bomb testing on Bikini Atoll between 1946 and 1958 was the deaths of, and ill-health effects for, Marshallese, American and Japanese people. In addition, all of the original coconut groves on the twenty three surviving atoll islands were destroyed.

On Sunday 1 August 2010, during the 34th session of the World Heritage Committee (in Brasília), *Bikini Atoll Nuclear Test Site*, Republic of the Marshall Islands (RMI), was inscribed on the World Heritage List as an outstanding example of a nuclear test site and as a source of globally significant cultural symbols and icons of the twentieth century (Decision 34COM 8B.20, World Heritage Committee 2011). This act, I suggest, was simultaneously poetic and political: the property was at once imagined as wasteland (a landscape ravaged and made dangerous by technologies of the modern era) and wonderland (a remote landscape conceptualised as deserted, exotic and earthly paradise), a subversive space coopted as 'heritage' by both global and local actors. The UNESCO World Heritage project, driven by a credible and representative world heritage ethos, sought out the remote atoll of Bikini for listing. The global attention listing brought was in turn conscripted by Bikinians displaced from their homeland to further their goals of being heard, recognised and supported.

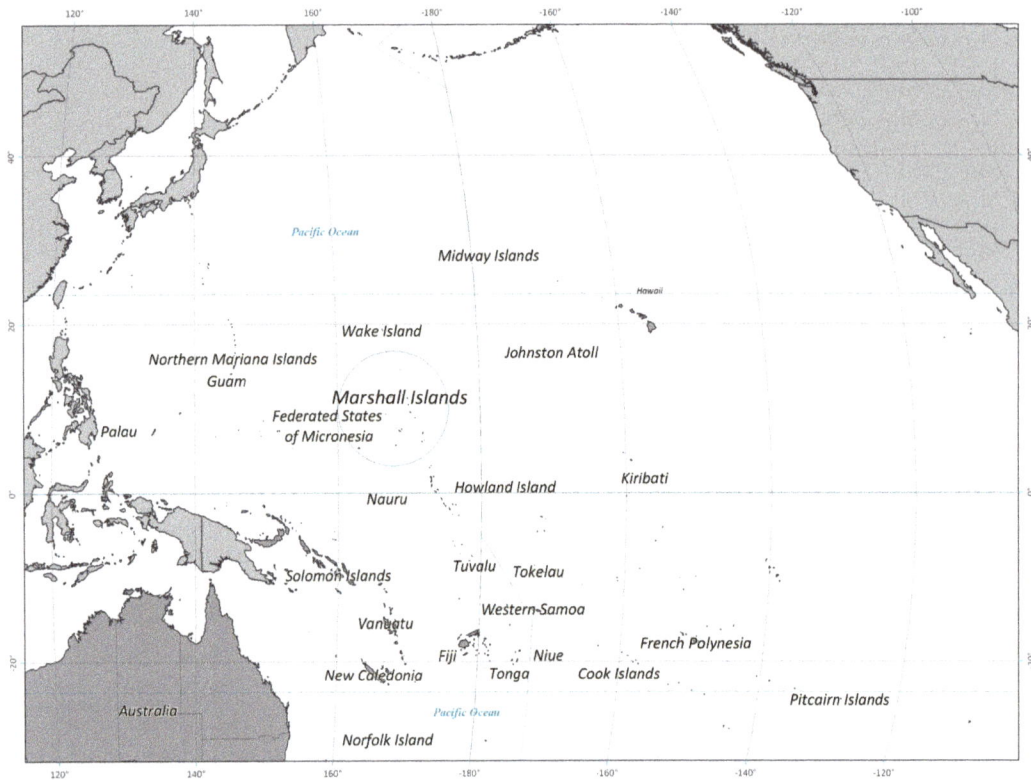

Figure 1. Map of Pacific region, showing location of the Marshall Islands.

Source: RMI 2009.

In the case of Bikini Atoll there is a cultural engagement between global and local in world heritage listing that comprises complex processes of creative cooptation in political, economic, and administrative practices (Hertzfeld 1997:3). There is an element of social poetics, a term used by the anthropologist Michael Herzfeld, in the way that local Bikinian people find advantage through engaging with the world heritage project and global heritage practice. The loss of homeland through nuclear testing is at the heart of a struggle to be supported; a struggle that is necessarily a part of Bikinian identity.

However, at its core, the history and heritage enlisted to raise the atoll to world heritage status – the 'technical ensemble' of bomb-testing along with the associations with globally significant cultural symbols and icons is not strictly the history and heritage of the Bikinian people. The Indigenous inhabitants of Bikini Atoll were shipped out to make way for the United States' nuclear test program of 1946 and 1954–1958. Radioactive contamination has meant that the Bikinians have never been able to permanently return to their homeland. Nuclear testing *became* Bikinian heritage, a heritage not of their making.

In this paper I explore the complexities of the nuclear test period heritage of Bikini Atoll and its management. I describe the parallel, yet slippery, universes of local and global in the recognition, protection and management of Bikini's cultural landscape/seascape. I argue that entangled local and global management regimes have initially established a mutually beneficial arrangement for UNESCO and for the Bikinian community. However, a possible future resettlement of the atoll has a potential to destabilise this relationship because of the tensions likely to be created between the rights of Bikinian people to reoccupy their homeland and the inevitable impacts on the universal values of the property.

Figure 2. Boat and groves of coconut trees, Bikini Atoll.

Source: US Government, 1946.

The historical context is that, after atomic bombs were dropped on Hiroshima and Nagasaki in 1945, nuclear testing in the Pacific took place almost continuously between 1946 and 1996 (Smith 2007). At each test location Indigenous populations were displaced. The USA chose Bikini Atoll as the site for the first major atomic experiments. Two atomic bomb tests ('Operation Crossroads') were undertaken at the atoll in July 1946 and a further twenty-one nuclear (hydrogen) bomb tests between 1954 and 1958 (RMI 2009:29–44).

The history of nuclear testing at Bikini Atoll, beyond the summary above, is not detailed in this paper. Anita Smith (2007) has written an overview of colonialism and the bomb in the Pacific. There are a number of descriptive narratives of the Marshall Islands tests (Kiste 1974; Weisgall 1994; Carucci 1997; Simon 1997; Barker 2004; de Groot 2006), several outstanding documentaries (e.g. Academy Award nominated *Radio Bikini* [Stone 1988]), and contemporary published accounts (e.g. Bradley 1946) as well as newspaper coverage analysis (Keever 2004) and popular magazine stories (Davis 2005a). There is also a huge archive of primary source material on the testing, including declassified US government documents and film. There are a number of accounts of the Bikinian relocation and forced migration story (Weisgall 1980; Niedenthal 1997), including oral histories with Bikinian Elders (Niedenthal 2001), anthropological studies (Mason 1950, 1954) and recent articles that contextualise the Bikinian's historical experience read from different, including geographic and folkloric, perspectives (Davis 2005a, 2005b; McArthur 2008). A description of the terrestrial material traces of nuclear testing on Bikini Atoll can be found in Brown (2013).

Indigenous peoples, heritage and free, prior and informed consent

> The common understanding of the right to free, prior and informed consent is that consent should be given freely, without coercion, intimidation or manipulation (free); sought sufficiently at all stages, including from inception to final authorization and implementation of activities (prior); based on an understanding of the full range of issues and implications entailed by the activity or decision in question (informed); and given by the legitimate representatives of the Indigenous peoples concerned. (UNPFII 2011:7[34])

Sunday 10 February 1946

On this particular Sunday after church, the United States military governor of the Marshall Islands, Commodore Ben H Wyatt, addressed the assembled Bikinian community. Wyatt asked if they were willing to leave their atoll *temporarily* so that the United States could begin testing atomic bombs 'for the good of mankind and to end all World Wars'. Juda, the *iroij* or leader of the Bikinian people ' … after much confused and sorrowful deliberation among his people' (Niedenthal 1997:29), announced: 'We will go believing that everything is in the hands of God.' There was no signed agreement. Commodore Wyatt reported by cable: '… their local chieftain, referred to as King Juda,[1] arose and said that the natives of Bikini were very proud to be part of this wonderful undertaking' (Weisgall 1994:107).

Figure 3. Filming of staged re-enactment of Commodore Wyatt addressing the assembled Bikinian community, 6 March 1946.

Source: US Government, 1946.

---

1   Tobin (1953:20–25) provides an account of how the use of the term 'King' from this time was read by Bikinians as official recognition of Juda's sovereignty on Bikini by the American government. This position strengthened the case for Juda as the 'rightful ruler' over a rival's claim to be paramount chief.

On 6 March 1946, Commodore Wyatt staged a filmed reenactment of the February 10 meeting in which world peace was exchanged for homeland (Figure 3). The next day, the 167 Bikinians inhabiting the atoll along with their belongings were shipped out (Figure 4) to Rongerick Atoll, a place believed to be inhabited by evil spirits. Two atomic bomb tests were undertaken as part of *Operation Crossroads* in July 1946. Juda was brought back to Bikini Atoll by the US to witness the explosion of the second and larger bomb test, the Baker Atomic Test on 25 July 1946 (Niedenthal 2001:3).

No one today would argue that the homeland for world peace transaction reflected free, prior and informed consent on the part of the Indigenous population. Anthropologist JA Tobin (1953:8) suggested that, in 1946, for the Bikinians there was an historical conditioning to 'obedience' after a quarter of a century of autocratic Japanese rule. Thus it seems reasonable to surmise that negotiating with men in military uniform like Commodore Wyatt would have been intimidating, regardless of the colonialist context or power dynamic of the engagement (cf. Smith 2007:56, 69).

Tobin's 1953 report emphasises the firm belief held amongst the Bikinian people that their removal was only temporary, and that they would be returning to their ancestral home, a return that the Bikinians believed (in 1953) was imminent. In this regard Tobin's findings reflected those expressed in anthropological reports previously provided to the US Trust Territory administration (e.g. Mason 1950; Drucker 1950 in Tobin 1953:13–15). Tobin had been part of a mission to Kili Island in January 1951 to negotiate 'the release of rights to Bikini', which the Bikinian Council 'unequivocally and unanimously refused to consider' (Tobin 1953:17).

Figure 4. Loading a boat onto US naval vessel for departure from Bikini Atoll, March 1946.

Source: US Government, 1946.

Bikinians today continue to believe that a return to their homeland will happen. The memory of Sunday 10 February 1946, now some 67 years distant, is forever inscribed on the flag of the people of Bikini – *Men Otemjej Rej Ilo Bein Anif* ('everything is in the hands of God') (Niedenthal 2001:175).

Wednesday 29 June 2011

On this day representatives of the United Nations Permanent Forum on Indigenous Issues (UNPFII) and of the International Work Group for Indigenous Affairs (IWGIA) addressed the 35th session of the UNESCO World Heritage Committee in Paris. The representatives voiced their concerns about making the implementation of the World Heritage Convention (UNESCO 1972) consistent with the United Nations Declaration on the Rights of Indigenous Peoples (United Nations 2007).

In May 2011, UNPFII called on UNESCO, IUCN, ICOMOS and ICROM ' … to review current procedures and capacity to ensure free, prior and informed consent, and the protection of Indigenous peoples' livelihoods, tangible and intangible heritage' with regard to World Heritage Listing (UNPFII 2011:8[36]). These concerns are expressed in a 'Joint Statement on Continuous violations of the principle of free, prior and informed consent in the context of UNESCO's World Heritage Convention' (IWGIA 2011). The joint statement was submitted to UNESCO's World Heritage Centre and the Bureau of the World Heritage Committee. The issues are encapsulated in the following extract from the joint statement:

> Last year, at the World Heritage Committee's 34th Session in Brasilia (25 July–3 August 2010), the Committee inscribed two sites on the World Heritage List although questions had been raised regarding Indigenous peoples' participation in the nomination processes and their free, prior and informed consent: the Northwest Hawaiian Islands Marine Monument ('Papahānaumokuākea Marine National Monument') and the Ngorongoro Conservation Area in Tanzania. The latter was re-inscribed as a cultural World Heritage site, because of its significance as an archaeological site, not because of the significance of the Maasai culture. We are concerned that the Committee's recognition of only the archaeological values, and not the living cultural values of the Indigenous residents, may exacerbate the already existing imbalances in the management framework for the Ngorongoro Conservation Area and lead to additional restrictions on the livelihoods of the Indigenous residents and further infringements on their rights.

In response to the joint statement and to the representations of the global Indigenous NGOs, the World Heritage Committee encouraged States Parties to respect the rights of Indigenous peoples and local communities and involve them in World Heritage nomination and management processes (Decision 35COM 12D, World Heritage Committee 2011).

Local rights and global interventions

The actions that took place on 10 February 1946 on Bikini Atoll and on 29 June 2011 in Paris illustrate aspects of the changing order of global politics with regard to the power of Indigenous people. The Cold War testing of nuclear devices in the Pacific in the second half of the twentieth century was part of a larger and longer history of imperialism in the region, though also a period when Pacific island nations gained independence (Smith 2007). In the case of the Marshall Islands, where the United States continues to occupy Kwajalein Atoll for military projects in return for rent payments, there is a complex interplay between independence from, and dependence on, the United States following the signing of the Compact of Free Association 1986 (US Department of Interior 1986), which was renewed in 2003 for 20 years.

Akin to the Compact of Free Association, the creation of the UNESCO World Heritage Convention (WHC) in 1972, and its subsequent operation (particularly through the 1970s and 1980s), is viewed by some commentators as a 'neo-colonial' project. This is illustrated for example in the way the WHC promotes the idea of 'universality', a concept deeply rooted in European cultural tradition (Cleere 2001:24), in its adoption of the notion of Outstanding Universal Value (OUV). Laurajane Smith, for example, argues that the concept of universality is 'deeply rooted in the process of colonization and imperial expansion and assumptions about the cultural and technological evolutionary achievements of the West' (Smith 2006:99). To give credit to the work of the WHC over forty years of its operation, it has responded actively to various challenges (e.g. a credible and representative world heritage list, discussed below) and evolution in understandings of the meanings of the idea of heritage (e.g. including intangible heritage – Blake 2009:66).

The global history of colonialism and neo-colonialism is inextricably enmeshed with issues of Indigenous rights. The story of Bikini Atoll and its inhabitants is a part of this history, a situation where the islanders were forced to leave their homeland, a place where Bikinian myth, history and genealogy, and identity are inscribed into landscape (McArthur 2008:264).

Local and State: Indigenous heritage and 'conserving' culture

Traditional system

Anthropologist Leonard Mason documented aspects of the Bikinian matrilineal social structure following their relocation of the community to Rongerick Atoll in March 1946 (Mason 1954; Mason in Niedenthal 2001:v–vi). At that time there were eleven Bikinian matrilineages (among a population of 196 people), each with an *alap* ('hereditary head') of which Juda was the head of the top matrilineage. The system changed after this time, and for example by 1957 land on the 81 ha island of Kili was divided amongst twenty kin or 'family groups' (*bamle*) (Mason in Niedenthal 2001:vi).

Concerned that knowledge of the traditional land system would be lost, as well as to get ready for a future contamination clean-up of the atoll prior to a re-settlement, a delegation of Bikinians returned in 1987 to re-establish the traditional land boundaries on Bikini (237 ha) and Enue (125 ha) islands. The 'parcels of land' (*weto*) run in 'strips' across the islands extending from the lagoon to the ocean sides (Holmes and Harver 1989) (Figure 5). Bikinian Jukwa Jakeo (in Niedenthal 2001:110) said of the re-inscribing process:

> The technical difficulties that we experienced in our attempts to re-establish the boundaries stem from the fact that all the natural surroundings and the markers that we used to delineate the land partitions are now gone. They were destroyed by the U.S. government and all of their atomic bomb testing. Today, when we draw the lines, we are using estimations only.

Within the context of the Indigenous customary land tenure on Bikini Atoll, it is clear that the terrestrial material traces of testing are located on land that is privately owned; and the impact of nuclear testing (land clearance, radioactive contamination) effects all land and landholders. Needless to say, the Bikinian customary land tenure system does not provide protective mechanisms for USA-created nuclear test features, heritage features that were not created by Bikinians but none-the-less are the property of customary landowners.[2]

---

2   There is a gendered element to this situation. The Bikinian social system is a matrilineal one yet the delegation to re-establish land boundaries on Bikini Island in 1987 comprised elderly men (Niedenthal 2001:110). The way gender roles work in land ownership, and in the ownership of the nuclear test structures, is beyond the scope of this paper. I confess to having little knowledge on this topic.

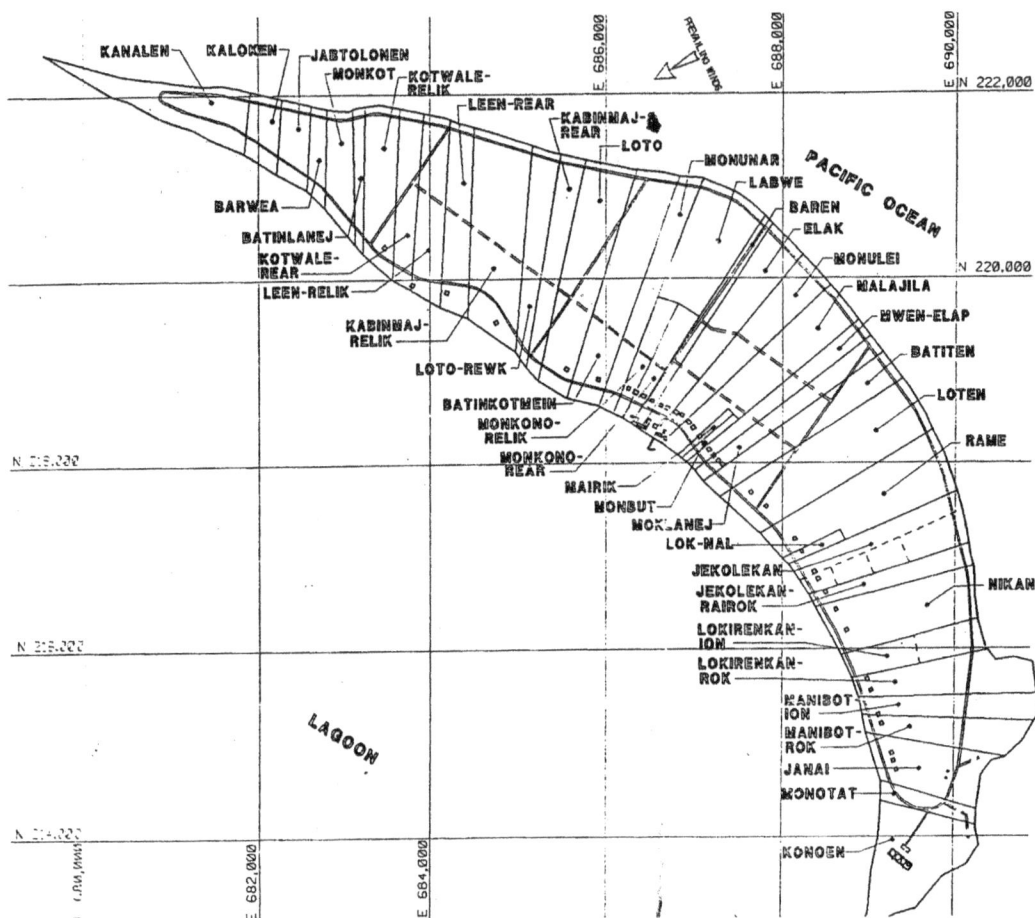

Figure 5. Diagram showing re-established land boundaries (weto) on Bikini Island.

Source: Holmes and Harver, 1988.

Legal system – State and local

The Republic of the Marshall Islands *Historic Preservation Act of 1991* and five regulations relating to this Act, largely based on USA laws, are the basis for managing cultural heritage at a national level in the RMI (Spennemann 2000). The legislation is administered through the Historic Preservation Office (HPO) which has the primary responsibility for identification, recording, protecting and preserving the tangible and intangible cultural properties of the RMI. No 'cultural and historic property' for Bikini Atoll is currently included in the RMI National Register of Historic Places. It would appear therefore that historic sites, both on-land and submerged, relating to nuclear testing at Bikini Atoll are not protected under the *Historic Preservation Act of 1991*. However, protection is afforded to 'artefacts' removed from the atoll which may be seized if they are believed to be prohibited for export (Spennemann 2000).

The constitution of the RMI establishes a system of local government [*Constitution of the Republic of the Marshall Islands*, Article IX, Section 1(2)]. The Kili-Bikini-Ejit Local Government (KBE LG) may make ordinances for the area in its jurisdiction, which extends to the sea and the seabed of the internal waters of Bikini Atoll and to the surrounding sea and seabed to a distance of five nautical miles. The KBE LG has made ordinances regarding access to Bikini Atoll, illegal fishing, fish harvesting levels, wreck diving requirements and prohibiting removal of objects from Bikini lagoon (Baker 2010:13).

Since Indigenous customary land tenure on Bikini Atoll, like much of the Pacific, is 'extremely privatised', archaeologist Richard Williamson (2001:44) argues that within the RMI, '(w)ith few exceptions, the landowners, and not the government, have absolute control of land'. Williamson further argues that preservation of 'historic sites' – sites relating to colonial and military occupation – must therefore come through education rather than legislation.

Nuclear fleet

The complex interplay of history, ownership and heritage management regimes is illustrated in the case of the 'nuclear fleet' at Bikini Atoll. As part of *Operation Crossroads* in 1946, eighty-eight 'target' vessels were laid out across Bikini's central lagoon by the US in order to assess the effects of two atomic tests – Able and Baker. Twenty-one vessels were sunk in the 60 m deep lagoon with most clustered in the shallow crater formed by the Crossroads Baker Test. The *Compact of Free Association between the Government of the United States of America and the Government of the Republic of the Marshall Islands 1985* (Article 177) transferred ownership of all 'sunken vessels' (as well as 'cable'), including liability resulting from salvage operations, from the USA government to the people of Bikini (Spennemann 1990:22; Delgado *et al.*1991). Under the agreement, studies of the sunken fleet were conducted by the USA to assess environmental hazards (leaking fuel, structural condition of the ships, safety of live bombs and residual radioactivity) as well as an assessment of historic significance.

The US National Park Service (NPS) undertook an historic and archaeological assessment of nine vessels in 1989–90, focussing on the aircraft carrier *USS-Saratoga*, though also documenting to a lesser degree the battleships *HIJMS Nagato* (Japan) and *USS-Arkansas* and the US submarines *Pilotfish* and *Apogon*, as well as the German cruiser *Prinz Eugen*, which had been scuttled at Kwajalein Atoll after *Operation Crossroads* (Delgado *et al.*1991; Delgado 1996). All vessels were assessed to be historically and archaeologically significant and the establishment of an 'archaeological park' was recommended. The survey report (Delgado *et al.*1991) also outlines the risks associated with the detonation of unexploded ordnance and fuel associated with the sunken ships.

Between 1996 and 2008, the KBE LG established and operated a commercial diving enterprise, Bikini Atoll Divers, in part to provide an economic base for a possible future resettlement of Bikini Atoll. Typically the tourism operation ran from late April to November and involved groups of up to 12 divers visiting the atoll for a week, with the annual number of recreational divers numbering between 200 and 250. In addition to the diving, fishing and a tour of the atoll, the tourism operation emphasised the telling of Bikinian history. Documentary films as well as lectures were presented to the participants in order to share and perpetuate a local nuclear story.

In order to safeguard the moveable traces of the wrecks, the KBE LG enacted laws that made the removal of 'artefacts' from the sunken vessels illegal. Enforcement of the KBE LG's Ordinance No. 2–1996, relating to the removal of artefacts from the sunken vessels, is undertaken with regard to visitors in a number of ways. Firstly, access to Bikini Atoll is controlled by the KBE LG. Secondly, the 'Liability Release' form that all divers were required to sign made it clear that the taking of artefacts from any of the vessels is illegal and subject to a fine of US $5,000 for each artefact taken. The divers were required to also give permission to have their belongings searched. Thirdly, Mr Jack Niedenthal (Trust Liaison Officer, KBE LG) monitors E-bay, using 'Bikini' as an identifier, for any items appearing for sale.

In 2008 the KBE LG Council closed the Bikini Atoll Divers operation due to local airline reliability issues, soaring energy costs and US stock market conditions that impacted the local government budget. In 2011 the Local Government signed an agreement with a commercial operator (Indies Trader Marine Adventures) that would permit the commercial operator to operate their live aboard vessel, the *Windward*, on Bikini Atoll. The diving operation recommenced in June 2011.

The engagement of the Bikinians with the sunken nuclear fleet has served a number of purposes. As well as the economic and political benefits (touched on above), the Bikini Atoll Divers enterprise has enabled Bikinians to continue to revisit the atoll and maintain physical and spiritual connections. The project has also built local Bikinian capacity in cultural heritage management, largely undertaken with the support and assistance of international partners. In addition to the US National Park Service documentation of the sunken vessels (Delgado *et al.*1991), James Delgado has provided ongoing management and interpretation advice. Partnerships are proposed to monitor the ships and artefacts on them (with James Cook University and the Western Australian Maritime Museum, Australia) and to develop interpretation materials for the sunken vessels (Indiana University, USA).

The engagement of Bikinians with the nuclear fleet has a dimension of social poetics with regard to the way in which cultural heritage is recognised and managed in the Marshall Islands. While the focus of the RMI national heritage regime is on documenting and preserving traditional Marshallese culture, the Bikinians have 'sacrificed' a considerable amount of traditional knowledge (e.g. lagoon fishing skills, local ecological familiarity) as a result of nuclear testing. Bikini Atoll Divers arose from a conscious engagement with the atoll's nuclear test history, an engagement that provides benefits that include re-telling the contemporary 'nuclear nomad' story (cf. Niedenthal 1997).

Conservation measures

The present state of conservation of the material traces of nuclear testing is variable. The sunken ships and most of the bunkers are generally intact and well preserved, though are gradually deteriorating. For example, the bridge on the *USS Saratoga* is gradually collapsing and subsiding (Jack Niedenthal pers. comm., 2009); the 'Assembly Building' on Eneu Island, a large (c. 32 m x 15 m) wood framed and iron clad structure, was demolished in the early 1990s as it was considered unsafe; and terrestrial structures are continually impacted by tropical vegetation (cf. Spenneman and Look 2006).

No active conservation measures, including maintenance work, are being undertaken on the sunken ships, bunkers and other nuclear test period physical remains. The World Heritage *Nomination Dossier* (RMI 2009:63) and *Bikini Atoll Conservation Management Plan* (Baker 2010) make the point that the deterioration of the sunken ships and bunkers contribute to the distinctive character of the elements. These documents take a position that physical conservation works are not required to retain the OUV of the property. In general the current management system of 'do nothing' is effective, in part because of the lack of threats, the robustness of the tangible cultural heritage and the remoteness of the nominated property. However, there remains a need to document and monitor the nuclear test features, most likely achieved with the support of international partners.

UNESCO's credible and representative world heritage ethos

The nomination of Bikini Atoll is in large part a consequence of a process to address the under-representation of properties in the Pacific Islands Region on the World Heritage List. A series of World Heritage Global Strategy meetings for the Pacific Islands Region, commencing in 1997 (Smith and Jones 2007:6–13), have been undertaken as part of a process to develop a representative, balanced and credible World Heritage List (World Heritage Committee 1994; UNESCO World Heritage Centre 2004, 2008).

In 2007, ICOMOS published a thematic study, *Cultural Landscapes of the Pacific Islands* (Smith and Jones 2007). The publication was prepared to support possible nominations for World Heritage status through a summary of available evidence. Bikini Atoll is presented as a case study, emphasising the nuclear testing history and in particular the ships sunk in the lagoon as part of the 1946 atomic bomb tests (Smith and Jones 2007:86–87). The January 2009 Nomination Dossier (RMI 2009) for Bikini Atoll was prepared with the support of an 'International Preparatory Assistance Grant' from the World Heritage Centre in 2006.

On Sunday 1 August 2010, during the 34th session of the World Heritage Committee (in Brasília), *Bikini Atoll Nuclear Test Site*, Republic of Marshall Islands, was inscribed on the World Heritage List as an *outstanding example of a nuclear test site* and as a *source of globally significant cultural symbols and icons of the 20th century* (Niedenthal 2011). The bikini swimming costume is one such 'pop culture icon' (RMI 2009:52).

World Heritage listing has (re)materialised and (re)imagined Bikini Atoll by privileging 12 years of the atoll's history (1946–1958). The UNESCO synthesis of OUV recognises the Bikinian people's experience as part of the nuclear narrative of Bikini Atoll (http://whc.unesco.org/en/list/1339/). However, to my reading with reference to the way the nomination dossier relates the Bikini Atoll story (RMI 2009: 42–44, 55), the UNESCO account of the World Heritage meta-narrative fixes the heritage values more firmly in the materiality of history and global symbols/icons. The materiality of history – the sunken fleet, the Bravo bomb crater, and the concrete bunkers as well as the documentary record, evidences the nuclear narrative. The symbols and icons, such as the image of the mushroom cloud, the bikini, Godzilla, and the rise of international movements advocating nuclear disarmament, are global rather than local. In addition the tangible cultural heritage, seen from the gaze of the global NGOs, constructs the material expression of nuclear testing on Bikini Atoll as belonging *within* this particular landscape/seascape rather than belonging *to* a people (cf. Greenfield 1989 in Gillman 2010:47).

The slipperiness of local and global, history and prehistory

Folklorist and anthropologist Phillip McArthur (2008) has examined how Marshall Islanders make sense of their tumultuous, 'ambivalent' and 'destructive-creative' relationship with the United States. He argues that the islanders alternative story to official history (and this includes the world heritage meta-narrative) prompts consideration of the slipperiness of the terms 'local' and 'global'.

To examine the US-RMI relationship, McArthur interrogates the local use of a Marshallese trickster narrative (*Letao* – 'The Sly One'), a narrative that positions the local at the centre of the global. He relates the episode of the trickster's travels to America to bestow his power on the Americans. McArthur describes this process in the following way (2008:263, 266–267):

> The Marshall Islanders playfully dramatize their ambivalent history with Americans through a trickster narrative. By analogy the Americans, with their destructive-creative capacities, draw upon the cosmological power of the Marshallese trickster to become the most powerful and dangerous world chiefs... Thus, instead of positioning the local as invaded and dominated by more centralised structures, ... the Marshallese reverse the order of centralization (which can include globalization) in order to comprehend these structures on their own terms and to place themselves at the centre (Stoeltje and Baumann 1989). They do so through their own prehistorical narrative about their trickster figure ... creative Marshallese draw upon significant meanings and narratives to bypass a national identity and to explore new kinds of transnational values and meanings.

An important aspect of the narrative of *Letao* is that it relates to a cosmology of the past, a 'prehistoric' time (in Western terms) before the arrival of Spanish, Germans, Japanese and Americans. This dynamic and mobile Marshallese past 'continues to inform the globally connected present' because it recontexualises the trickster narrative in terms of modern historical events – 'a pre-historical production of the present' (McArthur 2008:267–268, 289). The *Latao* narrative appears to do more than act as an explanatory framework that weaves together the cosmos and history – it utilises the Indigenous system of meanings to position power at a local level.

The Marshallese are not passive victims of history or ignorant of the realities of global power. As noted by McArthur (2008:281), ' ... the memory of World War II, atomic bomb testing, and continued missile testing at Kwajalein are an indelible part of their experience with Americans.' Marshallese have considerable experience of American culture since many people travel to and from the US in accord with agreed rights under the *Compact of Free Association* (1986 [2002]). Additionally, the Bikinians have constantly engaged with Americans since 1946 – initially with the US Naval administration (1946–1947), subsequently with the US Trust Territory administration (1947–1986), as well as with the US federal administration and courts from 1975 when a first lawsuit was filed (Niedenthal 2001:12–14). Bikinians are incredibly experienced in world affairs and have regularly visited the US from the 1970s to meet with administrators, lawyers and politicians in relation to seeking, negotiating, and administering compensation packages. In addition, delegations of Bikinians attended the Academy Awards (when the documentary *Radio Bikini* (Stone 1988) was nominated for best documentary feature) and, also in 1988, a delegation of Bikinians visited Aboriginal groups impacted by British-Australian testing at Maralinga, Australia.

Thus, while worldly lived experience enables Bikinians to recognise their own marginality on the international scene, the *Latao* narrative simultaneously positions them at the centre of the global. There is a social poetics in this positioning that underpins the status quo between local Bikinians and global political, including World Heritage, actors.

Heritage, rights and resettlement

I have argued in this paper that the World Heritage Listing of Bikini Atoll for its nuclear test history is based on a cultural engagement between an Indigenous population (through a State Party) and the global NGO UNESCO that mutually benefits and advantages both parties. While Bikini Atoll remains relatively uninhabited and tourism remains small in scale (i.e. the specialised diving enterprise), there is equilibrium in the World Heritage project for both parties derived from the benefits that each receives.

What would be the effect on the symmetry of this relationship if Bikinians were to resettle the atoll? I examine this hypothetical question in order to elicit issues between, and challenges for, the global heritage project and local Indigenous rights. Before doing this, I briefly mention the key challenges to future resettlement of the atoll: radiation and climate change.

Resettlement of Bikini Atoll has, and continues to be, a goal of the Bikinian community. The key obstacle to this occurring has been radiation levels considered to be not 'really safe' (Davis 2005b:213), principally because of the bio-accumulation of radioactive cesium (^{137}Cs) in terrestrial food sources such as coconuts and coconut crabs. The science and politics of radiation at Bikini Atoll (and the RMI generally) is complex (Robison *et al.*2009; Robison and Hamilton 2010). For the management of the heritage items this means that there are human health and safety considerations – nuclear test period fallout is essentially co-constituted with the fabric and landscape of nuclear testing.

Climate change and natural disasters are factors that have considerable potential to impact on an option for permanent re-settlement of Bikini Atoll. The long term impacts of climate change – including sea level rise, increased exposure to storm surge and rising tides, desertification and warmer sea temperatures resulting in erosion, salinisation, decreasing biodiversity and scarcer water supplies (Collett 2009: vii–viii) – have the potential to be catastrophic to the atoll system. Somewhat ironically, natural disasters in the form of storm surges, and more rarely typhoons, are unlikely to impact the sunken ships or Bravo Crater, the physical features most highlighted in the World Heritage nomination.

In March 1946, the number of Bikinians living on Bikini Atoll was 167 (of a total population of 196 – Niedenthal 2001:177). Today the population is over 4,000 (a 'Bikinian' is defined by custom as a person with Bikinian blood or a person married to a Bikinian), a more than 20 times population increase on the 1946 figure. Thus any future resettlement of the atoll could involve a large number of people relative to the past population number. There would likely be a considerable impact on the landscapes, terrestrial habitat, marine resources and nuclear test period heritage features resulting from occupation (e.g. housing, infrastructure, waste and sewage disposal, agriculture and husbandry) and local resource use. The impacts on the nuclear fleet and Bravo Crater resulting from resettlement would, as with climate change impacts, likely be negligible.

A conflict lies however in the conservation and management of Bikini Atoll's marine biodiversity (Pinca *et al.* 2002), birdlife (Vander Velde and Vander Velde 2003) and terrestrial fauna (e.g. coconut crabs and reptiles). Despite the considerable disturbance of the terrestrial environment caused by the construction of the test site (Fosberg 1988:2 observed that 'no unaltered vegetation has survived'), the bomb blasts, rehabilitation efforts and introduction of invasive species, the relative absence of humans for more than fifty years (following American abandonment after bomb testing ended in 1958) has allowed species to return in considerable number to the atolls. For example, stands of *Pisonia grandis* are a favourite nesting place for birds (Reimaanlok 2008); the rare and threatened giant clam *Tridacna gigas* is particularly abundant in Bikini lagoon; there is a high diversity of fish fauna with a high concentration of shark species considered threatened elsewhere; and the recovery of coral reefs within the Bravo Crater provide a rare example of early succession and the development of reef structure (Richards *et al.* 2008). In brief, Bikini Atoll is home to many species that are threatened or depleted in the rest of Micronesia and the world and the ecological values have global significance (Baker 2010:4).[3] Bikini Atoll is a living laboratory of island re-colonisation. However, this situation (i.e. the recovery of the 'natural environment' as a 'cultural artefact' deriving from nuclear testing and abandonment) highlights artificiality in the nature/culture dichotomy within the World Heritage Listing criteria.

The Bikinian local government has been active in the protection of the atoll's biodiversity. In July 1997, the KBE LG passed regulations making commercial fishing within 12 nautical miles of the atoll illegal (in line with RMI national regulations), banning the use of gill nets, regulating the taking of lobster, and making fishing for shark and turtles illegal without authorisation. Bird habitats are entirely protected on some islands (Enue to Aeomen islands) and heavily regulated on the remainder ('Bird Islands'). In addition, the World Heritage nomination documents prepared for Bikini Atoll (RMI 2009:26) foreshadow a future re-nomination of the atoll under 'natural' criteria[4] that would re-categorise the property as either a cultural landscape or mixed cultural and natural site.

---

3   Baker (2010:5) notes: 'The entirety of the Marshall Islands lies in the central-western part of the Conservation International Polynesia/Micronesia Hotspot and the northern Marshall Islands form the Key Biodiversity Area, Kabin Meto.' The 'Micronesian Challenge' is a region-wide initiative aimed at conserving at least 30% of near-shore marine resources and 20% of terrestrial resources across Micronesia by 2020 (Beger *et al.* 2008).

4   Criteria (ix): 'Be outstanding examples representing significant ongoing ecological and biological processes in the evolution and development of terrestrial, fresh water, coastal and marine ecosystems and communities of plants and animals.' The criterion refers to the recovery of the marine environment as a result of the bomb detonations and particularly within the Bravo Crater.

Thus the increasingly-recognised ecological importance and values of Bikini Atoll is bringing the place within the purview of regional and global nature conservation agencies and groups. Resettlement of Bikini Atoll would challenge global and regional conservation management goals for both natural and cultural values. There are Indigenous traditions of sustainable use of marine and terrestrial resources controlled by complex social rules in the Marshall Islands (Weissler 2001). However, these customary values and practices have substantially changed as a result of 150 years of colonial entanglement and a transition to cash-based lifestyles. For the Bikinians, a loss of traditional lagoon fishing skills was exacerbated by their removal to the isolated and lagoon-less island of Kili.

A Bikinian homeland resettlement would bring into sharp focus tensions between local Indigenous rights (e.g. Articles 3, 10 and 11 of the *United Nations Declaration on the Rights of* Indigenous *Peoples* – United Nations 2007) and the global conservation movement. Issues associated with the tensions between human use of the environment and notions of 'natural' are also evidenced in the case of the semi-nomadic Maasai pastoralists of the Ngorongoro Conservation Area in Tanzania (referenced above). The UN declaration recognises the right of Indigenous peoples to freely pursue economic, social and cultural development (Article 3) and the right to practice and revitalise cultural traditions (Article 11). However, resettlement and development on Bikini Atoll by a relatively large population (i.e. up to 4,000 people today compared to 200 in 1946) would inevitably have impacts on the atoll's marine and terrestrial ecology.

Since global recognition of the ecological values of Bikini Atoll has been foreshadowed and resettlement is a goal of the Bikinian community, there appear to be opportunities for global and local actors to work together to plan for mutually beneficial economic, cultural and conservation outcomes. Environmental and heritage goals might best be achieved through harnessing Indigenous cultural knowledge and western scientific approaches via an approach that supports the rights of Indigenous peoples.

Conclusion

When I visited Bikini Atoll in September 2009, I was struck by its extent, remoteness and incredible beauty – the colour of the water, the teeming marine life in the shallow lagoon edges, and the closeness to which I was able to get to the birdlife on the outer islands. I was deeply impressed by my Bikinian hosts, not only for their hospitality and generosity in showing me the homeland given up 'for the good of mankind', but also their skill and knowledge – for example, in navigating the atoll and the routes through the coral reefs to access landings on the shores of the many islands. It seemed paradoxical therefore that the reason that Bikini Atoll is inscribed on the world heritage list as centred on the 'monuments' and 'icons' of nuclear testing, including the Bravo crater, the nuclear fleet, the mushroom cloud motif and *le bikini*, and not for the atoll ecology or knowledge and contemporary lived experiences of the Bikinians.

World Heritage is a strange beast. It manufactures history and heritage at a global scale, and as Sandra Pannell (2006:76) notes: 'World Heritage – as both a global concept and, for some, an imposed process – is subject to negotiation, opposition and a range of accommodations at the local level.' Derek Gillman (2010:1–2, 41–43) describes how the idea of heritage encompasses 'two ways of thinking about cultural property' (Merryman 1986) ' … represented respectively by cultural cosmopolitans, who seek to promote the idea of 'the heritage of all mankind', and cultural nationalists for whom art, architecture … are always a part of someone's particular heritage.' Gillman argues there are deep antecedents for the binary created in Western thinking between *universalism* (the global) and cultural *particularism* (the local). The World Heritage project (as expressed through the 1972 World Heritage Convention) lodges together the 'two ways' (Gillman 2010:49), recognising that the 'World Heritage of mankind as a whole' also belongs to specific people ('to whatever people it may belong') (UNESCO 1972: Preamble).

In its early years, the practice of the World Heritage project had a tendency to work within a Western neo-colonial model (e.g. privileging monuments as well as particular constructions of nature and culture) and to centre power in the global actors (e.g. see Pannell's 2006 discussion of Komodo National Park and Gondwana Rainforest of Australia). In its adoption of the idea of cultural landscapes and striving for a balanced, credible and representative World Heritage List from the early 1990s, the idea of World Heritage saw a marked change in approach through its accommodation of local readings of heritage.

Concerns voiced by global Indigenous peoples NGOs about making the implementation of the World Heritage Convention consistent with the United Nations Declaration on the Rights of Indigenous Peoples is a further challenge to the way World Heritage is constructed and practiced. As at April 2012, the World Heritage List comprised 936 items categorised as cultural (725 of which 66 are cultural landscapes), natural (183) and mixed (28) properties.[5] The categories themselves still retain a Western construction of nature and culture and this construction is what is being challenged by Indigenous peoples – an idea that the heritage of humans, ecology, landscape, things and history are deeply connected in Indigenous cosmologies.

In this situation, Bikini Atoll Nuclear Test Site, a property which symbolises the dawn of the nuclear age, provides an example of cultural engagement between global/universal and local/ particularist, polarities that can obscure complex processes of creative cooptation (Herzfeld 1997:3) and binary slipperiness. World Heritage Listing has strings attached for those State Parties and local governments who have responsibility for the property (e.g. requirements for state of conservation reporting). But the local administration, such as the KBE LG, has and indeed actively takes the opportunities offered by World Heritage to act back on the global NGOs (UNESCO, ICOMOS, IUCN) and heritage framework. The push for the rights of Indigenous peoples to be foregrounded in the practice of world heritage listing challenges but ultimately can benefit the idea of heritage.

Bikinian people have a long history of political activism concerning restitution for events that can be traced back to Sunday 10 February 1946. Only recently, since the mid-2000s, have the Bikinians coopted the global heritage system as a political tool to continue to expose the local and continuing consequences of nuclear testing. Thus Bikinians have enmeshed World Heritage in the politics of Indigenous rights in a way that Indigenous rights, for example the living cultural values of the Maasai resident population in the Ngorongoro Conservation Area, has seldom been integrated into global heritage. Bikini Atoll makes the politics of Indigenous heritage rights visible in World Heritage, highlighting the tensions between official heritage practice and the lived experience of local actors. And that is social poetics.

Acknowledgements

Deputy Mayor Wilson Note, Councillor Banjo Joel, Jack Niedenthal, Lani Kramer, Edward Maddison and Jim McNutt were my generous hosts, guides and companions on Bikini Atoll. My visit to the atoll was organised through the World Heritage Unit, Paris. I thank the two referees for the paper for their insightful comments and feedback. Finally I thank the editors of the volume, Denis Byrne, Sally Brockwell and Sue O'Connor for their input and support.

---

5   The nuclear landscapes/places on the list comprise the Hiroshima Peace Memorial (Genbaku Dome), Japan (listed 1996), and Bikini Atoll Nuclear Test Site.

References

Baker, N. 2010. *Bikini Atoll conservation management plan*. Kili-Bikini-Ejit Local Government, Majuro.

Barker, H.M. 2004. *Bravo for the Marshallese: Regaining control in a post-nuclear, post-colonial world*. Thompson and Wadsworth, Belmont.

Beger, M., Jacobson, D., Pinca, S., Richards, Z., Hess, D., Harriss, F., Page, C., Peterson, E. and Baker, N. 2008. The state of coral reef ecosystems of the Republic of the Marshall Islands. In: Waddell, J.E. and Clarke, A.M. (eds), *The state of coral reef ecosystems of the United States and Pacific Freely Associated States*, pp. 387–418. NOAA/NCCOS Centre for Coastal Monitoring and Assessment's Biogeography Team, Silver Spring.

Blake, J. 2009. UNESCO's 2003 Convention on Intangible Cultural Heritage: The implications of community involvement in 'safeguarding'. In: Smith, L. and Akagawa, N. (eds), *Intangible heritage*, pp. 45–65. Routledge, New York.

Bradley, D. 1946/1984. *No place to hide*. University Press of New England, Armidale.

Brown, S. 2013. Archaeology of brutal encounter: Heritage and bomb testing on Bikini Atoll, Republic of the Marshall Islands. *Archaeology in Oceania*, 48(1):26-39.

Carucci, L.M. 1997. *Nuclear nativity: Rituals of renewal and empowerment in the Marshall Islands*. Northern Illinois University Press, Deklab.

Cleere, H. 2001. The uneasy bedfellows: Universality and cultural heritage. In: Layton, R., Stone, P.G., and Thomas, J. (eds), *Destruction and conservation of cultural property*, pp. 22–29. Routledge, London.

Collett, L. 2009. A fair-weather friend? Australia's relationship with a climate changed Pacific. Paper No.1. The Australian Institute, Canberra.

Davis, J.S. 2005a. Representing place: 'Deserted isles' and the reproduction of Bikini Atoll. *Annals of the Association of American Geographers* 95(3):607–625.

Davis, J.S. 2005b. 'Is it really safe? That's what we want to know': Science, stories and dangerous places. *The Professional Geographer* 57:213–221.

de Groot, G.J. 2006. *The bomb: A life*. Harvard University Press, Cambridge.

Delgado, J.P. 1996. *Ghost fleet: The sunken ships of Bikini Atoll*. University of Hawai'i Press, Honolulu.

Delgado, J.P., Lenihan, D.J. and Murphy, L.E. 1991. *The archaeology of the atomic bomb: A submerged cultural resources assessment of the sunken fleet of operation crossroads at Bikini and Kwajalein Atoll lagoons, Republic of the Marshall Islands*. Submerged Resources Centre Professional Report No. 11. US National Park Service, Santa Fe.

Fosberg, F.R. 1988. Vegetation of Bikini Atoll, 1985. *Atoll Research Bulletin* 315:1–28.

Gillman, D. 2010. *The idea of cultural heritage*. Cambridge University Press, New York.

Herzfeld, M. 1997. *Cultural intimacy: Social poetics in the Nation-State*. Routledge, London.

Holmes and Harver Inc. 1988. *Bikini Island Wato diagram, Bikini Atoll masterplan*. Unpublished report. Holmes and Harver Inc., Albuquerque.

International Work Group for Indigenous Affairs. 2011. *Joint Statement on continuous violations of the principle of free, prior and informed consent in the context of UNESCO's World Heritage Convention*.

Keever, B.A.D. 2004. *News zero: The New York Times and the bomb*. Common Courage Press, Monroe.

Kiste, R.C. 1974. *The Bikinians: A study in forced migration*. Cummings Publishing. Menlo Park.

Mason, L. 1950. The Bikinians: a transplanted population. *Human Organisation* 9(1):5–15.

Mason, L. 1954. *Relocation of the Bikinian Marshallese: A study in group migration.* University of Hawaii Press, Honolulu.

McArthur, P.H. 2008. Ambivalent fantasies: local prehistories and global dramas in the Marshall Islands. *Journal of Folklore Research* 45(3):263–298.

Merryman, J.H. 1986. Two ways of thinking about cultural property. *American Journal of International Law* 80:831–853.

Niedenthal, J. 1997. A history of the people of Bikini following nuclear weapons testing in the Marshall Islands: With recollections and views of Elders of Bikini Atoll. *Health Physics* 73(1):28–36.

Niedenthal, J. 2001. *For the good of mankind: A history of the people of Bikini and their islands.* Bravo Publishers, Majuro.

Niedenthal, J. 2011. World's debt to Bikini recognised: nuclear test site added to Heritage List. *Voices UNESCO in the Asia-Pacific* 25(1):10–11.

Pannell, S. 2006. *Reconciling nature and culture in a global context? Lessons from the World Heritage List.* Research Report. James Cook University, Cairns.

Pinca, S., Beger, M., Richards, Z. and Peterson, E. 2002. Coral reef biodiversity community-based assessment and conservation planning in the Marshall Islands: Baseline surveys, capacity building and natural protection and management of coral reefs of the atolls of Bikini and Rongelap. Report to the Rongelap Government, Republic of the Marshall Islands.

Reimaanlok National Planning Team. 2008. *Reimaanlok: National conservation area plan for the Marshall Islands.* N. Baker, Melbourne.

Republic of the Marshall Islands. 2009. *Bikini Atoll: Nomination by the Republic of the Marshall Islands for inscription on the World Heritage List 2010.* Republic of the Marshall Islands, Majuro.

Richards, Z., Beger, M., Pinca, S. and Wallace, C.C. 2008. Bikini Atoll coral biodiversity resilience revealed: five decades after nuclear testing. *Marine Pollution Bulletin* 56:503–515.

Robison, W.L., Brown, P.H., Stone, E.L., Hamilton, T.F., Conrado, C.L. and Kehl, S. 2009. Distribution and ratios of ^{137}Cs and K in control and K-treated coconut trees at Bikini Island where nuclear test fallout occurred: effects and implications. *Journal of Environmental Radioactivity* 100:76–83.

Robison, W.L. and Hamilton, T.F. 2010. Radiation doses for Marshall Island's atolls affected by U.S. nuclear testing: All exposure pathways, remedial measures, and environmental loss of ^{137}Cs. *Health Physics* 98(1):1–11.

Simon, S. 1997. A brief history of people and events related to atomic testing in the Marshall Islands. *Health Physics* 73(1):5–20.

Smith, A. 2007. Colonialism and the bomb in the Pacific. In: Schofield, J. and Cocroft, W. (eds), *A fearsome heritage: Diverse legacies of the Cold War*, pp. 51–72. Left Coast Press, Walnut Creek.

Smith, A. and Jones, K.L. 2007. *Cultural landscapes of the Pacific Islands: ICOMOS thematic study.* UNESCO World Heritage Centre, Paris.

Smith, L. 2006. *Uses of heritage.* Routledge, London.

Spennemann, D.H.R. 1990. *The ownership of cultural resources in the Marshall Islands: An essay in pertinent jurisprudence and legal history.* Alele Corporation, Majuro.

Spennemann, D.H.R. 2000. *Cultural heritage legislation in the Republic of the Marshall Islands. Historic Preservation Act of 1991, Republic of the Marshall Islands.*

Spennemann, D.H.R. and D. Look. 2006. Impact of tropical vegetation on World War II-era cultural resources in the Marshall Islands. *Micronesian Journal of the Humanities and Social Sciences* 5(1/2):440–462.

Stone, R. 1988. *Radio Bikini.* (Documentary Film, Director: Robert Stone).

Tobin, J.A. 1953. *The Bikini People, past and present.* Unpublished report. Marshall Islands District, Majuro.

United Nations. 2007. *United Nations Declaration on the Rights of Indigenous Peoples.*

UNESCO. 1972. *Convention Concerning the Protection of the World Cultural and Natural Heritage.* UNESCO, Paris.

UNESCO World Heritage Centre. 2004. *The State of World Heritage in the Asia-Pacific Region.* World Heritage Reports 12. UNESCO, Paris.

UNESCO World Heritage Centre. 2008. *Operational Guidelines for the Implementation of the World Heritage Convention.* UNESCO, Paris.

United Nations Permanent Forum on Indigenous Issues. 2011. *Report on the Tenth Session (16-27 May 2011).*

US Department of the Interior. 1986. *Compact of Free Association Act of 1985.* U.S. Public Law 99–239 – Jan.14, 1986.

Vander Velde, N. and Vander Velde, B. 2003. *A review of the birds of Bikini Atoll, Marshall Islands with recent observations.* Unpublished report. Bikini Atoll Local Government, Majuro.

Weisgall, J.M. 1980. The nuclear nomads of Bikini. *Foreign Affairs* 39:74–98.

Weisgall, J.M. 1994. *Operation crossroads: the atomic tests at Bikini Atoll.* Naval Institute Press, Annapolis.

Weissler, M.I. 2001. Precarious landscapes: prehistoric settlement in the Marshall Islands. *Antiquity* 75:31–32.

Williamson, R.V. 2001. The challenges of survey and site preservation in the Republic of the Marshall Islands. *Cultural Resource Management* 1:44–45.

World Heritage Committee. 1994. Expert Meeting on the 'Global Strategy' and thematic studies for a representative World Heritage List. UNESCO, Paris.

World Heritage Committee. 2011. *Decisions Adopted by the World Heritage Committee at its 35th Session (UNESCO 2011).*

Zinn, H. 2010. *The bomb.* City Lights Books, San Fransisco.

4

Nature and culture in a global context: A case study from World Heritage Listed Komodo National Park, eastern Indonesia

Sandra Pannell, James Cook University, Cairns, Australia

Introduction

As the title suggests, the 1972 UNESCO Convention Concerning the Protection of the World Cultural and Natural Heritage enshrines one of the most pervasive dualisms in Western thought – that of nature and culture (MacCormack and Strathern 1980).[1] The 936 properties currently inscribed on the World Heritage List are identified as either 'natural', 'cultural', or as 'mixed' heritage.[2] Although the Convention provides 'definitions' and 'guidelines' regarding natural and cultural properties, it is clear from a comparative analysis of a number of World Heritage sites that the values ascribed to nature and culture are not a global given. Nor is it necessarily the case that the invocation of the nature-culture distinction results in a set of universally uniform effects. What is clear, however, is that nature and culture are made visible in a range of inter-cultural and trans-political contexts. In the operationalisation of the Convention, member nation-states provide some of the localised venues for the invention of nature and culture, while science and other 'expert' disciplines provide some of the procedures for producing nature and culture in these contexts (Smith 1998; Lowe 2006).

In the case of Komodo National Park, we can readily track the historical impact of an 'international community' of experts upon a remote island in the eastern reaches of the Indonesian archipelago. In the same way that Celia Lowe tracked the story of biodiversity conservation in Indonesia in the period from 1988 to 1998 and, more specifically, the making of nature in the Togean Islands of northern Sulawesi, the example of Komodo National Park allows us to also follow the discursive transformation of the 'nature' scientifically monitored and managed by these experts – from nature as an evolutionary oddity to ideas about 'nature in balance', and more recently, to a bio-diverse vision of nature (see Lowe 2004, 2006). As I hope to demonstrate in this politico-

---

1 'Cultural' and 'natural heritage' are defined in Articles 1 and 2 of the Convention. In summary, 'monuments', 'groups of buildings' and 'sites', the latter being the 'works of man or the combined works of nature and of man' (UNESCO 2005:13), are considered as 'cultural heritage'. 'Natural heritage', on the other hand, refers to 'physical and biological formations', 'habitats of threatened species' and 'natural sites or natural areas', which are of 'outstanding universal value' from the point of view of science, conservation and/or aesthetics (UNESCO 2005:13.). A complete definition of 'cultural' and 'natural heritage' is given as part of the definition of World Heritage in the 2005 'Operational Guidelines for the Implementation of the World Heritage Convention' (UNESCO 2005).

2 As of March 2012, there are 725 cultural, 183 natural and 28 mixed listed World Heritage properties, located in 153 of the 189 state parties which have ratified the World Heritage Convention.

environmental history of Komodo Island spanning a hundred years or so, the globalisation of the nature-culture distinction, and the complex of values it engenders, often encourages the very threats and dangers *the Convention* seeks to ameliorate through listing.

'Here there be dragons': Discovering dragons and the creation of media-genic megafauna

The scientific 'discovery' of the world's largest terrestrial reptile, the 'Komodo dragon' (*Varanus komodoensis*), in 1912 set in motion a history of regional regulation, national legislation and international conservation measures aimed at protecting the dragon and its habitat, restricted to several islands in the Komodo archipelago.

Responding to both the problems created by scientific curiosity and the conservation concerns of scientists, during the course of the twentieth century, the islands of the Komodo archipelago were granted some form of reserve or protected area status, leading to the declaration of Komodo National Park in 1980 (PHKA and TNC 2000(1):36).

Informed by an IUCN evaluation, in 1991 the Park was inscribed on the World Heritage List as a 'natural' property, based upon its 'superlative natural features' (Criteria III) and as the 'habitat of a threatened species' (Criteria IV), the Komodo monitor. As indicated in the original nomination document, 'nature' not only includes the Park's 'rugged' landscape, but it also encompasses the endangered 'dragons', the only acknowledged endemic inhabitants of this 'dramatic' space'.

Up until the mid 1990s, the 'singular focus of the [P]ark', and that of the previous regulations and decrees, was on the 'very impressive and remarkable animal – *Varanus komodoensis*' (IUCN 1991:27). Up until this time the value of the Komodo dragon to the 'international community' and to the general public had largely been as a scientific curiosity.

The World Heritage listing of Komodo National Park in 1991 set in motion a re-evaluation and re-presentation of this dragon-focused view of nature. As I argue in this paper, changing 'scientific' ideas about 'nature' have serious social and cultural implications for the approximately 3,000 individuals living within the Park and for a further 17,000 people living in villages directly surrounding the Park (PHKA and TNC 2000 (1):5).

Culture Wars: Part I

In his account of the American Museum of Natural History scientific expedition to Komodo in 1926, Expedition Leader, W. Douglas Burden, characterised the villagers as: 'a degenerate lot of diseased people, that have reached such a degraded state that they don't seem capable of curiosity' (1927:103). In contrast, Burden valorised the 'wilderness of romantic Komodo' and he concluded that: 'Komodo is a place where every prospect pleases, and only man is vile' (1927:103). Burden's depiction of the Komodo Islanders arguably represents the first shot fired in the twentieth century in the ensuing 'culture wars' aimed at vilifying the local population. Burden's distinction between a 'vile' humanity and the 'beautiful scenes' afforded by nature on Komodo Island is a trope consistently reproduced in the history of this region as a protected area.

The 'discovery' of the Ata Modo

Up until 1982 it was popularly believed that the entire population of Komodo Island originated from elsewhere or that they were the descendants of 'convicts', while the language spoken on the island was commonly seen as a 'mixture of other tongues' (Needham 1986:54). However, in 1982, the Dutch missionary, Father Jilis A. J. Verheijen, published a monograph on Komodo Island, which alludes to a history of human occupation of Komodo Island spanning some 2,000 years (1982:256). The antiquity of human occupation of the island is supported by preliminary archaeology undertaken in the late 1960s (see Auffenberg 1981:350). Verheijen speaks of a distinctive Komodo people, the Ata Modo, with an independent language, in which the Komodo dragon is referred to as '*ora*' (Verheijen 1982:115). Verheijen's reporting of the existence of the Ata Modo and of their status as the original occupiers of Komodo Island is further supported by the work of a number of anthropologists (see Needham 1986; Forth 1988).

On the subject of *ora*, the people of Komodo believe that they and the dragons are descended from the same ancestral set of twins. According to these beliefs: 'if one of these animals is injured, then its relatives, who have taken the form of human beings, will also become ill' (Hitchcock [citing Bagus 1987] 1993:305). Until recently, after every hunt or fish catch, a portion of meat was left by the Ata Modo for the Komodo dragon (Ellis 1998:75–76).

It is clear from Verheijen's work that, prior to the establishment of the National Park, the Komodo Islanders enjoyed a mixed subsistence and cash economy, focused upon both terrestrial and marine environments, with edible starch from a number of palm species on the island constituting a staple food. Named garden and fishing areas, together with former settlements and clan territories are detailed on a map appended to the monograph. The map clearly indicates that Komodo Island is far from the barren landscape commonly depicted in National Park documents and in tourist guides.

'New nature'

In 1977, as part of the preparations for the 1980 establishment of Komodo National Park, a management plan was drawn up by a 'multinational team of experts under the auspices of the United Nations' (Hitchcock 1993:310). The authors of the plan were scientists interested in the zoology and botany of Komodo. The ideological emphasis in the 1977 management plan was upon restricting or eliminating the impact of humans on the environment and thus restoring nature to its proper balance. Notable among these proposals to restore 'prey/predator relationships' is the recommendation to cease baiting and the feeding of the dragons with goats purchased from the local community. Emphasising the 'wild' and 'dangerous' nature of the Komodo dragon is a critical dimension of these proposed restoration activities and associated scientific ideas about the 'normal state of nature' (Budiansky 1995:71). As such, these restoration activities were aimed at distancing dangerous dragons from defenceless human populations.

In line with the idea of nature 'in balance', and in keeping with attempts to restore it to this 'natural' state, the management plan made several recommendations to curtail and ultimately stop the hunting of deer in the Park, one of the key prey species for the Komodo monitor. Associated with the 'deer poaching' issue, the plan also identified local burning of grasslands on the island as a threat to the Komodo monitor and its primary prey species.

Notwithstanding the considerable impact of the 1977 management plan upon the local human communities within and adjoining the National Park, the plan portrays a somewhat confined view of nature, albeit one regarded as 'out of balance', but still focused upon the 'dragon' and its terrestrial habitat. However, since the 1990s, with the World Heritage listing of Komodo National Park, nature has not only been re-defined (once again), but it has also been privatised in the process.

UNESCO's 'Man and the biosphere programme', launched in 1970, signaled a new global vision of nature in terms of the idea of 'biodiversity', and the need to halt the loss of it. As Luke observes, by the 1970s the Enlightenment idea of 'nature' as 'untouched and undisturbed expanses' appeared obsolete, indeed, 'dead' (1995:12). In drawing public attention to the impact of humans on nature in the form of wholesale extinctions, 'industrial pollution, greenhouse gases, chemical contamination, and radioactive wastes' (Luke 1995:12), science played an important role in bringing about the 'end of nature' (McKibben 1989).

Confronted with the reality that few, if any, environments were undisturbed, nature has undergone an 'involution' (Katz 1998:46). Natural spaces have been re-worked to produce greater 'internal sub-divisions' (Katz 1998:46). Biodiversity, biosphere reserves, and biodiversity 'hotspots' are some of the products of this redefinition. Accordingly, conservation efforts have shifted from an emphasis upon nature as a quantity (i.e. untouched expanses 'hoarded' for their 'pristine appearances and organic presences') to focusing upon the 'quality of nature' in terms of biodiversity, and the idea of rare or endangered species.

The redefinition of nature in terms of biodiversity not only set in motion a global program aimed at documenting the earth's environmental riches but it also went hand-in-hand with the re-emergence of ideas about the preservation and restoration of nature. What Cindi Katz calls the 'new enclosure movement' (1998:47) entails setting aside 'discrete patches of nature' (Katz 1998:47) in the form of 'park enhancement districts', 'world wildlife zones', 'biosphere reserves', and so on. This strategy of 'bio-accumulation' and the idea of investing in nature for the future also encouraged new forms of corporate environmentalism and the 'increasing privatization of public environments' (Katz 1998:47). In Komodo National Park, these observations about the 'private productions of space and the preservation of nature' have been a reality for the past fifteen years.

In 1995, The Nature Conservancy (TNC), a US-based, private environmental organisation, joined with the Indonesian National Park Authority, PHKA,[3] to 'help' manage Komodo National Park (Michael 2001:35). Through its multi-million dollar portfolio of properties purchased around the globe, and underwritten by corporate donations and individual contributions, TNC operates 'the largest private system of nature sanctuaries in the world' (cited in Katz 1998:59). TNC is also the latest in a long list of international NGO's to be involved in management of the Park since 1980. Promoted as the first example of 'collaborative park management' in Indonesia, TNC's involvement in Park management is justified by the organisation in terms of its global mission to 'preserve plants, animals and natural communities that represent the diversity of life on earth by protecting the land and waters they need to survive' (TNC 2002:1).

While TNC presents its mission as 'transnational', conserving the biodiversity of Indonesia for the 'well-being of humankind' (TNC 2002:1), officials from the Indonesian Department of Agriculture and Forestry cite a 'lack of government funds' and the resulting 'no or poor park management' as the primary reason for this new arrangement.

As heralded in TNC promotional literature, Komodo National Park is also the 'pilot site to test new park financing mechanisms' (TNC n.d.:1). These new financing mechanisms refer to the Joint Venture company, PT Putri Naga Komodo (PNK), established between TNC and an Indonesian tourist entrepreneur (TNC 2002), to implement the 30 year concession to 'manage tourism in Komodo National Park' (TNC 2002).[4]

---

3   Management of national parks in Indonesia is the responsibility of the Director General for Nature Conservation and Forest Protection in the Department of Agriculture and Forestry (*Direktorat Jenderal Perlindungan Hutan dan Konservasi Alam, Departemen Kehutanan dan Perkebunan*).

4   According to a TNC-commissioned 'Environmental Assessment Study', undertaken as part of the compliance requirements for the World Bank's private sector financing arm, the International Finance Corporation, for a US $5 million bridging grant to 'kick-start' PT Putri Naga Komodo's new 'eco-tourism development enterprise' (TNC 2005a), Mr Hashim's tourism company is called PT Jaytasha Putrindo Utama (Singleton et al. 2002:3).

TNC's vision for 'local people' in the National Park is that they 'will be trained as tour and dive guides … and will be able to supplement their income by designing and selling handicrafts' (TNC n.d.*a*:1). TNC's vision for 'local people', 'eco-tourism' and 'environmental protection' constitutes the content of the latest management plan for the Park, *The 25 Year Master Plan for Management, 2000–2025, Komodo National Park* (PHKA and TNC 2000).

As stated in the 2000 Plan, biodiversity protection is now the primary goal of management, signaling a major shift in emphasis from the previous 'dragon focused' 'nature in balance' plans. In this more inclusive view of nature, the introduced animals and translocated native species, previously hunted when Komodo Island was a game reserve, are all regarded as part of the Park's 'rich biodiversity', and thus afforded protection (cf. Lowe 2004). Dogs and cats, both of which are linked to the local residents in the Park, are the only animals identified in the management plan as 'exotics' that pose a threat to the biodiversity of the region (PHKA and TNC 2000 (1):21,(2):67).[5]

While both endemic and introduced terrestrial species are regarded as part of the Park's biodiversity, the real focus of TNC's conservation program is upon the marine environment within the Park. In this remaking and refocusing of nature, the environmental significance of Komodo National Park has been extended to the point where WWF identified it as a 'global conservation priority area' (Singleton *et al.* 2002:7). This change in significance also introduces a new role for the National Park as a 'genetic/species storehouse with which to replenish and re-colonize devastated coral habitats elsewhere in the Indo-Pacific region' (Singleton *et al.* 2002:8). The re-invention of the Komodo archipelago as an area of 'rich biodiversity' with one of the "world's richest marine environments" (TNC n.d.*b*:1) has some serious consequences for local people.

Culture Wars II

The 2000 plan identifies the human population of the Park as 'already over the carrying capacity of the area' (PHKA and TNC 2000 (2):66) and states that 'human population pressure' is 'leading to degradation of the terrestrial resource base' and 'overharvesting of marine resources' (PHKA and TNC 2000 (2):67). These Malthusian scenarios of unchecked population growth and dramatic resource depletion pivot upon the construction of local people as both ignorant and immigrant.

In the management plan, the 'low level of education' of the villagers in the Park is seen as a major obstacle to 'economic diversification' among the local population (PHKA and TNC 2000 (2):66) and, as such, an impediment to the Government's attempts to financially attract people living in the Park to resettle elsewhere. There is also a strong suggestion in TNC's promotional and educational materials that the villagers' 'low level of education' lies behind their "destructive fishing practices and overfishing of the Park's marine resources" (TNC n.d.*c*:1). There is little or no realisation that the implementation of previous management regulations, particularly the prohibition upon harvesting terrestrially-based staple foods, has directly contributed to the scenario today where the villagers are 'wholly dependent upon marine resource utilization' for both their incomes and food sources (see Pet and Djohani 1998:18).

Indeed, the current management plan attempts to rewrite the long human occupation and economic history of the Park by stating that 'few are farmers and little land is used for agricultural purposes within the park'. As this statement indicates, the villagers' aboricultural productions and agroforestry practices do not fit into the official view of field/food-crop agriculture as the only form of landed productivity, while the 'barren' features of the Park do not conform with the

---

5   A similar situation emerged in the Central Kalahari Game Reserve, established in Botswana in 1961. As Adam Kuper reports, 'environmentalists complained that residents were keeping donkeys and goats that interfered with the game and that they were engaged in poaching' (2003:393).

ideas presented in the plan of fertile anthropogenic landscapes. The effect of these erasures and this kind of revisionism is to create an ahistorical landscape, populated by recent resource raiders, engaged in nothing more than 'extractive-economies' (Peluso 2003:212).

In the Master Plan for the Park, immigration is seen as the primary source of the 'exponential' population increase and resulting human pressure upon the Park's biodiversity. The alleged 'steady influx of migrants into the area' (PHKA and TNC 2000 (2):64) is also held responsible for the importation of destructive fishing methods and modern external influences, such as television and radio. Furthermore, immigration is directly linked to the loss of 'traditional customs', language and ethnic identity (PHKA and TNC 2000 (2):64-67). However, as Verheijen reported, the human history of Komodo Island is one of settlement, migration and immigration, reflected in the fact that the original language of Komodo incorporates vocabulary from languages spoken on nearby islands, particularly Manggarai, spoken in western Flores (Verheijen 1982; Needham 1986). Indeed, in origin narratives from Komodo, the historic arrival of people from Sumba, Sumbawa, Flores and Ambon is presented as an essential development in the foundation of the village of Modo on Komodo Island. When coupled with the written historical record, these origin narratives point to a long tradition of human movement and settlement in the region. As McWilliam (2002:18) and other anthropologists have observed, 'orders of precedence', based upon relative priority in time or place, constitute a fundamental form of social organisation throughout eastern Indonesia. Thus, contrary to the image presented in the management plan, migration in the area is not a recent response to a situation of scarce resources or growing population pressure elsewhere.

While the plan acknowledges that the Ata Modo are the 'original people of Komodo' (McWilliam 2002:18), it further states that "there are no pure blood people left and their culture and language is slowly being integrated with the recent migrants" (McWilliam 2002:18). It seems that for the people of Komodo no sooner had their culture come into view than it was deemed to have disappeared.

Reflecting The Nature Conservancy's preservationist view of nature as either valorised pristine areas or demonised expendable environments, the current plan thus separates the population of the Park into a near (if not already) extinct Indigenous minority and an endangering immigrant majority (see Pannell 1996).

The 2000 Master Plan for Komodo National Park also carves up the marine and terrestrial environment into seven new protection zones (with additional sub-zones) (see Figure 1).[6] The activities of villagers within the Park are now restricted to miniscule areas of land and sea designated as part of the 'traditional use zone'. Technology permitted in this zone is restricted to 'traditional tools', while the use of these tools is 'licensed by the Head of Komodo National Park' (Pannell 1996) In this respect, the Plan reflects an earlier ecological assessment, which concluded that 'limited eco-tourism and research' are the 'only true sustainable uses' of the terrestrial portions of the Park.

For local people, the new zoning system transforms what is an inhabited environment into a 'dartboard' of pristine 'natural' areas, which not only limits or prohibits their future use and occupation but also gives priority to tourist access and use of the Park. As this last point suggests, these 'dartboards of nature' are 'often constructed and overseen by non-residents whose livelihood is not dependent on the preserved environment' (Katz 1998:55). Indeed, in the management plan, tourism is depicted as an 'eco-friendly' activity and is associated with minimal environmental impact, as opposed to the high impact, 'destructive' practices of local people. Locking up biodiversity in a complex zoning and buffer system, it is clear that management of Komodo National Park is now focused upon preserving nature outside of culture, or in spite of it.

---

6   They are: core zone, wilderness zone with limited tourism, tourism use zone, traditional use zone, pelagic zone, special research and training zone and traditional settlement zone (PHKA and TNC 2000 (1):44).

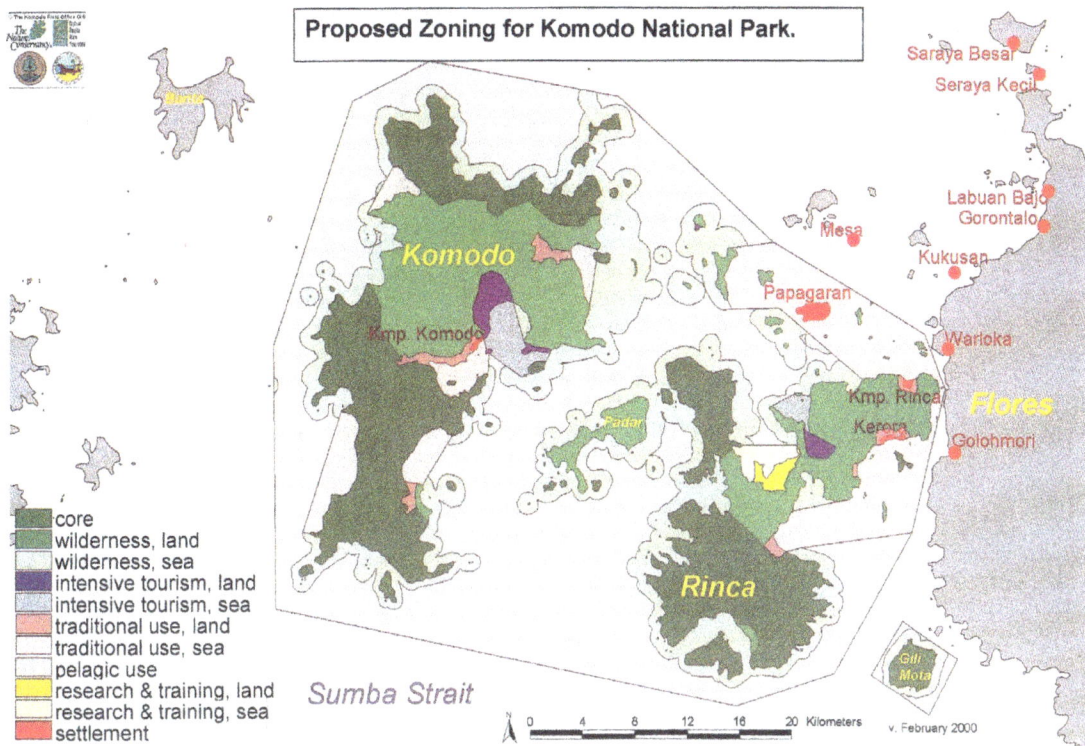

Figure 1. The proposed system of zoning for Komodo National Park.

Source: PHKA and TNC 2000.

The Good, the bad and the ugly: A postcard from Komodo National Park

While it might appear to visitors to the National Park that scientific 'management' of the World Heritage Listed natural values of the Park amount to nothing more than guiding fee-paying tourists to and from the official dragon viewing site for a couple of hours per day, for non-visitors, however, management has more serious consequences. In 2001, for example, 38 villagers were given jail sentences, ranging in length from 6 months to 3 years for 'deer poaching' and 'illegal fishing' activities. What is not reported in TNC-sponsored reports are the deaths and injuries resulting from the 'enforcement and protection' strategy, implemented as a result of the 2000 Management Plan. As reported by the peak Indonesian conservation NGO, the Indonesian Forum for the Environment (WALHI),[7] routine 'floating' patrols within the Park 'have operated in an extremely violent fashion, beating and shooting fisher folk they encounter' (WALHI 2004). According to WALHI, in a four-month period in 2002–2003, three fishermen were shot by Park and allied government officers on the patrol boat *Ora*, while 'at least 40 fishermen have been tortured and arrested [and] several fishermen and their families have been exiled from the National Park zone' (WALHI 2004).[8] In response to these deaths and beatings, members of the affected community destroyed the National Park post in Sape, Sumbawa, and took control of the local passenger ferry operating between Sape and Labuan Bajo, the gateway town to the Park.

The Islanders' battle for cultural survival is also apparent in the tension between them and the Park Authority regarding feeding of the Komodo lizards. As previously mentioned, from the

---

7   WALHI stands for 'Wahana Lingkungan Hidup Indonesia'.

8   It appears that this kind of extreme policing of the waters of the National Park continues into the present. On 29 March 2012, The Jakarta Post newspaper reported that a Komodo National Park ranger had shot and killed a fisherman from Sape, Sumbawa, and, in the same incident, the ranger also shot three other fishermen in the boat of the deceased, inflicting life-threatening injuries. According to the report, the four fishermen were accused of using explosives to catch fish in waters west of the National Park.

inception of the Park in 1980 and up until 1994, Park rangers regularly fed the 'dragons' with goats purchased from the local community. The 'ritual' feeding of the dragons for the benefit of tourists neatly dovetailed with the Islanders' belief that they have a kinship obligation to 'look after their twin brother' and provide *ora* with a portion of each catch (Ellis 1998:75). The Park Authority's decision to cease the 'goat-gobbling' (Ellis 1998:84) spectacle on Komodo was based on the scientific fashioning of nature as 'wild' and 'pristine'. From the point of view of the Komodo Islanders, however, rather than restoring the balance of nature, the intervention of science has threatened the very survival of this nature. Komodo Islanders state that the dramatic decline in the number of Komodo dragons on the island soon after regular feeding stopped is due to starvation (in 1997 the population of dragons on the island was recorded as 1,722, a year later only 1,061 lizards were recorded) (PHKA and TNC 2000 (2):21).[9] As these and other restrictions in the name of 'conservation' exemplify, the World Heritage listing of the Park and its current management plan ignore the ongoing connections between the local population and 'nature', and disregard the values that local people attach to the Park's so-called 'natural heritage'.

WALHI's allegations about TNC's management of Komodo National Park stand in stark contrast to the rhetoric of management produced by The Nature Conservancy through its internet website, its Bali-based, 'Coral Triangle Centre', and its reams of glossy information sheets and up-beat media releases. In this respect, TNC is highly successful in its management of the virtual environment of Komodo National Park. In the new millennium, we see that the role of international conservation organisations and national environmental agencies in managing nature increasingly revolves around image management and the manipulation of the eco-tourism experience. In order for the nature produced by these forms of political ecology to be perceived as environmental realism, the anthropogenic conditions of its production must be concealed or certainly back-grounded.

Based on the history and experience of Komodo National Park, serious questions need to be asked about the identities of the producers and consumers of this nature. For example, 'from whom' and 'for whom' is TNC 'saving the last great places on Earth' (TNC logo). Given the numerous references in TNC-produced material about the 'destructive fishing practices' of local people, the answer to the first part of this question seems obvious.

While the evidence strongly supports the claim that the preservation efforts of TNC 'insistently evict people from nature' (TNC logo), as the example of Komodo National Park illustrates, only certain kinds of people are affected. 'Eco-tourists' and the 'nature lovers' from around the world who underwrite TNC's international biodiversity investments are certainly not amongst those evicted from nature.

As evident in Komodo National Park, The Nature Conservancy and the World Heritage Committee are assisted in their efforts to 'save' nature for (certain kinds of) humanity by an international array of scientists and other experts. As governments and regional authorities increasingly look to World Heritage listing as a means of delivering economic benefits and a much-sought after international status from global tourism, expert input and advice is accorded greater political currency and leverage. In this commonly found scenario, as Harrison concludes: "'supervision' by experts sometimes comes to mean domination by experts" (2005:8–9).

As the example of Komodo National Park indicates, using nature preservation as the 'measure and arbiter' (Katz 1998:57) of rightfulness and what constitutes a global good produces an ugly environmental politics. It also produces a landscape of 'nature cemeteries' (Luke 1995:17), for

---

9   Walter Auffenberg (1981) recorded a population of 2,348 dragons on Komodo Island in 1970. The Park Authority states that 'the decline appears to be due to high mortality rates in the young and juvenile classes', but also suggests that fluctuations in population size may be an 'artifact of the methods employed' (PHKA and TNC 2000 (2):21).

the most part, off limits to the living. In the context of Komodo National Park, the production of these very un-natural spaces, wiped clean of culture, is further reified by the World Heritage listing of the Park solely for its natural values. Like so many other 'natural heritage' properties on the List, the inscribed values of Komodo National Park both conceal and reflect certain cultural assumptions, as well as a history of ideas, about nature.

The example of Komodo National Park, like so many others, also alerts us to the fact that the international presentation and 'protection' of 'nature' is 'linked to power: the power to impose a view of the world' (Harrison 2005:9). In Komodo National Park, the world view promulgated by TNC and the Park's management authority, together with the many local effects resulting from this imposition, stands in stark contrast to UNESCO's promotion of World Heritage as a global public good, in which we all 'join hands to protect and cherish the world's natural and cultural heritage'.

Acknowledgements

This study was made possible by a grant from the Rainforest Cooperative Research Centre, James Cook University and the Department of Environment and Heritage, Canberra. I am particularly indebted to Professor Nigel Stork, Director of the Rainforest CRC, and his unreserved support for this research project. I would also like to acknowledge the research assistance, intellectual input and on-going interest of Daniel Vachon in the preparation of this report. My interest in World Heritage, and particularly the nature-culture paradigm that underscores UNESCO's World Heritage Convention, was initially fueled by conversations with my colleague, Joseph Reser. I would like to take the opportunity here to thank Joe for our many stimulating exchanges. Particular thanks also go to my colleagues: Joan Bentrupperbaumer, Roger Wilkinson, and Dermot Smyth, for their respective contributions in the form of suggested reading material, photographic images of World Heritage sites visited in the course of this study, and coffee shop chats.

References

Auffenberg, W. 1981. *The behavioural ecology of the Komodo Monitor*. University Presses of Florida, Gainesville.

Bagus, I.G.N. 1987. Komodo National Park: Its role in tourism development in Indonesia. *Man and Culture in Oceania* (Special Issue) 3:169–176.

Budiansky, S. 1995. *Natures keepers: The new science of nature management*. Weidenfeld and Nicolson, London.

Burden, D.W. 1927. *Dragon lizards of Komodo: An expedition to the lost world of the Dutch East Indies*. G.P. Putnam's Sons, New York.

Ellis, C. 1998. *The land of the Komodo Dragon*. Times Editions, Singapore.

Forth, G. 1988. Komodo as Seen from Sumba: Comparative remarks on an Eastern Indonesian relationship terminology. *Bijdragen tot de Taal -, Land-en Volkenkunde*, 144:44–63.

Harrison, D. 2005. Contested narratives in the domain of World Heritage. In: Harrison, D. and Hitchcock, M. (eds), *The politics of World Heritage: Negotiating tourism and conservation*, pp. 1–10. Channel View Publications, Clevedon.

Hitchcock, M. 1993. Dragon tourism in Komodo, Eastern Indonesia. In: Hitchcock, M., Victor T., King, V.T. and Parnwell, M.J.G. (eds), *Tourism in South East Asia*, pp. 303–317. Routledge, London.

IUCN. 1991. *World Heritage Nomination – IUCN technical evaluation. 609. Komodo National Park (Indonesia)*. IUCN, Gland.

Katz, C. 1998. Whose nature, whose culture?: Private productions of space and the 'preservation' of nature. In: Braun, B. and Castree, N. (eds), *Remaking reality: Nature at the millenium*, pp. 46–64. Routledge, London.

Kuper, A. 2003. The return of the native. *Current Anthropology* 44(3):389–395.

Luke, T.W. 1995. The nature conservancy or the nature cemetery: Buying and selling 'perpetual care' as environmental resistance. *Capitalism Nature Socialism* 6:1–20.

Lowe, C. 2004. Making the monkey: How the Togean Macaque went from 'new form' to 'endemic species' in Indonesians' conservation biology. *Cultural Anthropology* 19(4):491–516.

Lowe, C. 2006. *Wild profusion: Biodiversity conservation in an Indonesian archipelago*. Princeton University Press, Princeton.

MacCormack, C. and Strathern, M. (eds). 1980. *Nature, culture, and gender*. Cambridge University Press, Cambridge.

McKibben, B. 1989. *The end of nature*. Doubleday, New York.

McWilliam, A. 2002. *Paths of origin, gates of life: A study of place and precedence in Southwest Timor*. KILV Press, Leiden.

Michael, A.W. 2001. Komodo Aria. *Asian Geographic* January:26–41.

Needham, R. 1986. Principles and variations in the social classification of Komodo. *Bijdragen tot de Taal -, Land-en Volkenkunde* 142(1):52–68.

Peluso, N.L. 2003. Fruit trees ad family trees in an anthropogenic forest: Property zones, resource access, and environmental change in Indonesia. In: Zerner, C. (ed.), *Culture and the question of rights: Forests, coasts, and sea in South East Asia*, pp. 184–219. Duke University Press, Durham.

Perlindungan, Hutan dan Konservasi Alam [Direktorat Jenderal] (PHKA) and The Nature Conservancy (TNC). 2000. *25 Year Master Plan for Management, 2000-2025, Komodo National Park*. 3 Volumes. Direktorat Jenderal Perlindungan Hutan dan Konservasi Alam, Jakarta.

Pet, J.S. and Djohani, R.H. 1998. Combating destructive fishing practices in Komodo National Park: Ban the hookah compressor! *SPC Live Reef Fish Information Bulletin* 4:17–28.

Pannell, S. 1996. *Homo nullius* or 'Where have all the people gone'? Refiguring marine management and conservation approaches. *The Australian Journal of Anthropology* 7(1):21–42.

Singleton, J., Sulaiman, R. and The Nature Conservancy-SE Asia Center for Marine Protected Areas Staff. 2002. *Environmental assessment study – Komodo National Park Indonesia*. TNC-SE Asia Center for Marine Protected Areas, Bali.

Smith, N. 1998. Nature at the millenium: Production and re-enchantment. In: Braun, B. and Castree, N (eds), *Remaking reality: Nature at the millenium*, pp. 271–285. Routledge, London.

The Nature Conservancy. n.d.*a. Collaborative park management: Partnerships, financing and ecotourism at Komodo National Park*. Information Sheet. The Nature Conservancy-Indonesia Program, Coastal and Marine Conservation Centre, Bali.

The Nature Conservancy. n.d.*b. Komodo National Park: Effective marine protected area management*. Information Sheet. The Nature Conservancy-Indonesia Program, Coastal and Marine Conservation Centre, Bali.

The Nature Conservancy. n.d.*c*. *Enforcement in Komodo National Park*. Information Sheet 10. The Nature Conservancy-Indonesia Program, Coastal and Marine Conservation Centre, Bali.

The Nature Conservancy. 2002. *Frequently-asked questions about The Nature Conservancy (TNC) in Komodo (August 1, 2002)*. The Nature Conservancy-Indonesia Program, Coastal and Marine Conservation Centre, Bali.

The Nature Conservancy. 2005. *Komodo collaborative management initiative: Transforming the Komodo Project into a responsible and sustainable Business*. The Nature Conservancy-Indonesia Program, Coastal and Marine Conservation Centre, Bali.

UNESCO. 1972. *Convention concerning the protection of the World Cultural and*

Natural Heritage. UNESCO, Paris.

UNESCO. 2005. *Operational guidelines for the implementation of the World Heritage Convention*. UNESCO World Heritage Centre, Paris.

Verheijen, J.A.J. 1982. *Komodo: Het eiland, Het volk en de Taal*. Martinus Nijhoff, The Hague.

5

Changing perspectives on the relationship between heritage, landscape and local communities: A lesson from Borobudur

Daud A. Tanudirjo, Jurusan Arkeologi, Fakultas Ilmu Budaya,
Universitas Gadjah Mada, Yogyakarta

Figure 1. The grandeur of the Borobudur World Heritage site has attracted visitors for its massive stone structure adorned with fabulous reliefs and stupas laid out in the configuration of a Buddhist Mandala.

Source: Daud Tanudirjo.

The grandeur of Borobudur has fascinated almost every visitor who views it. Situated in the heart of the island of Java in Indonesia, this remarkable stone structure is considered to be the most significant Buddhist monument in the Southern Hemisphere (Figure 1). In 1991, Borobudur was inscribed on the World Heritage List, together with two other smaller stone temples, Pawon and Mendut. These three stone temples are located over a straight line of about three kilometres on an east-west orientation, and are regarded as belonging to a single temple complex (Figure 2). Known as the Borobudur Temple Compound, this World Heritage Site meets at least three criteria of the Operational Guidelines for the Implementation of the World Heritage Convention: (i) to represent a masterpiece of human creative genius, (ii) to exhibit an important interchange of human values over a span of time or within cultural area of the world, on developments in architecture or technology, monumental arts, town planning or landscape design, and (iii) to be directly or tangibly associated with events or living traditions, with ideas, or with beliefs, with artistic and literacy works of outstanding universal value (see also Matsuura 2005). This is an affirmation of the Borobudur Temple Compound as a magnificent example of the heritage of humankind and a source of pride for Indonesian people, especially the local people of Borobudur who consider themselves direct descendants of the builders.

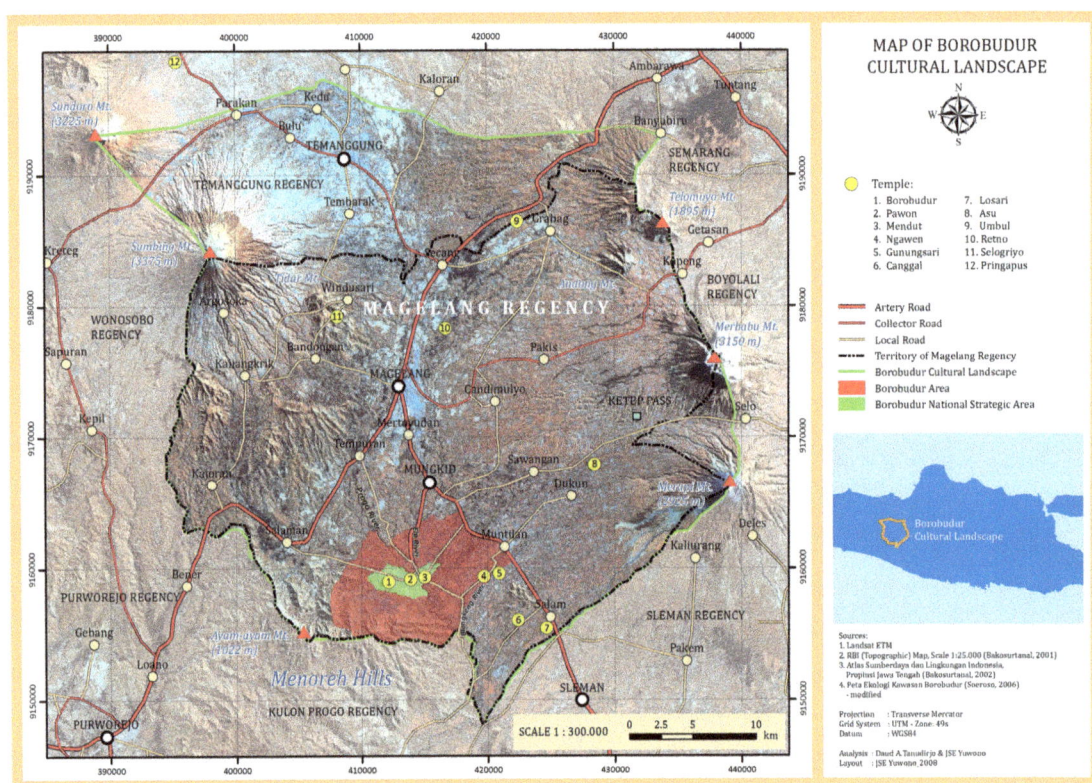

Figure 2. Map showing the Borobudur Temple Compound and the wider cultural landscape. As the local people were excluded from undertaking their traditional practices within the area of the Borobudur monument, they constructed a new wider Borobudur Cultural Landscape which includes the entire area of Kedu Plain encircled by the seven mountains (highlighted).

Source: Prepared by Daud Tanudirjo and J. Susetyo Edi Yuwono.

To maintain and conserve the Borobudur Temple Compound, the Government of Indonesia employs a centralised management policy in which local people are marginalised, having no role in management. This policy has triggered long-standing conflict between the local communities and the Government. The Borobudur case study allows us to evaluate the varied and changing perspectives of local people in relation to the monument Borobudur and the broader cultural landscape. This paper attempts to explicate how these changes occurred and the impact they have had on heritage management in Indonesia.

A brief history of Borobudur

Although the Borobudur Temple Compound is comprised of three stone temples – Mendut, Pawon, and Borobudur – it is the latter that is the exclusive focus of interest and most disputed in the Compound's heritage management. Borobudur is the main temple and the largest of the three. The monument was built on a hill that rises about 46 metres above the surrounding plain, with an areal measurement around the base of about 120 metres by 120 metres (Figure 1). More than 55,000 cubic metres of andesite stone was carved to build this massive structure in the form of a low step-pyramid. It consists of six rectangular terraces, three circular terraces, and a huge bell-shaped stupa on top. The wall and balustrade of the rectangular terraces are adorned with meticulously worked reliefs. A series of niches with Buddha statues rise above the lower balustrades and are crowned with small stupas. In total, 432 sitting Buddha statues with various hand-poses (*mudras*) were placed around the balustrades. On the three circular terraces, perforated larger stupas of 3.8 metres diameter at the base and 3.7 metres in height are arranged in circles containing 32, 24, and 16 stupas from lower to upper terraces respectively. Each stupa contains a sitting Buddha stone statue with *dharmacakramudra* ("turning the wheel of the law" hand-pose) representing a figure who has attained the highest level of enlightenment, variously known as the Highest Buddha, Vajrasattva (Krom, 1927), or Mahavairocana (Bernet-Kempers 1976).

The dome-shaped structure of Borobudur is commonly described as a replica of the universe and is comprised of four layers: the underworld (*kamadhatu*), material world (*rupadhatu*), spiritual world (*arupadhatu*), and the Eternal world (*Nirvana*). The first two layers are presented as five square terraces surrounded by walls with more than 1460 panels of relief, while the spiritual and the Eternal layers are symbolised as three round terraces and the main stupa respectively. The reliefs depict at least four themes: the punishment and reward for evil and good conduct (*Mahakarmavibhangga*), previous lives of the Buddhas or *Jatakamala-Avadana*, the life of Siddhartha Buddha Gautama or *Lalitavistara*, and the story of a young man Sudhana in search of the Highest Wisdom or *Gandavyuha* (Miksic 1990). Following the reliefs consecutively from lower to higher levels, the stories teach us the spiritual path to attain the ultimate Enlightenment. This may also be interpreted as a representation of the essential path of human progression from the physical towards the spiritual (Tanudirjo 2008). The existence of reliefs that could guide viewers on the path to obtain Enlightenment has led many scholars to suggest that Borobudur was built as a place of pilgrimage.

Borobudur can also be viewed as a Mandala, a symbolic diagram consisting of various geometric shapes, such as triangles, squares, and diamonds which are arranged in a circle. Usually a Mandala is used during meditation to help people to obtain better concentration (Walker 1983). In Buddhism, a Mandala is a diagram showing images of gods in specific positions. Each god represents a certain power. The design of Borobudur, especially when seen from above, clearly resembles a Mandala diagram. The placement of Buddha statues with particular hand-poses in specific directions supports this notion (Miksic 1990). However, the unique design of Borobudur has also been interpreted by some scholars as evidence for syncretism between Buddhism and the pre-Hinduistic Javanese beliefs based on ancestor worship. This interpretation is partly based on a phrase in an inscription related to the building of Borobudur, dated to 824 AD, which mentions the builder's appreciation of the merit that had been made by his/her predecessors (de Casparis 1950).

Borobudur Temple Compound was constructed during the reign of Syailendra dynasty around the ninth century AD and took more than 50 years to complete (Dumarcay 1978). Its location was selected in relation to a number of considerations (Voute 2005). Located amidst the bowl-like Kedu Plain of Central Java, Borobudur is very close to the geographic centre of Java Island. The area is surrounded by mountains and hills, namely: Merapi, Merbabu, Andong, Tidar,

Sindoro, Sumbing, and the Menoreh Hills respectively (Figures 2 and 3). This configuration references the location of Mount Mahameru, the centre of the Universe and the abode of the gods. According to Hindu-Buddhist cosmology, this sacred mountain is encircled by concentric rings of sea and land. Recent geological research in the vicinity of Borobudur has discovered evidence of an ancient lake that once existed around Borobudur (Murwanto *et al.* 2004). The occurrence of an ancient lake surrounding Borobudur was hypothesised by W.O.J. Niewenkamp more than 75 years ago (de Casparis 1981:70, 83). Its verification is indeed a strong indication that Borobudur was constructed to represent Mahameru. What is more, astronomical studies show that Borobudur, Mendut, and Pawon are arranged along an east-west axis pointing directly to the Merapi volcano to the east (Figure 3). The alignment is considered to symbolise the birth of a New World Age (Voute 2005; see also Tanudirjo 2011). On certain days of the year, the sun rises behind the volcano and sheds rays of light on Borobudur before touching the surrounding plain. Such a scene may have been imagined and configured in the construction of Borobodur by its architect as a symbol of Enlightenment. Clearly the Borobudur Temple Compound was designed to reference and illustrate the cultural and spiritual world of the community that constructed it.

The glorious age of Borobudur was not long-lasting. The monuments declined in use shortly after the centre of Old Javanese Kingdom shifted from Central Java to East Java around the end of tenth century AD, possibly due to the continuous eruption of Merapi and other volcanoes around this area. However, Borobudur may not have been totally abandoned. Its existence is noted in *Nagarakrtagama*, a fourteenth century Old Javanese manuscript. The temples appear to have been totally abandoned around the sixteenth century when the influence of Islam grew stronger in Central Java (Tanudirjo 2011). By the eighteenth to nineteenth century AD, some Javanese manuscripts describe Borobudur as merely a sacred mound, or even as a dangerous place to visit. Most of Java was ruled by the Islamic Kingdom and the majority of its population had been converted to Islam by this time. There was no sign of Buddhism existing in Central Java and Borobudur had been almost forgotten.

Figure 3. The Borobudur cultural landscape is surrounded by mountains and hills. Merapi and Merbabu volcanoes are among the mountains that mark the boundary to the east.

Source: Daud Tanudirjo.

The rediscovery and restoration of Borobudur

Borobudur was rediscovered in 1814 when Java was under the authority of Governor General Sir Thomas Stamford Raffles (1811-18). Knowing of the existence of temple ruins in the Borobudur area, Raffles sent H.C. Cornelius to examine the region (Soekmono 1976). With the help of around 200 local labourers, Cornelius cleaned away the vegetation and soil covering the ruins and revealed the ancient monument (Raffles 1817). Subsequently, Borobudur was often visited by Europeans for research and documentation, as well as for recreation (Figure 4). After the exposition, the monument's condition gradually deteriorated and this raised concern among scholars. Deformation of the stone structure, especially the stupas and the three round terraces, endangered the monument. In the early 1900s, the Dutch Government appointed a military officer van Erp to lead the restoration of Borobudur. Van Erp decided to focus the restoration on the three round terraces and the stupas as he considered that the square terraces were still in relatively good condition. After four years of hard work, from 1907 to 1911, van Erp managed to restore all the stupas, including the main stupa, and the three upper terraces. This restoration saved Borobudur from collapse, at least temporarily (Tanudirjo *et al.* 1994).

Figure 4. Since it was discovered in 1814, Borobudur has attracted visitors for recreation as well as research. The Dutch officials built a viewing hut on the top of the Main Stupa.

Source: Courtesy of KITLV/Royal Netherlands Institute of Southeast Asian and Caribbean Studies.

Twenty years after van Erp's restoration, another problem emerged: the square terraces began to deform due to water seepage. No significant action was taken by the Dutch Government to address this. It was not until the 1950s, after the independence of Indonesia, that Borobudur was declared an endangered monument. The new Indonesian Government initiated an international campaign to save it but this was unsuccessful. Although lacking sufficient resources, in the 1960s the Government strove to save the monument from ruin. It was not until 1972 when UNESCO agreed to take part in the restoration that real progress was made. With financial support from Japan, West Germany, Britain, Belgium, and Australia, UNESCO began an international safeguarding campaign for the monument. Following this, the Indonesian Government and UNESCO decided to carry out a more comprehensive restoration, executed from 1973 to 1983. Since 1978 the Japan International Cooperation Agency (JICA) has assisted the Indonesian Government to establish a master plan for managing the Borobudur-Mendut-Pawon complex. According to the plan, the Borobudur area should be developed as the Borobudur National Archaeological Park and an integrated zoning system should be implemented. The zonation consists of five zones (JICA Study Team 1979):

Zone I   Zone for protection and prevention of destruction of the physical environment of the archaeological monuments

Zone II   Zone for provision of park facilities for the convenience of visitors and preservation of historical environment

Zone III   Zone for regulation of land use around the parks and preservation of the environment while controlling development in areas surrounding the parks

Zone IV   Zone for maintenance of historical scenery and prevention of destruction of the scenery

Zone V   Zone for undertaking archaeological survey over a wide area and prevention of destruction of undiscovered archaeological monuments

At the conception stage of the JICA masterplan, the cultural landscape of Borobudur was considered as important and meriting protection. A leading Indonesian archaeologist, R. Soekmono, who was project manager of the Borobudur restoration (1971-1983), had expressed his concern regarding the potential impact of the tremendous changes to the Borobudur cultural landscape that were occurring as a result of activities attracted by the newly restored monument, especially the increase in visitation due to its prominence as a tourist destination. Even before the restoration officially concluded, he strongly emphasised the need to establish a plan for the protection of the cultural landscape of Borobudur (Soekmono 1983; see also Tanudirjo 2007). Soekmono's concern was partly accommodated in the JICA masterplan. Zone IV was designed to protect the historic scenery and cultural landscape. However, the plan for Zone IV has never been implemented.

Since the rediscovery of Borobudur, the local population living within the surrounding landscape has embraced the monument and expressed pride in residing in its proximity. They made offerings there and performed traditional ceremonies, including *wayang* (shadow) puppet show. To celebrate Idul Fitri, the end of Muslim fasting season, they gathered at Borobudur to greet their relatives and friends. They also provided food, souvenirs, and services for visitors to the monument. These actions engendered a feeling of ownership among the local people. They perceived Borobudur to be their heritage, although most of them had no historically informed knowledge of its history. It became part of their life and their cultural identity. The local people considered themselves the guardians of the cultural complex.

To establish Zones I and II, a protected area of about 90 hectares had to be freed up from human settlement. At least 381 households inhabiting five villages were resettled. These local people had to give away their land, mosques, kiosks, cemeteries, and gardens to the Government with very minimal compensation that amounted to only about one fifth of the market value at that time. Although the Government promised to provide jobs for these people during the development stage of the park, only small numbers of local people worked on the project. Some of the local people resisted giving up their properties, but as commonly happened at that time, the New Order regime in power used oppressive measures to force them to relinquish their homes and lands and move to other villages or the resettlement locations provided for them. Since then they have had no access to the monuments of Borobodur and consequently they feel disenfranchised from their heritage. Meanwhile, the Government founded a state-owned corporate body, PT Taman Wisata Candi Borobudur and Prambanan, to manage the monuments and the surrounding area which then became Borobudur Tourism Park rather than a National Archaeological Park. The name chosen by management reflects the stance of the Indonesian Government that perceives the Borobudur Temple Compound more as economic resource than a cultural and educational resource.

Although the establishment of the Integrated Zoning System by JICA was aimed at protecting the broader Borobudur landscape (JICA 1979), removing the villagers who had lived there for centuries was actually an act of landscape clearance. A landscape is not only composed of its topography and physical elements, it includes the people who live there, their acts, memories, relations, and the histories they have in that place (see Smith 2008; Thomas 2001). From this perspective, the residents of these villages, with their long-standing relationship with Borobudur, are an integral part of the Borobudur landscape. Landscape is always meaningful for people who have long and close relationships to it. It is often seen as part of their identity and involves a strong sense of belonging (Smith 2008). Hence, the eviction of the local people from their villages and land around Borobudur has predictably caused a terrible impact upon them. It is thus understandable that they resisted the Government edict to move to other places.

On 21 January 1985 a high-explosive bomb exploded on the circle terraces of Arupadhatu. At least nine perforated stupas were damaged (Tanudirjo *et al.* 1994). Local people were made the scapegoats. A number of prominent leaders who had resisted government control of the monument were brought to the military camp, interrogated and even tortured. They were forced into confessions admitting responsibility for the blast. Fortunately, just a couple days afterward, the real bomber was arrested. He was a Muslim fundamentalist who was afraid of Buddhist revivalism and considered the restoration of the Buddhist monument to be a return to paganism. This traumatic incident terrified the local people of Borobudur and increased their suffering. However, even in the face of such oppression many of them maintained the struggle for greater access to the monument and to be involved in the management of their heritage, although they confronted the Government less overtly. After this time, they exercised different strategies to take hold of their rights.

The management of Borobudur Temple Compound

In 1992, the Central Government issued Presidential Decree Number 1/1992 which assigned different institutions to manage the five zones of the Borobudur area. Zone I is managed by the Conservation Office of Borobudur which focuses on the conservation of the monument. Zone II is under the management of PT Taman Wisata Candi Borobudur and Prambanan, especially for the tourist activities. Zones III, IV, and V are managed by the local Government (Magelang Residency), which administers the whole area. Strangely, the conservation of Mendut and Pawon is conducted by another office, the Conservation Office of Central Java. The reason for this is that these two monuments are situated in Zone III and are under the management of the Magelang local Government, as a part of the Provincial Government of Central Java. Therefore, at least four institutions are assigned to manage the Borobudur Temple Compound.

Its distribution among different institutions has made management more complicated. The lack of coordination between the institutions jeopardises the implementation of the JICA masterplan and generates conflict. Every party pursues its own goals without considering the others and they impose different, and often conflicting, values and meaning on the heritage. The one thing all management bodies have in common is that they barely involve local people in their planning or implementation. The Borobudur Conservation Office is most interested in the academic value and conservation of the monument. PT Taman Wisata Candi Borobudur and Prambanan aims to gain maximum profit or financial value. Similarly, the local Government attempts to raise income from their authorised zones and thus often 'forgets' to enforce necessary regulations. Meanwhile, the local people struggle for their right to have more access to obtain cultural as well as economic benefits from the monuments. The latter becomes essential as more and more local people depend for their subsistence on Borobodur's value as a tourism attraction.

Among the four management bodies, PT Taman Wisata Candi Borobudur and Prambanan is the only profit-oriented body. It is a state-owned corporation which was established to gain as much profit as possible from Borobudur, especially through tourism. Sometime its policies and conduct are not in accord with the conservation principles pursued by other bodies. To obtain more profit, Zone II is often used for activities irrelevant to the heritage, such as to house a flying-fox adventure game and merry-go-round. The protection embankment to the south of the monument was modified to enable the performance of a newly choreographed opera narrating the interpretive history of the construction of Borobudur. Housing facilities formerly constructed for the Borobudur Research Centre were converted to an exclusive hotel. While sometimes it was claimed that the local people were given the opportunity to perform their traditional ceremonies and festivals within Zone II, in fact it was groups of people from outside Borobudur who were invited to perform there. Most of the profit obtained by PT Taman Wisata Candi Borobudur and Prambanan is rendered to the Central Government. Only a very small portion is shared with the Borobudur Conservation Office to support conservation programs, and even less is contributed to local community development programs. The Magelang Local Government gains a considerable sum of money from various taxation schemes and parking fees. Obviously, in using this management model, the Indonesian Government was implementing the spirit of the so-called "archaeology in the service of the state" (Kohl and Fawcett 1995: 3) where "archaeology has served the needs of the nation-state and those in a position of power and privilege. Archaeologists often work as technicians of the state, under a system of 'Governmentality'... In actuality, public access to the material and intellectual results of archaeological research remains limited, and in most contexts 'held in the public trust' means 'owned and managed by the state' for particular state interests and purposes" (Nicholas and Hollowell 2006: 60). This is exactly what happened at Borobudur.

With such a management system, it is obvious local people obtain almost no financial or cultural benefits from Borobudur's heritage. An official Government statistic shows that Borobudur is the poorest village in Magelang Residency (Biro Pusat Statistik 2006). As they have no land, gardens, or paddy-fields, the local people have been increasingly forced to rely on tourism related activities for their subsistence. But as most of them do not have any skills base, the simplest way for them to make a living is to become vendors or street hawkers. Everyday more than 3000 hawkers swarm the monuments around Borobudur temple (Figure 5). This has caused problems for the tourists as well as religious pilgrims. Cultural Resource Management (CRM) expert Myra Shackley (2001) commented that Borobudur would no longer be a good place for pilgrimage since "any spiritual experience is militated against by crowds of persistent hawkers". A significant decline in the number of foreign visitors caused tension between local people and PT Taman Wisata Candi Borobudur and Prambanan to be amplified. The latter institution blamed local people, especially vendors and street hawkers, for the decline. For their part, the local people accused the Government of monopolising the management of Borobudur heritage. Local people felt betrayed by the Government. They argued that their predecessors had been forced to forfeit their property and heritage and the Government failed to fulfill its promise to improve the quality of life for the local communities of Borobudur. They also complained that PT Taman Wisata Candi Borobudur and Prambanan was allowed to carry out profit-oriented activities in Zone II, which violated the regulations, while local vendors and hawkers were strictly prohibited from entering this zone.

Figure 5. Local people have turned to tourism dependent subsistence since their lands were converted to the Borobudur Tourism Park. Thousands of them have become vendors or street hawkers who often chase visitors to buy their merchandise causing much inconvenience.

Source: Daud Tanudirjo.

Soon after the New Order regime collapsed in 1998, local people felt they had the freedom to access their Borobudur heritage. They acted as if the Government had no control of management anymore and that the extant regulations had been relinquished. They held several public meetings and rallies to protest the Government "ownership" of Borobudur. Vendors and hawkers behaved uncontrollably. They freely entered and offered their merchandise in Zone II, even up to the top of the monuments. PT Taman Wisata Borobudur and Prambanan lacked the courage to stop them as this was an expression of the people's euphoria after more than 30 years under the New Order authoritarian Government. But, this lapse in Government control was only temporary. Through mediation, gradually Government agencies regained their control of Borobudur.

Unable to have greater access to Borobudur, the local people pursued a different strategy. They were aware of their inadequate power to change Government policy on Borobudur so they shifted from a focus on access to the monuments to building greater integrity among the local communities. They revitalised their traditional culture by more intensively performing their traditional ceremonies and art festivals outside the protected area. Through such activities, they engaged communities living outside the resettlement areas, as far as the western slopes of Mount Merapi and Merbabu (Figure 3). It was apparent that the local people impacted by the establishment of the Borobudur Tourism Park were not only attempting to reinforce social cohesion among themselves but also to build more intimate relationships with communities outside them. Through this effort they were creating wider social networks and solidarity for their cause. Interestingly, the local people then started to identify themselves not only with Borobudur, but also with the broader landscape surrounding it and even with the Kedu Plain in general. They fostered a new awareness among the wider communities that the Borobudur landscape covers not only the Borobudur-Pawon-Mendut temples and the nearby villages but the entire area encircled by the seven mountains (Merapi, Merbabu, Andong, Tidar, Sindoro, Sumbing and the Menoreh Hills) (Figures 2 and 6) and extended their cultural landscape. This is evident in the transformation from what was initially the Borobudur Folk Festival to the Five Mountains Festival, and then Seven Mountains Festival (Figure 7). And, now even the concept of Borobudur as a Mandala (see above) has been embraced by the younger generation of Borobudur.

Figure 6. Menoreh Hills. The undulating Menoreh Hills have become the boundary of the Borobudur cultural landscape to the south.

Source: Daud Tanudirjo.

It cannot be denied that the construction of a new, wider Borobudur landscape was partly influenced by the academics who carry out research, teach field schools, and encourage community empowerment in the Borobudur area. Most of them work closely with and build good relationships with the local people. It was through the academics that the notion of the original Borobudur cultural landscape was introduced. In fact, most prominent leaders within the local community who actively engaged in the construction of the new Borobudur cultural landscape were generally closely associated with the academics. It was from academics that the local people learned about the Mandala concept, the cosmology of interrelations between the Borobudur monuments and the surrounding mountains and hills, the existence of an extinct lake around Borobudur, and even a basic knowledge of Cultural Resource Management (CRM). What they learned from the academics played a vital role in triggering their understanding of the wider cultural landscape. It also created a kind of new outward looking mind-set in which they framed their activities.

The ways in which the people of Borobudur attempted to construct a new and wider landscape were various. Sucoro, one of the prominent local leaders who was expelled from his land and even tortured by the military for his resistance, strives to revitalise traditional ceremonies that had almost died out in villages around Borobudur. He has travelled from village to village, some of them in isolated remote parts of the Menoreh Hills, to encourage the inhabitants to maintain their traditional ceremonies, arts, crafts and other cultural events (Figures 6 and 7). Once every two or three months, he gathers a considerable number of people from different villages around Borobudur to conduct the so-called *Pitutur* meeting in which some of the attendants sing traditional songs offering good advice (*Pitutur* literally means 'a good say'). Inspired by field schools held by academics at Borobudur, he started to organise his own style of field school for local people, especially the younger generation, in which they study many subjects, from practical skills (such as how to make certain kinds of craft items and videos) to more conceptual ones (such as CRM, social participation and heritage conservation). He often invites scholars to share their knowledge with local participants at the events he organises (Sucoro pers. comm., 2006).

Figure 7. As the local people had no access to Borobudur anymore, their traditional art festivals were annually conducted outside the protected zone to keep their social solidarity and integrity intact. Traditional cultural events played an important role in constructing the wider Borobudur Cultural Landscape.

Source: Photo courtesy Waring Info Jagad Cleguk.

A progressive young leader, Jack Priyatna, took a different approach. With some of his colleagues he initiated a tourism organisation network. Although he sometimes guided tourists visiting Borobudur, his focus is actually to develop tourism outside the protected areas as a means to empower not only local people who live close to Borobudur but also in the surrounding areas. As one of a younger generation who is fully aware of the injustice of the Government's treatment of his people, he is sceptical that the present management system of Borobudur will enhance the welfare of local people. He assumed that his people would not be able to rely on their heritage claims to the Borobudur monuments for their livelihoods. Therefore, he turned his attention to exploring other aspects of the broader heritage of the Borobudur region, such as the traditional villages with their own unique cultural as well as natural assets. He designated several villages each with a special attraction, such as a good view towards Borobudur, or traditional manufacture of pottery, vermicelli, tofu, or arts and crafts, and around these he created a Borobudur ecotourism package. Tourists travel around the Borobudur area by traditional horse-cart and visit villages with specialties. Along the road, they can view Borobudur from different aspects (Figure 8), appreciate an alternative experience and "visit and enjoy the real and comprehensive Borobudur heritage". The tourism organisation network also initiated geological tourism around Borobudur to see the traces of Borobudur's ancient lake (Jack Priyatna pers. comm., 2009) (Figure 9).

Figure 8. In response to their limited access to Borobudur, young people in local communities have taken the initiative to organise alternative tours which introduce visitors to the wider area around Borobudur, takes them to villages and provides views of the monuments from outside the protected area. This initiative was part of their effort to create a wider cultural landscape for Borobudur.

Source: Daud Tanudirjo.

Figure 9. Evidence of the existence of an ancient lake around Borobudur has been incorporated into alternative tours organised by the young local people of Borobudur.

Source: Daud Tanudirjo.

Futhermore, Priyatna promoted the Mandala concept for the management of Borobudur. It means that the Borobudur monument can be seen as the centre of the Mandala which attracts a great energy that can then be distributed to surrounding areas so that they in turn have the energy to protect Borobudur. The basic principle is that the management of Borobudur should not be focused on the ancient monuments alone, but also on the villages in the broader landscape. Hence, the more than two millions visitors who come every year to Borobudur would not solely spend their time in the Tourism Park, but would be encouraged to visit the rural areas around Borobudur. This strategy encourages PT Taman Wisata Candi Borobudur and Prambanan to work hand in hand with the local people and not monopolise the Borobudur Temple Compound. Such an approach has the benefit of reducing the burden on the monuments caused by large numbers of visitors, at the same time as dispersing prosperity to surrounding areas. Priyatna believes this Mandala concept will build up synergy among the stakeholders of Borobudur, between central and local Governments, between local people and the management authorities of Borobudur, and even between visitors and local people. All the stakeholders have their own position and power within a particular area, but they share responsibility and rights to make the system balanced. This is how the Mandala concept works.

Sucoro's and Priyatna's initiatives are indeed only a few examples of many other efforts by the local people of Borobudur in response to Government control of the existing management. All the initiatives made by the local people have played a crucial role in changing perspectives on relations between local people, heritage, and the landscape. They mean that Borobudur heritage is not only contained in the ancient monuments, but intertwines with all natural, demographic, and cultural aspects that exist within the wider area of Borobudur. And this eventually leads to the construction of a wider Borobudur cultural landscape that includes almost the entire Kedu Plain encircled by the seven mountains and hills. This changing perspective generates greater solidarity and stronger ties among the local people that significantly increases their bargaining power with the Government. Hence, the Borobudur cultural landscape has become a negotiating arena for a range of stakeholders.

The pressure brought to bear on the Central Government has ultimately encouraged it to change its policy on the management of the Borobudur Temple Compound. Of course, such pressure came not solely from the local people but also from academics, as well as UNESCO via the World Heritage Centre. Since the 1990s, academics have expressed their concern about Government management of Borobudur Temple Compound that focused almost exclusively on the built monuments and created inequality and conflict among the stakeholders. Since 2000, more academics have become part of the worldwide paradigm change in cultural resource management (see Tanudirjo 2006) that gives a greater role to local communities in the management of their heritage. This paradigm promotes the view that heritage should not be viewed as static and pertaining only to the past, but rather to be an integral part of the wider landscape, including the people who live around it and/or own the heritage (Shackley 2001; Smith 2004). The emergence of this new concept of cultural landscapes (Taylor 2003) also influenced Indonesian academics working in heritage management and has become one of the significant aspects of the Indonesian Charter for Heritage Conservation declared by heritage management activists and academics in 2003. With this new perspective, academics have tried to convince Government of the necessity to manage Borobudur Temple Compound as a cultural landscape in which local people and their traditional life style are integral, and to consult and involve local communities in planning processes and execution of CRM.

Lack of community participation in the management of Borobudur Temple Compound has become UNESCO's concern as well. This was partly brought about by escalating conflict due to the Government's intention to build the so-called *Pasar Seni Jagad Jawa* (The Spirit World of Java Art Mall) in Zone III of Borobudur in 2002. This project aimed to solve the problems caused by the large numbers of vendors and street hawkers at Borobudur. According to this plan, they would be concentrated in one big market and not allowed to sell their merchandise outside. However, this was not done in consultation with local communities and it instigated nation-wide protest. In April the following year, UNESCO-ICOMOS carried out Reactive Monitoring, followed by the Fourth International Experts Meeting at Borobudur organised by the Indonesian Government and UNESCO in July. On both occasions, the aspirations of the local people were collected and discussed. As a result, UNESCO recommended that the Government set up a conservation and management policy that would benefit local people. The policy should not only focus on the Borobudur temples but should embrace the wider cultural context of the monuments (Engelhardt *et al.* 2003; Adhisakti 2003).

The Indonesian Government anticipated these pressures by promoting a new catch phrase "the second stage of Borobudur restoration". In essence, over the next 20 years, this concept aims to restore within communities the socio-cultural relationships that are jeopardised by the current management system. In 2004, the Minister of Culture and Tourism appointed steering and organising committees to operationalise the concept. Most of the members were Government officials. Unfortunately, when another Reactive Monitoring was carried out by UNESCO-ICOMOS in 2006, it was found that the committees had achieved nothing. The report noted that an Action Plan drafted by the commitees to improve community roles in the management of Borobudur World Heritage was apparently only formulated as objectives, and there was no clear indication of responsibility to implement the plan. Hence, it was stressed again in the recommendation the need to strengthen the management system by establishment of a single management authority which includes representatives of local and Central Governments as well as local communities. The new management authority would ensure the protection of the monuments and their wider setting and the provision of benefits to the local communities (Boccardi *et al.* 2006).

The pressure brought to bear by local communities, academics, and the international bodies has prompted the Government to reconsider seriously the management policy of the Borobudur area. More and more high-level officials now realise the policy to exclude local people and focus only on the monuments is counter-productive. There is growing awareness that the significance of Borobudur World Heritage lies not only with the monuments but also with their natural setting or landscape, as well as with the communities living within the Borobudur area. In the last three years this changing perspective has been recognised through the review process of Presidential Decree no. 1/1992 that identified the source of conflict. In 2011, a draft of an integrated management plan for the Borobudur Temple Compound and surrounding areas was submitted to the Minister of Culture and Tourism. It is now being considered at the Inter-Ministerial Forum. Furthermore, this new perspective is being considered in the management plans of other World Heritage sites in Indonesia.

Epilogue

From Borobudur, we learn that the relationships between local people, heritage and landscape has changed over the course of time. Interestingly, this relates closely to the way its heritage has been managed. When Borobudur was still buried by earth and vegetation, it was considered a dangerous sacred place that was part of an unknown and different world. As the splendour of the monuments were revealed by the efforts of H.C. Cornelius in 1814, the local people started to identify themselves with them. They positioned themselves as the descendants of the great monument builders. Borobudur then became a source of pride and the local people were eager to be the guardians of this heritage. Through their various activities, such as village parades, art festivals, Idul Fitri celebrations and traditional ceremonies conducted around the monuments, the local people tried to always to renew and strengthen their traditional ties with Borobudur.

Under colonial Government administration at that time, the local people still had relatively open access to their heritage. However, since the 1970s the traditional ties between the local people and the monuments were disrupted by the Indonesian Government's policy of management, which began ironically as a desire to safeguard them. To facilitate extensive restoration, further protection and preservation of Borobudur-Pawon-Mendut complex, the Indonesian Government resettled local people who had lived nearby the monuments for hundreds of years. These local communities were then denied access to their heritage. Having lost their lands, their traditional cultural ties with the monuments gradually became economic ones as they now relied on tourism for subsistence. On the Indonesian Government side, the perspective shifted from academic to economic interests when they replaced the Borobudur National Archaeological Parks with the Borobudur Tourism Park. This is also evident in the Government establishment of a profit-oriented state corporation, PT Taman Wisata Candi Borobudur and Prambanan, to manage the heritage.

The desire of the local people to reclaim a relationship with their heritage led to the creation of the Borobudur cultural landscape. They identified themselves as part of a wider society that included communities living in the area encircled by the seven mountains, as well as with the ancient communities who built the monuments, thereby uniting the past and the present. These new relationships seem to have engendered stronger solidarity and ties among them.

By embracing the broader landscape and its population, they gain broader alliances and more power to negotiate with the Government, which is now seriously reconsidering the management policy of Borobudur World Heritage. A new policy is being prepared which will accommodate greater involvement of local people and comprehensive conservation of the wider Borobudur cultural landscape.

The Borobudur case study encapsulates a situation that is quite common in countries with a colonial history, one where the post-colonial Government claims a sole right to manage 'its' heritage. Similarly, this will tend to happen in countries with an "archaeology in the service of the state" perspective (Kohl and Fawcett 1995: 3). In these situations, the Government acknowledges no relationship between heritage and the living cultural traditions of local people. Consequently, it also disregards the landscape as a contextual setting for heritage that also requires management, as its focus is solely on the built environment. It is this perspective that allowed the Government to justify its decision to expel the local population from Borobodur in order to control the heritage. Consequently, heritage becomes isolated from its socio-cultural environment. It becomes an island in the past, static and disconnected from living tradition and practice, where the interpretation of meaning is imposed by the authority that manages it and is biased by that perspective. It is clear then that the establishment of a protected area by means of eviction and disenfranchisement of local people is not an effective measure for protecting heritage. Potentially, it will trigger long-lasting conflict among the stakeholders, which is counter-productive to the aims of heritage management.

References

Adhisakti, L.T. 2003. *Community Participation and Future Development of Borobudur Temple and its Environment*. A paper presented in the Fourth International Experts Meeting on Borobudur in Borobudur 4-8 July 2003.

Bernet-Kempers, A.J. 1976. *Ageless Borobudur*. Wassenaar, Servire.

Biro Pusat Statistik. 2006. *Kabupaten Magelang dalam Angka 2006*. Magelang: Bappeda – Biro Pusat Statistik Kabupaten Magelang.

Boccardi, G., Brooks, G. and Gurung, H. 2006. *Mission Report: Reactive Monitoring Mission to the Borobudur Temple Compounds*, World Heritage Property, Indonesia 18-25 February 2006.

de Casparis, J.G. 1950. *Prasasti Indonesia I*. Bandung: A.C. Nix.

de Casparis, J.G. 1981. The dual nature of Barabudur. In: Gómez, L.O and Woodward, H.W. (eds), *Barabudur: History and Significance of a Buddhist Monument*, pp. 47–83. Berkeley: Asian Humanities Press.

Dumarcay, J. 1978. *Borobudur*. Kuala Lumpur, Oxford University Press.

Engelhardt, R., G. Brooks, and Schorlfemer, A. 2003. *Mission Report: The Borobudur Temple Compound*, Central Java, Indonesia UNESCO-ICOMOS Reactive Monitoring Mission, 16-20 April 2003.

JICA Study Team. 1979. *Borobudur – Prambanan National Archaeological Parks*. Japan International Cooperation Agency.

Kohl. P.L. and Fawcett, C. 1995. Archaeology in the Service of the State. In: Kohl, P.L. and Fawcett, C. *Nationalism, politics, and the practice of archaeology*, pp. 3–18, London: Cambridge University Press.

Krom, N.J. 1927. *Barabudur: Archaeological Description*. The Hague: Martinus Nijhoff. 2 volumes.

Matsuura, K. 2005. Foreword. In: Anom, I.G.N. (ed.), *The Restoration of Borobudur*, p. 9. Paris: UNESCO.

Miksic, J.N. 1990. *Borobudur: Golden Tales of the Buddhas.* Periplus.

Murwanto, H., Gunnell, Y., Suharsono, S., Sutikno S., and Lavigne, F. 2004. The Borobudur monument (Java, Indonesia) stood by a natural lake: chronostratigraphic evidence and historical implications. *The Holocene*, 14 (3): 459–463.

Nicholas, G. and Hollowell, J. 2006. Ethical Challenges to a Postcolonial Archaeology: the Legacy of Scientific Colonialism. In: Hamilakis, Y. and Duke, P. (eds.), *Archaeology and Capitalism: From Ethics to Politics*, pp. 59–82, California, Left Coast Press.

Raffles, T.S. 1817. *The History of Java.* Oxford: Oxford University Press. 1978 edition.

Shackley, M. 2001. *Managing Sacred Sites: Service Provision and Visitor Experience.* London: Thomson Learning.

Smith, A. 2008. Landscapes of Clearance: Archaeological and Anthropological Perspectives. In: Smith, A. and Gazin-Schwartz, A. (eds). *Landscapes of Clearance: Archaeological and Anthropological Perspectives,* pp. 13–45, California, Left Coast Press.

Smith, L. 2004. *Archaeological Theory and the Politics of Culture Heritage.* London, Routledge.

Soekmono, R. 1976. *Chandi Borobudur: A Monument of Mankind.* Amsterdam: Van Gorcum.

Soekmono, R. 1983. Usaha Demi Usaha Menyelamatkan Candi Borobudur, in Menyingkap Tabir Misteri Borobudur. Prambanan: PT Taman Wisata Candi Borobudur dan Prambanan, pp. 16–17. (Originally published in daily newspaper Sinar Harapan on 17 February1983).

Tanudirjo, D.A. 2006. Heritage for All, Changing Perspective on Heritage Management in Indonesia. A paper presented in the *Rethinking Conference on Cultural Resource Management in Southeast Asia, National University Singapore* , 25-27 July 2006.

Tanudirjo, D.A. 2007. Cultural Landscape Heritage Management in Indonesia: an Archaeological Perspective. A paper presented in the *First International Symposium on Borobudur Cultural Landscape Heritage.* Centre for Heritage Conservation Universitas Gadjah Mada, 20 April 2007.

Tanudirjo, D.A. 2008. Inspirational Borobudur. A paper presented in a Cultural Dialogue "Borobudur dalam Jalinan Religi, Seni, dan Sejarah". Jogja Gallery, 6 May 2007.

Tanudirjo, D.A. 2011. Dampak Erupsi Gunung Merapi Terhadap Candi Borobudur. In: Sutopo, M. (ed.), *Menyelamatkan Candi Borobudur dari Erupsi Merapi.* Magelang, Balai Konservasi Peninggalan Borobudur.

Tanudirjo, D.A., Prasodjo, T., Yuwono, J.S.E. and Nugrahani, D.S. 1994. *Kualitas Penyajian Warisan Budaya kepada Masyarakat, Studi Kasus Manajemen Sumberdaya Budaya Candi Borobudur.* Research Report. Yogyakarta: Pusat Antar Universitas Studi Sosial, Universitas Gadjah Mada.

Taylor, K. 2003. Cultural Landscape as Open Air Museum: Borobudur World Heritage Site and its Setting. *Humanities Research* Vol. 10 No. 2, 2003, pp. 51–62.

Thomas, J. 2001. Archaeologies of Place and Landscape. In: I. Hodder (ed.), *Archaeological Theory Today*, pp. 165–186. Cambridge, Polity.

Voute, C. 2005. A New Perspective on Some Old Questions Pertaining to Borobudur. In: Anom, I.G.N. (ed.), *The Restoration of Borobudur*, pp. 213–249. Paris, UNESCO.

Walker, B. 1983. *Hindu World, an Encyclopedic Survey of Hinduism.* Vol. II. Munshiram Manoharlal Pusblisher.

World Heritage Center UNESCO. 2005. Operational Guidelines for the Implementation of the World Heritage Convention.

6

Being on Country: Githabul approaches to mapping culture

Nick McClean, The Australian National University, Canberra, Australia

With a general rise in interest in cultural mapping over recent decades, this chapter[1] discusses two approaches to recording and mapping Indigenous cultural relations to land employed by Githabul people in NE NSW. Mapping scope and methods in this project were developed with participants, revealing their particular views and situations. These approaches are explored, and I argue that the idea of being on Country, a common element of the two approaches, forms a key aspect of maintaining a shared cultural identity within the Githabul community. This influences perceptions of the value of mapping culture, while also opening heritage work up to include intangible aspects of culture.

Introduction – charting a brief history of Indigenous mapping

In an early and eloquent portrayal, Howard Morphy (1983) called attention to the complex interface of Indigenous culture and land politics in the post-protection era by describing Yolngu methods of representing 'sacred knowledge' in the public sphere as a means of demonstrating traditional ties to land. During the 1950s and 60s, Yolngu elders revealed to Australian courts and the world at large traditional bark paintings, usually kept hidden from the view of those uninitiated in their culture, and thereby demonstrated the basis of the laws that had in their eyes regulated a system of land ownership since time immemorial. This was a significant moment not only for its influence on the land rights movements of the 1970s, but it also represented a shift in the dynamics of Indigenous cultural politics in Australia, with Yolngu elders engaging directly, and to considerable effect, with the European legal system in terms that were rooted in Yolngu ceremony and culture. Morphy found it useful to think of the paintings as maps of the culture and politics in Yolngu society, maps which assisted Yolngu in maintaining political and cultural autonomy in a world that demanded their assimilation into the mainstream of white Australian culture.

In somewhat parallel developments, the last 20 years has seen an increasing trend of Indigenous communities across the globe mapping these relationships, increasingly in Western cartographic forms (Sirait *et al.* 1994; Rundstrom 1995; Fox 1998; Gambold 2001; Chapin *et al.* 2005; Pramono *et. al.* 2006; Byrne 2008; Tobius 2009). Peluso (1995) was the first to capture a precise image of this recent trend by coining the term 'counter-mapping'. In this case the Penan people

---

1   Fieldwork for this paper was undertaken during 2010 as part of my PhD research, and was jointly supported by the Department of Archaeology and Natural History, The Australian National University and the Country, Culture and Heritage Division, NSW National Parks and Wildlife Service through the 'Rethinking Cultural Heritage' ARC Research Project. I am grateful to all at Githabul Working on Country for their input and collaboration.

with whom she worked viewed their relationship with the forests of central Borneo as culturally embedded and unique, and chose to develop their own maps that provided a view of the landscape that ran 'counter' to the maps being developed by foresters and conservationists. Both of these ignored Penan relationships to land, instead focusing on either economic or scientific views of the forest, views that were proving influential in deciding the fate of those forests.

Peluso's account reflected some shifts in thinking since Morphy (1983) which would prove influential, widening the scope of these endeavours beyond the confines of the sacred, considering these issues within the context of conservation, and acknowledging the increased influence of mapping in bureaucratic land management and planning. Since then the idea of counter-mapping as a practical tool in land management has grown significantly, the successes of the land rights movement in Australia have delivered greater control over land to many Aboriginal communities, and the popularity of cultural mapping has remained.

A practical context for mapping

In New South Wales, this trend has been reflected in an increasing number of publications providing located explorations of mapping techniques, particularly in cultural heritage management contexts. English (2002), Byrne and Nugent (2004), and English and Gay (2005) have undertaken in-depth mapping projects in collaboration with Indigenous communities in NSW[2] and the opportunities and demands of the co-management system instituted within the NSW National Park and Wildlife Service in 1998 are frequently reflected in these studies, being a key local driver of community based heritage research.[3]

This paper explores mapping undertaken with Githabul people as part of a recent national parks co-management agreement in the Border Ranges in central eastern Australia. The Githabul are traditional owners of an area that encompasses the towns of Kyogle, Woodenbong and Bonalbo in the Northern Rivers region of NSW and extending north and west to Killarney and Warwick in southeast Queensland (see Figure 1). While some scholars, most notably linguist Margaret Sharpe (1985a, 1985b), have viewed Githabul as part of the Bundjalung group of tribes, many Githabul reject this categorisation and instead assert an identity as a separate cultural group with sovereign ties to their land.[4] In recognition of this, the Federal Court in 2007 recognised Githabul native title rights over 1120 km² of public land, stretching from the Border Ranges in the east, north to the NSW-QLD border and as far west as the Great Dividing Range. This land is managed as either timber production forest or as National Park, and following the determination, Githabul have negotiated agreements with the NSW Government for involvement in its management. In 2009 the Githabul Working on Country (GWC) organisation was established,[5] and since then they have been undertaking land management and conservation works inside the National Park lands, and have also recently begun developing programs with Forests NSW, who manage the remaining timber production forests.

---

2   See also Tobius (2009).

3   Adams and English (2005) provide a useful reflection on the co-management system in terms of negotiating diverging value systems and views of nature between Indigenous and non-Indigenous stakeholders.

4   When necessary Githabul point to anthropological studies since the 1950s which have identified their community as having a connection to traditional culture that is unusually strong in the context of southeast Australia (e.g. Hausfeld 1960; Creamer 1974; Native Title Services 2005).

5   The Working on Country program is an Australian Federal Government initiative designed to support the establishment of Indigenous land management organisations. In this case it has provided funding for the establishment of the Githabul Working on Country organisation to undertake nature conservation, natural resource, and cultural heritage management programs in the areas over which Githabul people hold Native Title rights. The principle means by which GWC achieves this is through their Ranger group, who undertake all on-ground works under the direction of the Githabul Native Title representative body.

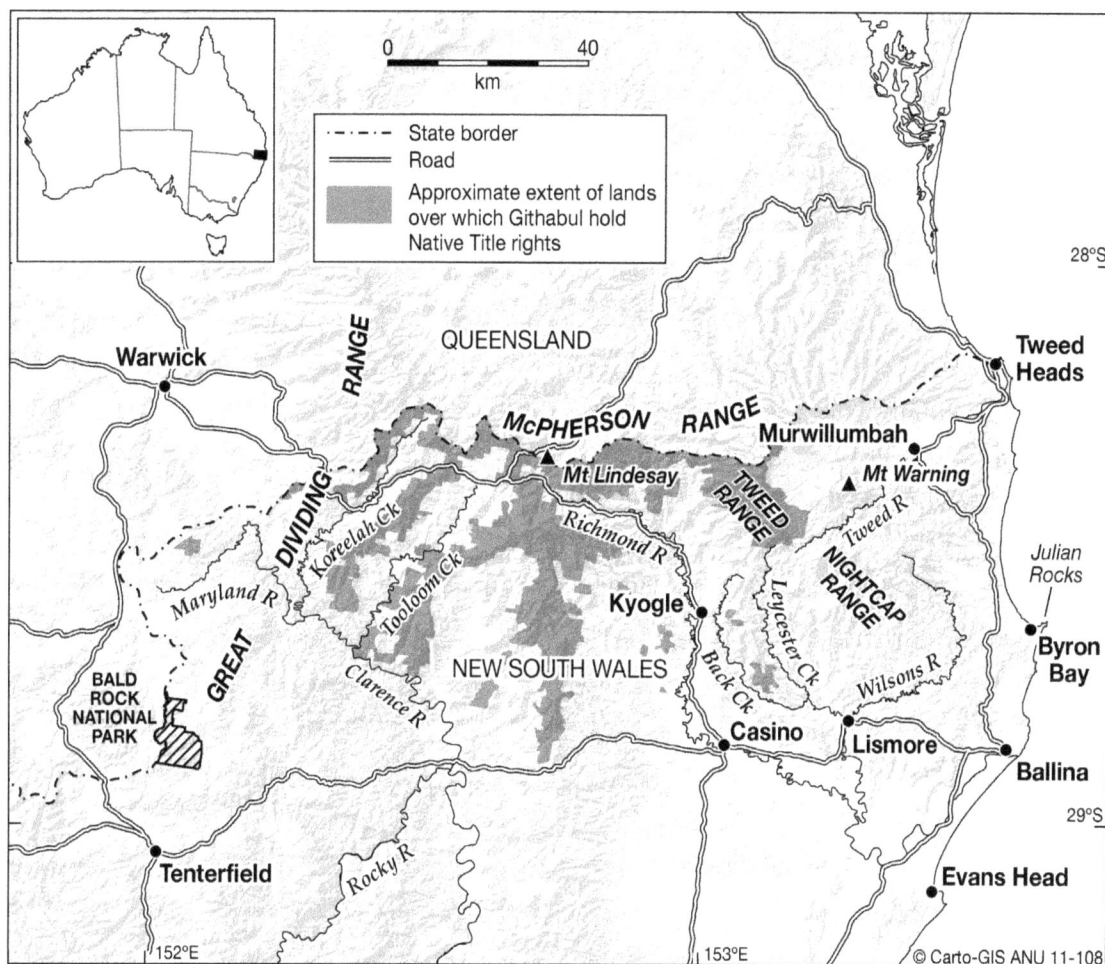

Figure 1. The Far North Coast of NSW.

Source: CartoGIS, College of Asia and the Pacific, ANU.

In this paper I will focus on two approaches to mapping Country[6] that have been utilised by Githabul community members within the context of this new co-management agreement —that of mapping traditional culture, such as stories, sites and language in traditional forms, and that of mapping everyday cultural practices, such as fishing places and hunting grounds, in Western cartographic forms. By describing the use of these contrasting methods, and considering these choices in the contexts of the Githabul community itself and the broader currents of Indigenous cultural politics in Australia, I hope to shed some light on the role that representing culture through mapping can play in a community context, as well as identify the limits of its value. In doing so, I also hope to reveal some of the basic aspects of cultural life for Indigenous people in rural New South Wales, and it is my view that this is a strong influence on determining perceptions of the value of such projects within the Githabul community, as well as the appropriateness of different forms of mapping.

As is common in many Indigenous communities, concepts of tradition and cultural continuity are central to individuals developing an understanding of their place within the Githabul community, and I initially want to show that the relationship to traditional Githabul culture, particularly as it relates to sacred knowledge, was the key factor in determining the approaches to mapping work

---

6   As is now widely recognised, *Country* is a term used by Aboriginal people throughout Australia to refer to their traditional lands. Used in this paper, it implies the culturally embedded nature of land ownership, and therefore includes in its scope of meaning the full range of cultural relationships that Indigenous people maintain with their lands. For an in depth discussion of the term see Rose (1996).

in each case. In the case of those in the community who held traditional knowledge and were considered its rightful spokespersons, the mapping process was similar to the Yolngu case, being determined and developed in ways seen to reflect and promote Githabul culture in its own terms. For those who were still learning about traditional culture, particularly younger participants, the work relied more heavily on the counter-mapping model. In the latter case the process was more oriented towards exploring the value of these techniques within the context of co-management, whereas in the former case the mapping was developed as an 'authentic representation' of traditional culture—its value was seen as both instrumental and intrinsic.

While the differences between the two approaches provide insight into the variation of knowledge and status within the Githabul community, the similarities reveal what can be seen as a fundamental aspect of a shared Githabul identity—that being on Country is seen as a central part of Githabul life. Similar connections can be found throughout NSW, however in the case of Githabul, the express linking of such activities to the survival of both traditional cultural forms and everyday cultural practices today provides a valuable insight into the views they hold about the relationship of their culture to that of mainstream Australia, and therefore the value they place on such planning and documentation activities as cultural mapping.

In this paper I seek to show that by insisting on the connection between being on Country and the documentation of cultural practices and values, Githabul are seeking to maintain their sense of tradition and continuity, while also resisting what Merlan has described as mainstream society's 'demands [for Aboriginal people] to demonstrate traditionality, or continuity with practices and forms considered traditional' (2006:85). In this way discussions of cultural heritage were taken beyond what might be considered the conventional archaeological view of heritage that focuses on sites to encompass issues of community health and well being, intangible values and spiritual philosophies, and contemporary visions of land management (Byrne 2008).[7] Such well-established themes in anthropological discussions and, increasingly, literature surrounding Indigenous land management in Australia are here extended to consideration of cultural heritage management and mapping in particular as a tool in program development with Indigenous organisations.[8]

Before proceeding, it is important to note some practical background of this project. This paper does not present results of a full mapping project and it does not necessarily represent the full range of views and expressions of cultural attachments to Githabul Country that exist within that community. Working in collaboration with Githabul Working on Country, in early 2010 a pilot project for developing cultural heritage programs was established within the organisation.[9] In the initial phase, this project focused on working with Doug Williams, a senior Githabul custodian, and in the second phase of the project, the work focused on involving the Githabul Rangers. As this was the first cultural project undertaken by the organisation, there was an identified need for the scope of the work to be negotiated on an ongoing basis, to allow for a period where I could be introduced to Githabul culture and Country, and also to allow for GWC to identify the best ways for collaborating researchers and consultants to be involved in cultural projects. With his in mind, this paper is an exploration primarily of the process of mapping Indigenous cultural heritage, through a case study of the issues that emerged in the course of conducting a pilot study with one Githabul community organisation.

---

7   See Maddock (1991) for a thorough examination of the rise of a sites-based discourse in Northern Territory land rights law. Many of Maddock's insights can be equally applied to subsequent developments since Mabo.

8   Rose (1996) e.g. in her discussion of the term 'Country' made the express link between community health and well being and connections to Country and this is becoming an increasingly accepted notion (see Albrecht *et al.*; Altman 2003; Johnston *et al.* 2007).

9   As noted in the acknowledgements, my capacity was as a PhD researcher, with our collaboration having the dual aims of contributing to a) a government funded research project exploring new directions in Indigenous cultural heritage, and b) the ongoing program of developing cultural heritage management programs within GWC.

One aspect that bears particular note here is that all of the active participants in this project were men, and while there was informal input and discussion with some Githabul women in the course of the project, the material presented here can be best understood as reflecting male perspectives on cultural life. Additionally, the nature of informed consent and the initial agreement brokered between myself and GWC required collaborative development of project themes with participants, and with this in mind I have chosen to present extensive explanatory material in the form of quotes from participants—where possible I have allowed their own words to speak for themselves. I have also included my own observations of the process itself where this has been required, and while this paper was cleared with project participants prior to its publication, the conclusions offered nonetheless remain my own.

Mapping traditional culture

In the initial field trips with Doug Williams, his status as an elder and fluent language speaker meant that the project focused on recording interviews about Githabul history and culture with a distinct emphasis on traditional cultural sites, stories and spirituality. He is regarded as an authority within his community and so to a large extent the work was about me learning and recording the different ways in which Doug has been asserting and maintaining the continuity of that culture. The central elements in this learning process were, firstly, a kangaroo skin map of Githabul tribal boundaries, place names and songlines that Doug has been developing over years, which was shown to me at our first meeting (see Figure 2), and secondly, a series of trips along the routes depicted on the map, to show me the Country it describes and relate stories, both cultural and historical, associated with those routes.

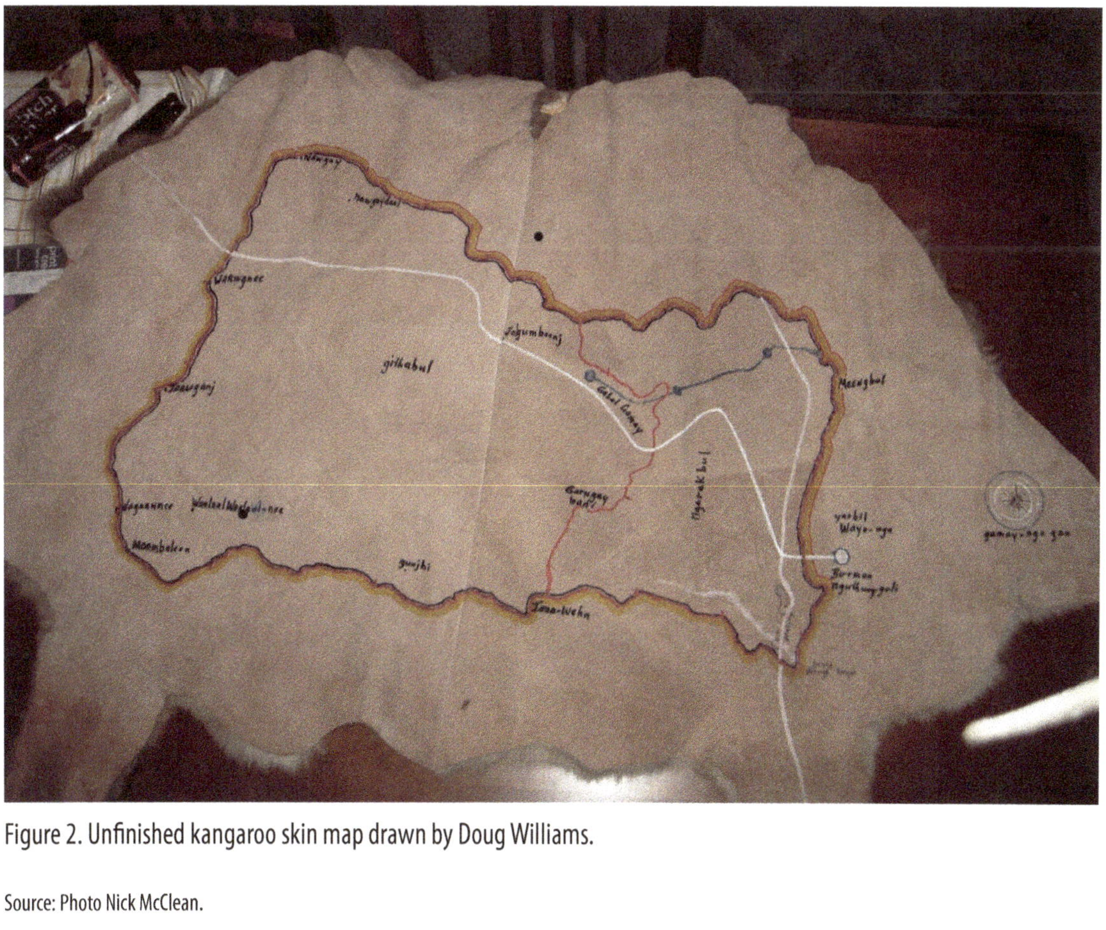

Figure 2. Unfinished kangaroo skin map drawn by Doug Williams.

Source: Photo Nick McClean.

To consider first the kangaroo skin map, Doug gave this description of its significance:

> First of all the two skins are tied or knitted together, one on the left is the Githabul one, it's a male kangaroo, it's Githabul, and that's all my grandfather's tribal boundary, his Country, and all the history, the *butheram*, the stories, *ngumbule*, ceremony. And on the right side's my grandmum, my grandmother, Ngarakbul nation, sovereign tribe. And there's that skin marriage, clans that were joined together of the two tribes, and everything pertaining to the laws within those two tribes were gelled together, and gave us free passage, access both ways. So they're my blood. I'm a blood connected sovereign owner of land, my grandfather and my grandmother.
>
> And why is it on the kangaroo skin? Well this is our way of recording and transcribing, and there are sacred stories and sacred songs that are associated with this skin that detail everything about both tribes, and primarily it starts with *gamay-nga gan*, Venus, the morning star, which is over on the right hand side. And the songline that started from Julian Rock, just out of Byron Bay, *wayo jalgumboonj,* the fairy-emu that my grandmother talks about, goes right through the central desert, right out through the west coast, out to sea out there.
>
> And so the kangaroo skin, to do a comparison to white man's history, they've documented everything, supposed to be on paper, and indexed and archived on paper. But this here, obviously the kangaroo, they use it now as a national symbol, but to us it's more sacred than all of that, because this here predates everything about federation, everything about the invasion.
>
> So this skin, it outlines the boundary of both tribes and it tells us about where our boundary is, and also the stories and the songline that connects up to neighbouring tribes. But primarily, all the names that I have here are names of some of the current towns and villages that have been named after a lot of the white history, I've changed and given the proper name in Githabul, the real name which has meaning and substance. (Kyogle 3/12/10)

Once Doug had shown me the kangaroo skin, the next task was for us to go on a series of trips across the Country he had mapped out as a way of familiarising me with it and relating different aspects of what he described above.

Throughout the Country, there are many places that Githabul often make oblique references to as powerful places. Some of these are referred to as *juraveels*, and form a key aspect of Githabul cosmology and beliefs, being seen as places where powerful spirits reside. They are also a key aspect of Githabul political organisation, with different families in the community having responsibilities to 'look after' particular *juraveels* and therefore are considered to have rights to 'speak for' those places. Today this translates across to practical issues of consultation over their use, be it for access, potential developments or for cultural recording projects (see Hausfeld 1960 and Creamer 1974 for more in depth discussion of *juraveels*).

One of the interesting things to observe was that as we'd travel to different places, Doug would often begin speaking in Githabul in a low voice, and on at least two occasions he sang out loudly. Once he stopped at the point where Mt Lindsay (see Figure 3) is first visible from its southern approach to 'sing out' to the mountain, and once from a large rock above a valley high up in the Bald Rock Range. When I asked him about the significance of language and song, he had this to say:

> There are two important factors that we consider very sacred, and of paramount importance in the history, and the families that are living in even *guri guriarba,* long before my time, *guriarba* days. That language, story and song are inseparable, they are an integral part of our whole culture, and its connected to the land

and also *gumanyargan*, the star laws, these are very prominent and very important, where the stars, the heavens connect up to the waters, and comes through the land, it goes through the cycle, it continues on, it lives on. And wherever we go there's stories and songs, we can walk our songlines. And there are certain places where we must sing to have access, to appease the *juraveel*, wherever, that mountain, that waterfall, to have safe passage. There's a whole lot of aspects to these different stories, the *butheram,* about certain significant and sacred sites. (Kyogle, 3/12/10)

The idea that Githabul must sing to Country as they enter sacred places extends further—to sing or speak about them while in another place is forbidden, you must be in a place to be able to talk about it. As described, this is a practical process of ensuring safety and prosperity while in that place; however, it also serves to maintain a highly localised cultural form, where separation from Country prevents the activities that directly maintain the connection of Githabul with their sacred places, a process that connects Githabul into their Country and its continuing cycles. Importantly, this was one of the primary reasons why Doug would not record specific information about spirits associated with places and kept our discussions largely in the realm of general ideas of Githabul culture, history and law – it can be viewed as a means by which the story could be related beyond the confines of its appropriate use, it opened the culture up to misuse. It was also the reason he would insist on visiting places as a central aspect of our work.

Figure 3. Mt Lindsay near Woodenbong, a prominent peak in the Border Ranges and one of the central places in Githabul cosmology.

Source: Photo Nick McClean.

We did however visit five areas as important places in the Githabul cultural landscape —Mt Lindsay (see Figure 3), Julian Rocks (see Figure 4), Bald Rock, Bulls Head Mountain and Toloom Falls. Of these, Mt Lindsay, Bulls Head Mountain and Toloom Falls are in the immediate vicinity of Woodenbong and are well documented as being associated with important *juraveels* (see Hausfeld 1960 and Creamer 1974). Julian Rocks is associated with Githabul culture by virtue of the alliance Doug claims between Githabul and Ngarakbul and by it being the beginning of a large songline that passes through Githabul Country. Bald Rock Range is said by Doug to form the Western extent of Githabul Country, and he chose to go to Bald Rock to re-establish a relationship that he says Githabul have been prevented from maintaining since white people came to Githabul Country.

Figure 4. Doug Williams at Julian Rocks, Byron Bay, the beginning of the fairy emu songline that passes through Githabul Country.

Source: Photo Nick McClean.

Of interest in this discussion are Doug's explanations of how he came to learn about traditional Githabul culture, which he related as follows:

> As recorded by my great uncle Dhuroom, that's my Dad's eldest brother, the family genealogy was done by him, taken way back to a man named Yaguy. Many occasions when I was growing up I'd listen to my dad, my grandma, the old men, who would always talk, and there were continuous and constant ceremonies and relating to the times and the sequence of events that used to relate to the land and to the different parts of the land, and particularly the two major factors was having the language and the songs, and the *butheram*, the stories. *Yaribil*, the singing of those songs, and *yawarr*, the ceremonies. And they were handed down, knowledge, and even how to avoid confrontation where necessary, and perhaps even suppress our own knowledge and culture, and then it would be brought out at an opportune time.
>
> And that has been the case, our people were very wise, they knew there were elements, there were people that was employed, engaged by the government and the policy makers, and all that, very disruptive through the Aboriginal welfare board policies and legislations that were forced on our people to suppress and to try and erase… many attempts, we thought they were managers that were commissioned by the government at the times, in the early 50s and 60s to come on and manage our people on the communities, on the missions, and they were anthropologists in disguise. And I guess some of them done good in a sense. It was a benefit in disguise I guess. They were able to engage in discussion with our older people to continue the history and the survival of our people. It has happened in a lot of other tribes that lost their language and their culture. But I'm grateful it didn't happen to us, cos we were here among the mountains and our people knew their Country very very well, they would steal away from any onslaught the attempts to suppress and forbid our people even speaking language. So there are many things there that I can't even mention perhaps. (Wiangaree 24/4/10)

By way of brief explanation, Doug's great uncle Dhuroom was also known as Stan Williams, and from this we know he was one of the chief informants of both Hausfeld and Creamer in their anthropological studies, and Geytenbeek (1971) in his linguistic study. The anthropologist in disguise is a reference to Hausfeld, who was the manager of Muli Muli Aboriginal station near Woodenbong and employed by the Aborigines Welfare Board at the time of writing his anthropological thesis on Githabul culture.

Mapping hunting and food gathering places

For the Ranger group, the focus of the work explicitly avoided dealing with traditional cultural sites and spiritual stories. Instead we focused on everyday cultural practices, such as food gathering and hunting places, which towards the end of our field season we began recording onto topographic maps.

Initially I put it to them that I was interested in learning about Githabul culture and finding a way for us to put that learning process to a practical use. However this caused some unease among the group, as they took this to mean that I wanted them to tell me about the *juraveels* and other traditional cultural matters. A number of the group were young men who were still learning the knowledge about spiritual places that elders like Doug held, and while some felt that our project could provide a useful means for them to go out on Country with elders and learn some of this knowledge, others were wary. The main reason offered related to the issue of speaking about places with spiritual associations described above, with some of the Rangers again expressing the view that these were powerful places and shouldn't be taken lightly.

After some discussion however we agreed to begin pursuing the process of documenting *bing-ging* (freshwater turtle) diving places in the rivers around Kyogle. *Bing-ging* diving is a popular pass time for almost every young Githabul person I met, and an activity that the Rangers saw as being an important part of being Githabul. Some related stories of all day dives of 30 or more people, walking up the Richmond River from Wiangaree into its upper tributaries and bringing the spoils back to the Muli Muli Aboriginal station at Woodenbong to share out among the community. As well as describing a wealth of hunting and gathering activities he practiced as a teenager, GWC manager Rob Boota gave this description of *bing-ging* diving:

> We used to hunt turtles, just sit in the water all day and wait for them to come up, when they came up we'd duck under and we'd swim up as close as possible to them. I guess it's a bit of an art, you've got to know when to duck under and swim up and grab them from beneath. But we wouldn't use nothing, you'd just hold your breath and come up underneath. You've got to do it in full sunlight, you couldn't do it on a cloudy day because you couldn't see too good, the waters got to be clear. (Kyogle 4/12/10)

This idea of *bing-ging* diving as a Githabul cultural practice is reinforced by the fact that it is a practice almost non-existent among the non-Indigenous population in the area. As we progressed, however, our discussions became broader and we began including the range of hunting and gathering places that the Rangers frequented.

After these discussions, the Rangers decided the next step was to take me out on a series of trips through their Country to show me some of the areas they would map, and also how to dive for turtles. Our main *bing-ging* diving trips were in the Richmond River (see Figure 5) and one of its tributaries Fingal Creek. We spent an afternoon chasing perch in Toloom Creek and they also invited me to join them for a day at a men's cultural camp they organised at Paddy's Flat on the Clarence River (see Figure 8).

Back at the GWC depot in Kyogle, large prints of topographic maps were used to mark the areas where people regularly fished, dived and hunted (see Figures 6 and 7). This started slowly, with me asking the Rangers to show them where they went so I could mark in the areas, but pretty soon they got the idea, marker pens were seized from my grasp and the mapping began.

Interviews were also carried out to provide background to the mapping. The maps and interviews indicate the extent to which Githabul have been able to maintain their day to day relationships with their Country, and show the depth of knowledge about the Country that has been retained through this.

It's worth briefly noting two details of the map which I will come back to in later discussion. Firstly, there was a high concentration of areas around Upper Toloom Creek and areas easily accessed by road from Woodenbong and Kyogle, as far down the Richmond River as Stoney Gulley. Secondly, while there were areas marked in all over Githabul Country, there were no sites recorded by the Rangers in the Duck Creek catchment or in the vicinity of Bonalbo, despite this being recognised as Githabul Country, and despite the Rangers having recalled visiting there numerous times.

Figure 5. *Bing-ging* diving, Richmond River.

Source: Photo Nick McClean.

Figure 6. Githabul Rangers mapping fishing and hunting places.

Source: Photo Nick McClean.

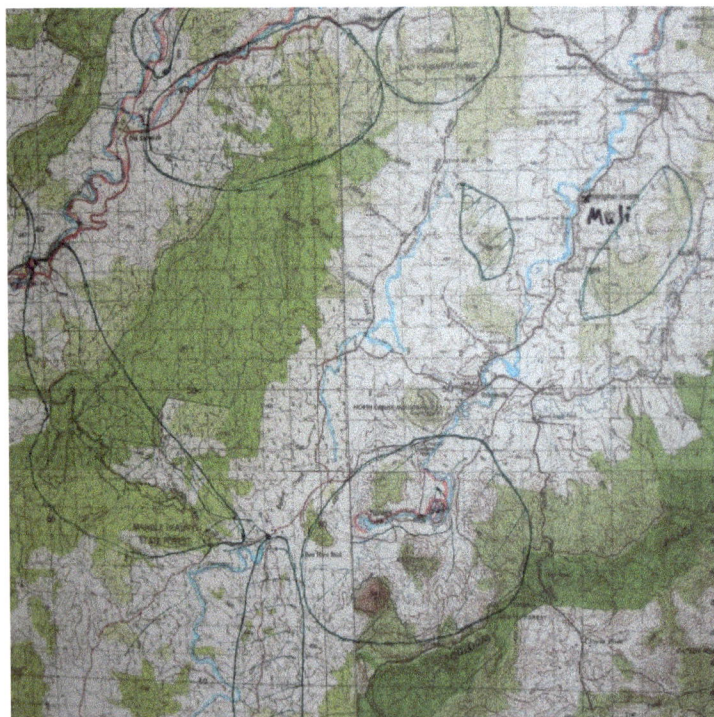

Figure 7. Detail of map.

Source: Photo Nick McClean.

Mapping and knowledge – comparing approaches

In terms of understanding approaches to mapping here, the first point to make is that the scope of mapping was directly influenced by the status of group members as bearers of Githabul cultural knowledge. Most obviously, Doug Williams is in a position of recognised authority about spiritual aspects of the culture, which gives him the right to determine how and when this knowledge should be represented publicly, while the Rangers, who as individuals have varying ages and levels of cultural knowledge, did not feel they were in a position to directly deal with spiritual matters when engaging with people outside their community. The reasons for these differing levels of knowledge are possibly quite complex, however it is clear that Doug Williams is closely related to people who spent much time documenting and recording their cultural knowledge in the past, so we know that he grew up directly in an environment where he could have access to that knowledge. For the Rangers, while some are in their early 20s and state they have not yet had the chance to learn the deeper knowledge of people like Doug, others are older and do have some of this knowledge. Even so they were not willing to discuss this in much detail, seeing this as Doug's particular role within the community.[10]

To consider the patterns of what each process recorded, there are two significant points to be made. Firstly, that there was a concentration of sites, both spiritual and hunting/fishing, in the immediate vicinity of Woodenbong. This may seem unsurprising given that all the participants are residents of Muli Muli, just outside Woodenbong, or have lived there for a significant portion of their lives. However, this pattern is also revealed in Creamer's (1974) site recordings and sketch maps in the early 1970s, so an initial comparison of these two studies suggests there has been a significant level of knowledge retention in the Githabul community, reflecting longstanding cultural affiliations and practices in that area. Conversely, there is a notable absence of any sites recorded by participants in the catchment of Duck Creek. This was explained by the fact that the families who have the right to speak for Country in that catchment were not participants in this project, so that even if participants did have knowledge or experience of that area, they explicitly refused to discuss it, or to take me there on trips to show me the Country.

A final point to make here is that for Doug, the mapping process was wholly initiated by him and using methods rooted in Githabul cultural forms. He saw this as an expression of the uniqueness and autonomy of Githabul culture, and a direct statement against the writing of history in white terms. For the Rangers however, the mapping was initially suggested by me, the rationale was discussed as being largely within the terms of co-management (that the maps might provide useful for developing programs and projects within GWC), and that the methods used largely reflected cultural forms rooted in non-Indigenous traditions (topographic maps, GIS truthing of data, English language names). This is not to diminish the value of the maps or the information they record but it is a relevant aspect of considering the process.

Mapping and culture – understanding the terms of engagement

The most significant common factor in these mapping processes was that participants insisted that they take me out on trips to show me the Country. The centrality of being on Country in traditional spiritual practices and relations to land has been described above, and that the discussions of contemporary, everyday culture were also linked to such a process indicates a shared cultural practice amongst Githabul, regardless of cultural status. This echoes English's observation that through 'fishing, hunting, plant food collecting, camping, walking along pathways and seemingly innocuous activities like swimming or sitting round a camp fire …[Aboriginal] people express and 'activate' their associations with place' (2002:4). This aspect of the project indicated that for Githabul, such associations with place remain an important part of their daily lives, and discussions even extended this idea, with connections to Country being seen as important for

---

10   Williams, pers. comm., 2010.

community health and wellbeing. While on the men's cultural camp, the Rangers encouraged me to interview one young Githabul community member (who chose to remain anonymous), and he described being on Country in this way:

> Yeah I've just come out here to get back into our cultural ways, hunting and bonding with the family again. Really too, just get away from society too, where there's drugs and alcohol. Really I've come out here to focus on culture, you know? Keep bringin' the young fellas through what the old people taught us, so we can bring them up and show them how to hunt, about the culture, about the land we're on, this is Githabul land, yeah, what can I say? We're just doing a bit of turtle divin', a bit of fishin', getting some wood, like doing chores every day, getting wood for the night and that. Oh yeah having a camp like this with the family, it's good you know, having family is good, you can speak more open to your family than other people you know? Yeah good, just to realise how our old people lived. We take that in too, take it in while I'm out here, give us remembrance of our old people, how they used to live, you know, how the atmosphere
>
> is you know? It's beautiful country, wouldn't think it'd be like this eh? (Paddys Flat 01/12/11)

So the practice of being on Country is seen as central to being Githabul in a way that goes beyond ideas of cultural knowledge, cultural forms or discussions of mapping, it's even seen as a basic aspect of being a healthy, happy person in that community. For the Rangers, the mapping process we undertook was an interesting project that they were exploring for its value in their working lives, but one of its most meaningful aspects, from my observation, was that it was linked to the things they do to stay connected to their Country. As a process connected to the world of funding agencies, organisations for land management and the world of Native Title, it was something they wanted to explore to gauge its value for them. The conclusion I came to was that if they felt it didn't contribute to their being able to continue connecting to Country, to continue being out on Country, its proper value would be within their working lives and their interactions with the world outside their community, rather than as a process that maintains cultural knowledge and practices within their community.[11]

In a similar way, Doug Williams took our discussions beyond the realm of documentation by discussing at length the connection he felt this knowledge had to Indigenous sovereignty, the idea that Indigenous people's legal title to land has never been extinguished due to the illegal nature of the occupation of the Australian continent by the British crown. While the legal debate relating to British settlement is not part of the scope of this paper, Doug described the nature of Indigenous sovereignty in the following way:

> We're free people. Before *dhagay*, long long time ago, our people had ownership, title, to land, water, everything about the environment. Because of our existence, our connection to our Country. And we can describe our land, our boundaries, our stories, our law. Githabul-Ngarakbul law. Very very sacred. (Wiangaree, 24/4/10)

So in this context, showing songlines, the telling of the stories, and Doug's recording these on the kangaroo skin are not a symbolic act or a cultural display divorced from a meaningful political outcome, but are seen as the assertion of an autonomous and continuing law that underpins land ownership in Githabul-Ngarakbul Country. As a result, Doug saw the value of present day arrangements coming out of Native Title as having a positive yet ultimately limited value, and continues to pursue recognition

---

11   Again I wish to point out that this does not diminish the value of the process, but it is relevant to understanding how and in what ways this sort of mapping can be useful. I found the Rangers to be overall quite enthusiastic about the project, while also interested to communicate the precise value of the work I proposed to them.

of sovereign ownership of Githabul lands based on the argument briefly outlined above. He states that the capacity to act as sovereigns, which includes the right to develop treaties in the political realm, will provide the basis for cultural renewal on Githabul Country.[12]

Figure 8. Githabul Men's Culture Camp at Paddys Flat, NSW.

Source: Photo Nick McClean.

Discussion and some conclusions

To turn to the issue of how the process of mapping relates to issues of cultural identity, in this case mapping and heritage recording undertaken here was assessed by Githabul participants for its value in promoting and maintaining a core aspect of Githabul life, being on Country.

To note first is that this assessment process (informal as it may seem) certainly doesn't seem unreasonable when we consider the experience of cultural mapping in other places. While Tobius (2009) presents the mapping experience for First Nations in Canada as being on the whole an empowering experience, Pramono *et al.* (2006) are less positive about its value in Indonesia, where Indigenous groups have mapped over 1,000,000 ha of traditional lands in the manner Peluso (1995) has described, and a number of other authors are interested to promote cautious and well planned approaches to mapping given the possible pitfalls of the process (e.g. Peluso 1995; Fox 1998; Deddy 2005; Chapin 2006). Some authors, such as Rundstrom (1995), even take the view that any mapping of Indigenous cultures done in Western cartographic forms represents a fundamental compromise of cultural autonomy.

---

12   See also Morphy (2007:31) for a discussion of the perceived link between cultural relationships to land and the continued existence of sovereignty among Yolngu. She observes that 'the Native Title process is an arena in which, among other things, the sovereignty of a colonising society over its colonised subjects is enacted'—a view which would not be without supporters among the Githabul.

The second is that for cultural documentation in general, the demands of the Native Title era in Australia have proven influential. Here, as Merlan (2006 see also 1995, Povinelli 2002) describes, Aboriginal people must demonstrate 'traditionality' in terms approved of in relation to the liberal multicultural values espoused by the state in order to have formal rights to land recognised. She argues that we must move 'beyond tradition' as it has been conceived in Australian land rights law and instead consider fresh means of working with divergent modes of culture and land relatedness. In a similar way the Githabul response to this dynamic has been a restatement of the importance of tradition in such a way that the power to define its meaning is not yielded wholly to the state, regardless of the value system it maintains. In doing so, it also opens up its meaning to a range of issues not often associated with traditional views of cultural heritage, in this case philosophies of land, spiritual beliefs and practice, community well-being.

Mapping here appears to exist at the watershed of this divide. Such mapping techniques as counter-mapping are essentially grounded in Western terms, and their principle value exists in articulating culture in ways that are conversant with the world outside the Githabul community. Such mapping techniques as recording spiritual knowledge onto kangaroo skins serve to assert a system of law that is seen as entirely autonomous from mainstream Australian law, reminiscent of the Yolngu episode described by Morphy. That both these forms are assessed for their value through the context of maintaining a connection to Githabul Country in my opinion serves to proscribe a space in which Githabul cultural life can exist, conceived by them as functionally and experientially separate from the processes that seek to define its official limits.

References

Adams, M. and English, A. 2005. Biodiversity is a whitefella word: Changing relationships between Aboriginal people and the NSW National Parks and Wildlife Service. In Taylor, L., Ward, G., Henderson, R. and Wallis, L.A. (eds), *The power of knowledge: The resonance of tradition*, pp. 86-97. Aboriginal Studies Press, Canberra.

Altman, J. 2003. People on Country, healthy landscapes and sustainable Indigenous economic futures: The Arnhem Land case. *The Drawing Board: An Australian Review of Public Affairs* 4(2):65–82.

Byrne, D. 2008. Countermapping: New South Wales and South-East Asia. *Transforming Cultures Journal* 3(1).

Byrne, D. and Nugent, M. 2004. *Mapping attachment: A spatial approach to Aboriginal post-contact heritage*. Department of Environment and Conservation, Sydney.

Chapin, M. 2006. Mapping Indigenous lands: issues and considerations. www.globetrotter.berkeley.edu/bwep/colloquium/papers/Chapin2006.pdf [accessed: 25/4/11]

Chapin, M., Lamb, Z., and Threlkeld, B. 2005. Mapping Indigenous lands. *Annual Review of Anthropology* 34:619–638.

Creamer, H. 1974. *A gift and a dreaming: the NSW Aboriginal sites survey.* National Parks and Wildlife Service, Hurstville.

Deddy, K. 2006. Community mapping, tenurial rights and conflict resolution in Kalimantan, Indonesia. In: Majid-Cooke, F. (ed.), *State, communities and forests in contemporary Borneo*, pp. 89–110. ANU e-Press, Canberra.

English, A. 2002. *The sea and rock gives us a feed: Mapping and managing Gumbaingirr wild resource use places*. National Parks and Wildlife Service, Hurstville.

English, A. and Gay, L. 2005. *Living land living culture: Aboriginal heritage and salinity*. National Parks and Wildlife Service, Hurstville.

Fox, J. 1998. Mapping the commons: the social context of spatial information technologies. *The Common Property Resource Digest* 45:1–4.

Gambold, N. 2001. Participatory land assessment: Integrating perceptions of Country through mapping. In: Baker, R., Davies, J. and Young, E. (eds), *Working on Country: Contemporary Indigenous management of Australia's lands and coastal regions.* Oxford University Press, Oxford.

Geytenbeek, B. 1971. *Gidabal grammar and dictionary.* Australian Institute of Aboriginal Studies, Canberra.

Hausfeld, R. 1963. Dissembled culture: An essay on method. *Mankind* 6(2):47–51.

Hausfeld, R. 1960. *Aspects of Aboriginal station management.* Unpublished MS thesis, University of Sydney.

Johnston, F., Jacups, S., Vickery, A. and Bowman, D. 2007. Ecohealth and Aboriginal testimony of the nexus between human health and place. *Ecohealth* 4:489–499.

Maddock, K. 1991. Metamorphosing the sacred in Australia. *Australian Journal of Anthropology* 2(2):213–232.

Merlan, F. 2006. Beyond tradition. *Asia-Pacific Journal of Anthropology* 7(1):85–104.

Merlan, F. 1995. The regimentation of customary practice: From Northern Territory land claims to Mabo. *The Australian Journal of Anthropology* 6(1):64–82.

Morphy, F. 2007. Performing law: the Yolngu of Blue Mud Bay meet the Native Title process. In: Smith, B.R. and Morphy, F. (eds), *Social effects of Native Title: Recognition, translation, coexistence.* CAEPR, Canberra.

Morphy, H. 1983. Now you understand: An analysis of how Yolngu have used sacred knowledge to retain their autonomy. In: Peterson, N. and Langton, M. (eds), *Aborigines, land and land rights,* pp.110–133. AIAS, Canberra.

Native Title Services. 2005. Unpublished anthropological report. Native Title Services, Sydney.

Peluso, N.L. 1995. Whose woods are these? Counter-mapping forest territories in Kalimantan, Indonesia. *Antipode* 27(4):383–406.

Pramono, A., Natalia, I., Janting, Y. 2006. Ten years after: Counter-mapping and the Dayak lands in West Kalimantan, Indonesia. http://dlc.dlib.indiana.edu/dlc/handle/10535/1997 [accessed: 25/4/2011]

Rose, D. 1996. *Nourishing terrains: Australian Aboriginal views of landscape and wilderness.* Australian Heritage Commission, Canberra.

Rundstrom, R.A. 1995. GIS, Indigenous peoples, and epistemological diversity. *Cartography and GIS* 22(1):45–57.

Sharpe, M. 1985a. *An introduction to Bundjalung and its dialects.* Armidale College of Advanced Education, Armidale.

Sharpe, M. 1985b. Bundjalung Settlement and Migration. *Aboriginal History.* 9(1):101–124.

Sirait, M., Prasodjo, S., Podger, N., Flavelle, A., and Fox, J. 1994. Mapping customary land in East Kalimantan, Indonesia: a tool for forest management. *Ambio* 23(7):411–417.

Tobius, T. 2009. *Living proof: The essential data collection guide to Indigenous use and occupancy map surveys.* Ecotrust Canada, Vancouver.

7

Exploring the role of archaeology within Indigenous natural resource management: A case study from Western Australia

David Guilfoyle, Applied Archaeology, Albany, Australia

Myles Mitchell, Applied Archaeology, Albany, The Australian National University, Australia

Cat Morgan, Applied Archaeology, Albany, Australia, University of Leicester, United Kingdom

Harley Coyne, Department of Indigenous Affairs, Southern Regional Office, WA, Australia

Vernice Gillies, Albany Heritage Reference Group Aboriginal Corporation, Australia

Introduction and background context

For Indigenous archaeology, an important measure of 'success' within any project is the level of control and ownership embedded with the local Traditional Owner community. If control/ownership is tokenistic, short-term, or undeveloped, archaeological research outcomes remain limited by default – in the understanding that Indigenous heritage management is linked to community identity and wellbeing, and requires delivery under customary practice/protocols. Any level of archaeological research – whether community, research or commercial – requires systems to ensure Traditional Owners are in control of all facets of project development, implementation, and reporting, at the level and context that they demand.

This paper outlines the results of a community heritage management project centred on a property bordering Lake Pleasant View, at Many Peaks, southwestern Western Australia (Figure 1). The focus here is on the role of archaeology and archaeologists in the design and implementation of the project as led and guided by the local Traditional Owners. The project was originally established as an Indigenous Cultural Heritage Management (ICHM) plan within a broader Natural Resource Management (NRM) project for a wetland under threat from natural and human processes (Guilfoyle *et al.* 2009). An outcome of this programme has been the acquisition of resources in securing a property vested with a Traditional Owner community organisation, that includes a long-term research programme examining the wider archaeological landscape centred around this wetland, and delivering an integrated conservation and management plan. For each stage of this project, an NRM/ICHM model was developed that drew heavily from, and integrated, a diverse body of theory within community, Indigenous and landscape archaeology.

This project was based on the recognition that effective Indigenous cultural heritage management requires protecting and managing both the physical fabric of places and landscapes, as well as the associated values related to community-identified social and cultural activity (Byrne *et al.* 2003). In so doing, structures are required to enable these activities to take place, and this necessarily involves moving beyond the assessment or identification of values, and to embedding a mechanism that allows social and cultural activity to take place. Although there are many facets of this dynamic project, this paper focuses on the role of archaeology in embedding community ownership and control of an area for effective management. It is suggested that a significant mechanism for linking diverse fields of archaeology may be found in the integration with methods and objectives that lie within NRM.

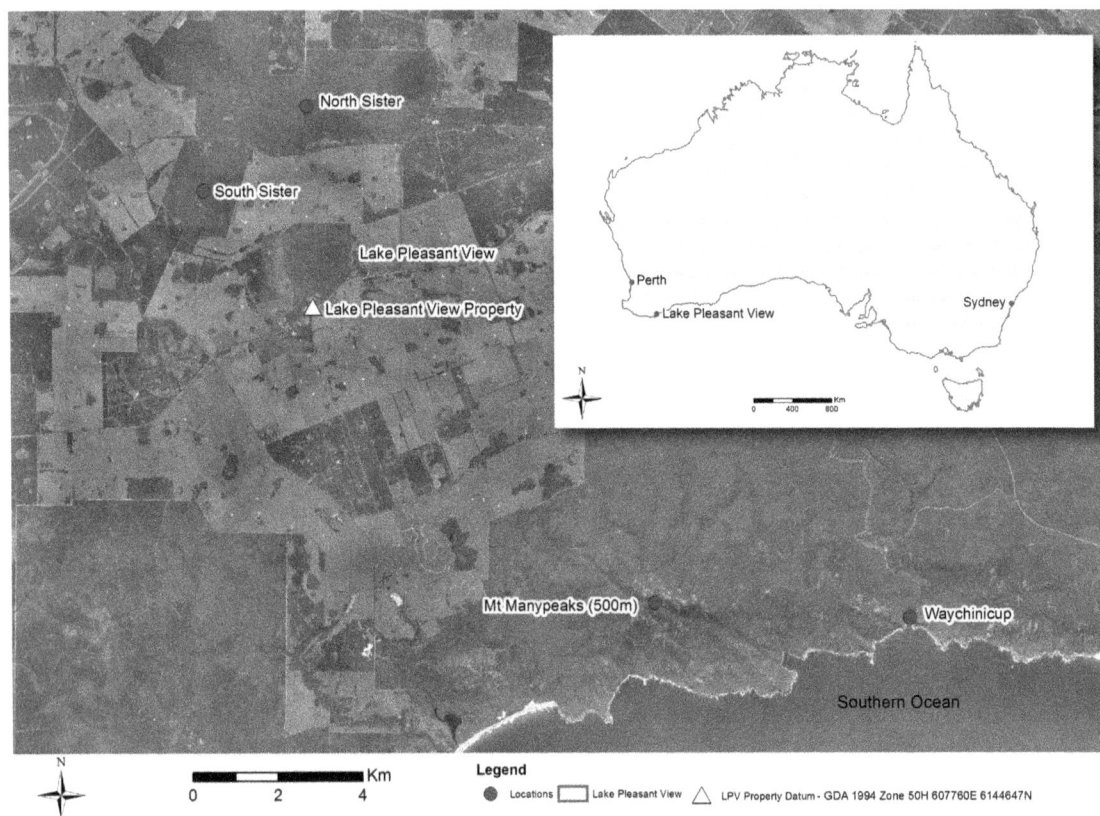

Figure 1. Location map.

Source: Image supplied by Gondwana Link.

The place

The area is dominated by a wetland surrounded by a narrow nature reserve managed by the Department of Environment and Conservation (DEC), and is listed as a Wetland of National Significance (Environment Australia 2001). The moderate-sized sedge lake (201 ha) comprises small open areas and extensive areas of sparse sedge. The wetland lies within the Albany-Fraser Orogen geological formation, bounded in the north by the older Yilgarn Craton, somewhat neatly divided by the Stirling Ranges. The major south coast river systems of the Albany area (King and Kalgan) formed during a southward down-tilting of the Yilgarn Craton, creating an extensive slope known as the Ravensthorpe Ramp, and much of the region was inundated by higher sea levels during the Eocene Age. As a result of these processes, the short rivers were formed and vast sand plains filled with Tertiary sediments were created, forming a low relief plain dotted with small wetlands, with the basement geology visible in various areas as impressive granite domes, such as Mount Clarence, Mount Lindsay, the Porongurup Range and the North and South Sister hills that surround the wetland.

The area is an important cultural area for the Traditional Owners and creates a 'sense of place' because of its associated oral histories, cultural features, and aesthetic values provided by the striking landform configurations. A Dreaming Story describes the creation of the wetlands – Lake Pleasant View – and surrounding wetlands and associations with the prominent, surrounding hills (Figure 2).

> This place here, we call Moolyiup, the hill that you see behind me is Moolyiup, and the one behind her is Twertup, they are two sisters. They belong to the kangaroo people – which are the Stirling's (ranges to the north). We call all this area Moolyiup. They were promised to the same man, that came from down Cape Riche way (south coast), and he travelled up to where the kangaroo people were, picked his two brides out, which happened to be the two sisters, and taken them back down to Warriup, and when they got so far they realised that the fellow that picked them, the bloke that married them was the devil himself Chunuk. The Chunuk lived over at Warriup and they stopped in their tracks when they realised, and they had a bit of a chat and said look we gotta make a run for it. So they ran straight up the foot range over there – they represent the devils footsteps as he followed them along, and he caught them here and their mother is Yoolberup, which is the peak behind you – Mount Manypeaks, her name is Yoolberup. Before they got there he caught them here and turned them to stone. The lake systems themselves were what the Devil left behind, his footsteps – and the water that you see in the lakes system, the entire lakes system, are the tears from those two girls. (Aunty Lynette Knapp, pers. comm., 2010)

The area has continued to be used by the local Traditional Owners (Menang People) throughout the post-contact period. In the earliest days of settlement, such areas provided a 'refuge' for accessing traditional resources and maintaining traditions in a context of increasing segmentation of the landscape into farming properties, as the region's European population increased and restricted patterns of traditional movement and settlement. During the 1950s members of nearly all families from the region utilised this area on a regular basis and well-known individuals from the community set up camps along the edges of Lake Pleasant View, using the wetland's resources. Today, individuals and families use the area for spiritual reflection and passing of their knowledge to their youngsters. Thus, the area is of important historical association and enhances cultural connections today.

There is a small area of private land that borders the southern edge of the wetland. On this property, a large granite dome dominates, providing 360 degree views of the vast sedge land, tall woodland, majestic hills, and surrounding open pasture. The associated archaeological landscape includes artefact scatters that occur throughout the property, the adjacent nature reserve, and exposed areas along various access tracks. *Gnamma* (water) holes were also located and recorded on the granite hill that extends to the neighbouring reserve to the west. Loose slabs of granite (now removed from this area) presumably functioned as lizard traps. Indeed lizard traps recorded previously can no longer be found, and so have presumably been destroyed by quarrying (discussed below). A putative modified (scarred) tree is also located on the site. The property and reserve also contain historical significance as a place where Menang people camped and hunted throughout the 1950s and 1960s.

The granite dome that dominates the property in question has been subject to major destruction by the previous landholder, as part of past quarrying activities. This includes the complete destruction of a rockshelter that was an important component of the archaeological and cultural landscape (Figure 2). These destructive activities represent past failures of compliance-based heritage legislation in protecting a significant cultural place that is a component part in an extended cultural landscape. While the past legislative failures to protect these cultural resources are not the focus of this paper, they provide an important background as the impetus to effect positive outcomes for cultural heritage values at this place.

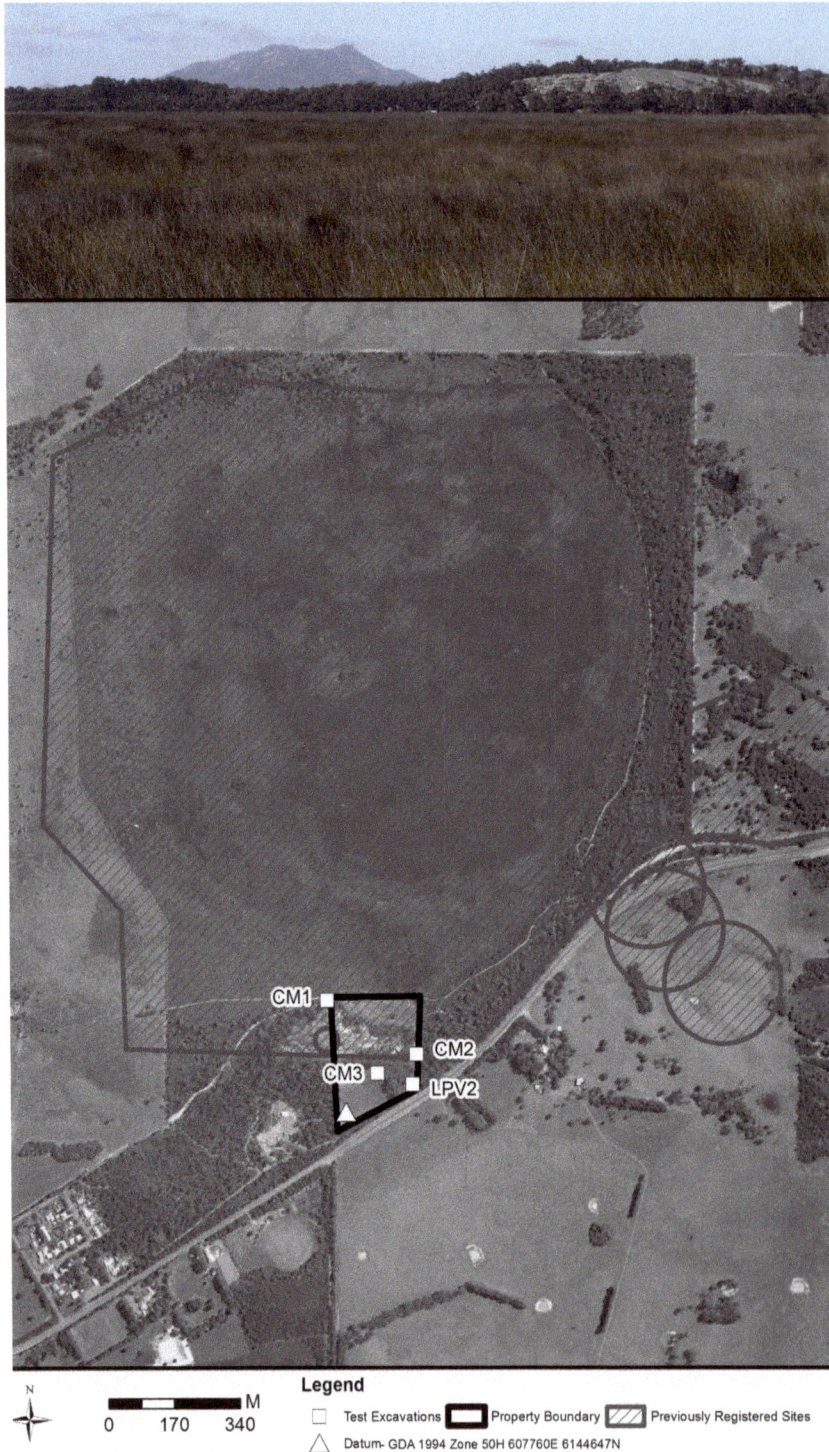

Figure 2. Map of the property, Lake Pleasant View, registered sites, and test excavations. Photo insert is a view south from the middle of the Lake to the granite dome (within the community-owned property) and Mount Many Peaks in the background.

Source: Image supplied by Gondwana Link.

The property has contemporary social significance as an area that should be restored and used for cultural activity and education. Active conservation continues to be carried out on the property as it is now managed by the local Menang community and a variety of partners as outlined below. The Menang Traditional Owners articulated the importance of caring for Country, as part of this project:

> The 'mountains' and the 'hills' you see, we call them kart, we also call our 'heads' kart, so that's the relationship. And 'water' is beelia, beel when it is still water like this (pointing to Lake Pleasant View) or beelia is running water. We also call our mothers, when they nurse us and their milk is running, that is beelia, so that is a part of us. We cherish and respect, when we are on Country, particularly walking in the bush. Kart, 'the hills', they've got eyes like we have, so they watch our every move. Which is why we respect our Country, we can't do anything wrong on our Country, we got to respect it and we got to look after our Country (Aunty Lynette Knapp, pers. comm., 2010).

Thus, from the outset, this project involved establishing a structure that integrated an assessment of the values associated with this area and also for ongoing management of the area under traditional concepts of 'caring for Country'. Both the natural and cultural values of this area have been degraded and were under threat from neglect, poor land-planning, and direct vandalism. The lake system is threatened by *eutrophication* resulting from agricultural and plantation fertilisers. The spread of non-native flora and fauna species was uncontrolled. Illegal rubbish dumping has taken place along the eastern edge of the lake. The previous landowner has damaged many heritage features associated with the granite outcrop. Sheep have grazed over the property and across culturally sensitive areas. Natural bush land has been cleared, including cultural plants. Quarrying and clearing have damaged natural and cultural heritage features.

The next section discusses the integrated archaeological framework that served to structure the way this project was developed and implemented, based on a review of the key stages/components of the project. This precedes a discussion that examines the wider implications drawn from this analysis in terms of the role of archaeology in the dynamic field of Indigenous cultural heritage management and NRM.

Cultural heritage ownership and the integration of archaeology

This project was implemented under an integrated model that served to overcome some of the limitations of narrowly-defined compliance based CRM processes, within a philosophy to work beyond compliance to a more integrated community archaeology model, as reflected in this statement:

> There is a genuine desire by many heritage professionals, despite legislative constraints, to work towards a cultural heritage practice that supports the integration of archaeology, cultural heritage and Indigenous knowledge. (Pragnell *et al.* 2010:152)

The limitations within compliance-based CRM processes are well documented; with reference to inadequate integration of traditional owner values and knowledge (Brown 2008; Hemming and Rigney 2010; Guilfoyle *et al.* 2011), a lack of integration into meaningful research paradigms (Morse 2009), limitations of relevant legislative mechanisms including the Australian National Heritage listing process (Sullivan 2008) and the *Western Australian Aboriginal Heritage Act 1972* (Anaya 2010), and relevant inadequacies in teaching and learning practices with regard to professionalism in Australian archaeology and CRM (Colley 2007). It is particularly noteworthy that James Anaya – United Nations (UN) Special Rapporteur on the Rights of Indigenous Peoples – made specific mention of the *Western Australia Aboriginal Heritage Act* as a point of concern in a report to the UN General Assembly, 1st June 2010 (Anaya 2010). Within this context of identified weaknesses in CRM processes, this project aimed to raise the bar above minimum standards and achieve outcomes for the protection and management of an important cultural place, attaining tangible community/social outcomes through all stages of the project. This required consideration of the need to integrate diverse archaeological frameworks and methods with Aboriginal values and knowledge, adopting a framework similar to that espoused by Prangell *et al.* (2010), in which power relations are renegotiated through a community-led, landscape based approach. This section examines each of the main stages/components of the project in relation to the adopted archaeological framework.

Land access and land ownership

Following community meetings set up initially to address the impacts caused to the registered heritage site by the previous landowner, the local Elders expressed their desire to obtain access to actively use, restore and manage the entire wetland area. These suggestions became the impetus for management that necessarily involved legal access to the area, and a subsequent land purchase application. The purchase of the 6 ha parcel of land was funded by the Indigenous Land Corporation (ILC) in 2005 for the Albany Heritage Reference Group Aboriginal Corporation (AHRGAC). The AHRGAC is comprised of Elders, Traditional Owners, and representatives of all Noongar families with historical, social, cultural and spiritual attachments to the Albany region and surrounding hinterland. The main purpose of the AHRGAC is to provide guidance on all matters affecting Noongar cultural heritage in the region. The Department of Indigenous Affairs (Southern Region, WA) works closely with the AHRGAC and provides support to facilitate the cultural use of this and other areas of land.

The land acquisition application was developed by (co-authors) Guilfoyle and Coyne (see Guilfoyle *et al.* 2009), following a site disturbance investigation involving quarrying of the large granite outcrop (and registered archaeological site) by the previous landowner. Because the property had been privately owned for many years, it had been difficult to access for cultural activities. Thus, following several field visits and planning sessions, one of the first actions was to draft a purchase application, which was submitted on behalf of the Reference Group. The result was a successful submission to the Indigenous Land Corporation in 2005 enabling the Reference Group to purchase the property.

This acquisition was the most direct form of ownership and control in the short-term but the integrated team also established a mechanism for ongoing community control of the management processes, and through all facets of this 'project'. Aplin (2002) has argued that ownership of heritage must move beyond tokenistic recognition, and also involve 'ownership' of the process that includes: "research, the development of listing proposals, the preparation of conservation and management plans, interpretation, and sharing financial rewards where they occur" (Aplin 2002:140).

In order to ensure these objectives, and for the land acquisition and subsequent management programme to be successful, a well-planned structure was required. This necessarily involves extensive planning, and in this case, the development of the incorporated body made up of community members that could manage the property – the AHRGAC. From here, a three-way collaborative agreement was established between AHRGAC, representatives from the *Restoring Connections Project* (South Coast Natural Resource Management Inc.), and the Department of Indigenous Affairs (Albany region). This process and partnership structure was set up to ensure that there were sufficient integrative mechanisms in place to deliver the ongoing management requirements of the property. It has been well argued that collaborative partnerships are critical to long-term sustainability of any community archaeology or CRM project (Moser *et al.* 2002:229) and it was partnerships that underpinned the ultimate success of this project. Restrictions in national parks, reserves, and private land mean that the land area for local Traditional Owners to continue cultural practices, such as obtaining wild resources and being on land in a traditional manner, is quite low. This creates an impetus to work towards attaining access, and ultimately ownership of land for cultural purposes, which was the basis of this project.

Management action plan: Heritage as social action

Another critical component centered on the need to ensure that the management process for the wetland and heritage landscape would be driven by the local community, to 'manage and protect country'. The premise was that heritage should be seen as a cultural and social activity, and it is only through such activity that vibrant cultural connections can be maintained (Byrne *et al.* 2003). The archaeologist, heritage officers and Traditional Owners involved in this project knew that the project must address this understanding in order to achieve any level of success, and so each component must serve to address this aspect of cultural heritage management. There is increasing recognition of the importance of heritage in matters of Indigenous identity and wellbeing, for example:

> Our acceptance that heritage helps to define identity and express values and aspirations, means we must be concerned about the conservation of Australia's Indigenous heritage. There is a fragile grasp on the management of Indigenous culture, which is integral to the identity and well-being of Australian Indigenous peoples. (Open Mind Research Group 2006:3)

Direct links have been drawn between the maintenance of culture, heritage and traditional lands, with Indigenous health in Australia and elsewhere; and NRM is posited as a key mechanism for delivering direct and indirect health-related outcomes for Indigenous communities (Burgess *et al.* 2005):

> Effective interventions in Indigenous health will require trans-disciplinary, holistic approaches that explicitly incorporate Indigenous health beliefs and engage with the social and cultural drivers of health. Aboriginal peoples maintain a strong belief that continued association with and caring for ancestral lands is a key determinant of health. Individual engagement with 'Country' provides opportunities for physical activity and improved diet as well as boosting individual autonomy and self-esteem. Internationally, such culturally congruent health promotion activities have been successful in programs targeting substance abuse and chronic diseases. (Burgess *et al.* 2005:117)

This community archaeology program aimed to achieve outcomes in identity, well-being and health, through the promotion and effective management of cultural heritage values. As such, a heritage management plan was developed which resulted in securing of funds for community members to carry out 'management actions' that included a range of components – including social gatherings, weed control, educational trips, and training courses. This plan was endorsed by the Indigenous Land Corporation and the DIA (Guilfoyle 2010). It is interesting to note that it was the incorporation of the cultural heritage values associated with this area that led to the development of the management plan for the wetlands. Despite its ecological and cultural heritage significance, no management plan existed for the Lake Pleasant View Nature Reserve. At the same time, much of the CRM activity implemented in the past in this region, and at this site in particular, has been reactive rather than strategic – focused on identifying options to protect a site (such as the now destroyed rockshelter) from imminent threat. Thus, it was clear that an effective heritage management system for this place must explore management at a landscape level, and in contexts outside of development-driven (or site-threat) proposals. In so doing it was necessary to work beyond the basic legislative requirements focused on assessment, and to link both NRM and CRM objectives. These principles are vital if CRM is to contribute to any sense of community wellbeing.

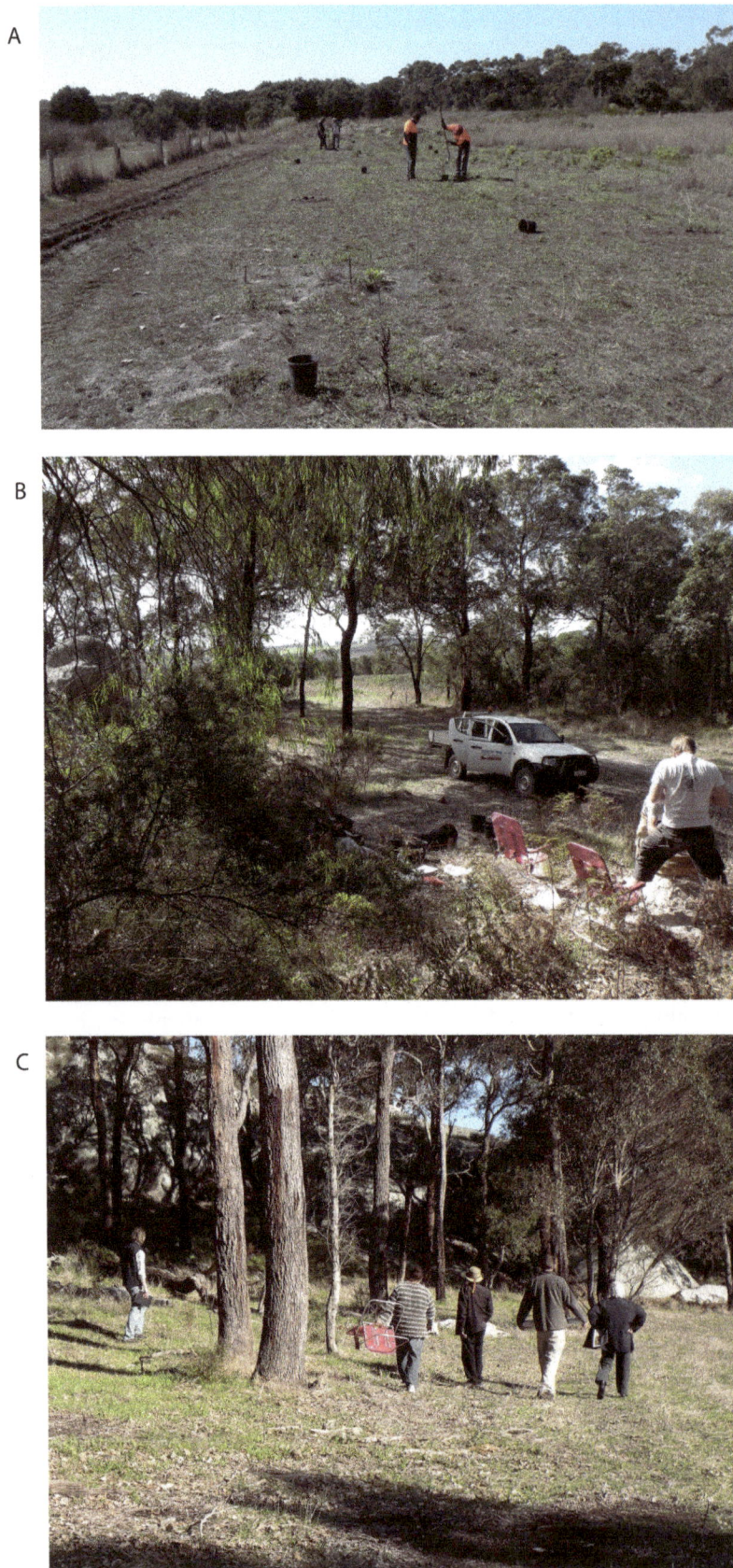

Figure 3. Indigenous conservation and land management training team undertaking (a) revegetation and weed control around the wetland (b) archaeological investigations, and (c) cultural/ecological mapping - occurring concurrently.

Source: Photographs (a) and (b) by David Guilfoyle; photograph (c) by Myles Mitchell.

At a broad level, the processes to be met in achieving the management plan targets were articulated within three main objectives that served to link NRM and CRM processes:

1.   facilitate sustainable property management and traditional use;
2.   maintain and enhance conservation values of the area; and,
3.   foster community ownership of and responsibility for the area.

At a practical level, the AHRGAC approves and facilitates short heritage and natural resource management training courses on the property (delivered by a registered training provider), and works with local schools in conducting educational trips to examine the natural and cultural values. A weed control programme includes a re-vegetation component, often resourced by the regional NRM body. Restoration work is carried out by community members at the wetland, including seed collection/propagation, weed eradication, and revegetation (Figure 3). Another method of securing ownership and control was through access arrangements, whereby any person wishing to visit or work in the area (including consultants and land managers) is required to obtain approval from the Reference Group. Thus, there was not only a focus on achieving environmental and cultural heritage management outcomes, but the approach ensured the methods contributed to community ownership and sustainable development:

> It's also about leadership, we are actually demonstrating some leadership in this regional community to say this is how we can do it, and these are the benefits for all of us. (Harley Coyne, pers. comm., 2010)

An additional component of the management plan centered on ongoing cultural mapping and monitoring, in order to integrate cultural and natural heritage management actions in a more strategic, regional manner. Its implementation is guided by the Traditional Owners and supported through partnership agreements linking local land management bodies, and centered on cultural mapping, landscape archaeology, and Indigenous knowledge. As the management plan continues to be implemented, it provides an ideal study area to develop and refine understandings of the integrated landscape. The next section explores this aspect of the project and develops an argument for embedding archaeological projects within NRM processes.

The integration of landscape archaeology, Indigenous knowledge, and NRM

A major aspect of the discussion with Traditional Owners was the suggestion of a landscape-level approach to management in order to ensure that the intrinsic links between cultural heritage and the wetland ecosystem were recognised. Quite often there is a discord between archaeological methods of site recording and analysis, and the Traditional Owner concept of the integrated cultural landscapes (Byrne 2008; Bradley 2008; Prangnell *et al.* 2010). Thus, this project required a method to explicitly examine the various components of this integrated cultural landscape and embed a process for assessing, documenting, and managing the associated values.

Natural resource management is the link in protecting and restoring cultural heritage places and values at a landscape level, or in Aboriginal terms 'caring for Country'. Just as Sullivan advocates for a cultural heritage management system at a national level, that "protects cultural landscapes at their fullest expression [or] in other words protects '*Country*'" (Sullivan 2008); the approach adopted here seeks to do so at a local level, under the cultural leadership of Traditional Owners. Despite the applicability of NRM for such landscape approaches, quite often NRM strategies are developed without adequate strategic planning with regard to integration of CRM and NRM. This results in lost opportunities that can be achieved when these strategic connections are made:

> While it might be difficult to truly understand the complexity of Aboriginal culture and its interconnectedness

with the landscape, the holistic nature of Aboriginal culture means that issues relating to natural resource management are interlinked with those of cultural heritage management. (Windle and Rolfe 2002:35)

Thus, while the immediate actions of this project focused on the area of private land bordering the wetland – when it was purchased for the local community group arising from threat to specific heritage features – the strategic development of the project led to the implementation of a community management structure that extended across the entire wetland and surrounding areas, enacting NRM actions within a CRM framework. In this way the project attained a landscape focus within a strategically linked integration of CRM and NRM.

At a methodological level, this component involved cultural mapping occurring concurrently with on-ground restoration work. This also required additional assessment of both the ecological and cultural places and values throughout the wider area, within a cultural landscape methodology. A landscape approach allowed for the management of ecological values while assessing and protecting aspects of the cultural landscape as defined by the Traditional Owners. This approach prompts local knowledge about how the landscape has changed, and ideas on what should be done for dual natural and cultural heritage conservation. As Byrne and Nugent (2004:73) wrote: 'it is the landscape themselves that ought to be considered heritage, rather than discrete and dispersed 'sites' within them'.

Thus, the objective was not to analyse and document individual sites, but to explore networks of places to reveal interconnections that reflect movement of people and materials around the landscape. Traditional Owners and heritage specialists worked together to map the granite landform system surrounding the wetlands, while identifying appropriate mechanisms to ensure the restoration and protection of both the natural and cultural resources. This process also had direct outcomes for the wetland system and associated biodiversity values. For one example, the local knowledge pertaining to wild resource use and interconnections of places and landforms facilitates archaeological inquiry, as Green *et al.* have noted:

> The environmental skills of people who have learned from the knowledge of many generations can add significantly to the way one reads a site and oral tradition can greatly enrich understanding of the meanings of places. (Green *et al.* 2003:369)

For this project, an ecological mapping and traditional ecological knowledge assessment was undertaken (Janicke 2010) at the same time as a landscape archaeology research project that began with the development of a CRM plan that included on-ground conservation work (Figure 3). The archaeological research component has since developed into a Masters level research project. This study aims to:

1. Examine the inter-relationships between the oral histories and archaeological landscape; and,
2. Establish a chronological framework for the area that examines both the formation of the wetland and changes in regional settlement patterns.

The investigation contributes to existing regional models which suggest the late Holocene period involved structured movements based on social and territorial boundaries, with seasonal movement between the coast (focused on the summer months) and inland areas (more dispersive use of the inland woodlands during winter). For instance, Dortch (1984:2002) suggests that past land use patterns involved seasonal movements linked to the availability of key resources, whereby groups most likely congregated on the coast during summer and autumn, largely dispersing inland during the winter months. The hypothesis here is that the Lake Pleasant View area was used as a congregative zone, whereby people would live in small family groups for most of the year, and seasonal gatherings of larger family and tribal groups would take place at certain times of the year when certain resources were available and abundant.

The archaeological investigations included surveys and test excavations on the community owned/ managed property and archaeological surveys around the wetland and the surrounding region. These surveys resulted in the identification of an extensive archaeological signature associated with Lake Pleasant View and wetlands to the north. Surface stone artefact scatters were recorded and analysed on site across the property and wetland. Three test excavations were undertaken revealing relatively intensive use of this area over the last three thousand years (Figure 2, Table 1), with an assemblage somewhat equally divided between tool maintenance, manufacture and core reduction, with a number of formal implements (backed artefacts). The preliminary data supports the view that there was a relatively intensive past history of human occupation in this area, and the settlement-subsistence patterns were somewhat structured by the wetland, and serving as a movement corridor between the upland hills and coastal resources to the south. The wetland sites likely represent a larger occupation area with a range of integrated smaller sites in the surrounding hinterland used as part of logistical forays, with larger congregative sites at specific, more distant locations, such as Waychinicup (to the south) and the Three Sisters Nature Reserve (to the north). This is how the Traditional Owners describe the place (as briefly mentioned above).

Table 1. Radiocarbon dates obtained from three test excavations.

Laboratory code	Test pit/spit	Depth (cm) below mgs	14C age (years BP)	δ13C	Calibrated age BP (95%)
Wk-28164	CM2_L2_S3.4	27	2562 ± 30 BP	-21.9 +/- 0.2‰	2592 cal BP – 2532 cal BP
Wk-28165	CM2_L2_S4.5	28-33	2550 ± 30 BP	-24.8 +/- 0.2‰	2580 cal BP – 2520 cal BP
Wk-28494	CM3_L2_S3.4	31	3140 ± 30 BP	-23.9 +/- 0.2‰	3170 cal BP – 3110 cal BP
Wk-27271	LPV_2_L4_S3	29	3009 ± 30 BP	-3.9 +/- 0.2‰	3039 cal BP – 2979 cal BP

Source: Author's research.

Here, the archaeological investigations complement, and are complemented by, the ethnographic and local knowledge of the area, and also an ecological assessment. Thus, the heritage complex is of important archaeological significance given that current knowledge largely pertains to the coastal zone and Albany area, and so more detailed investigations throughout this corridor are shedding light on patterns of the regional heritage landscape. It is also providing baseline data that is required for effective heritage landscape management, in the analysis of wetland formation, human impacts such as firing, and patterns of human-environmental relations over time and across space. Most importantly, the study links archaeological data to oral histories that also feed into the ongoing management actions. This mirrors other similarly deigned community heritage and NRM projects, such as that reflected in this statement from Ross *et al.*:

> From our field experiences, we have learned that Indigenous know-how is a form of 'local knowledge', intricately bound to particular communities and places as well as to whole ways of life. (Ross *et al.* 2010:34)

Discussion

The heritage place of specific focus here is valued as a significant cultural place embedded with the actions and continued presence of the Traditional Owners and the actions of their ancestors. In this way it maintains and displays the Traditional Owner's connection to the past and present, and contains meaning at a number of different levels. From a management perspective, these intangible values are included within the landscape-level protection and maintenance of the integrity of the features and places associated with this property, the wetland, and surrounding landscape. The place is also a central component to community identity as it provides a sense of belonging, historical association, and is part of the ongoing cultural practice of 'caring for Country'. This practice is an important aspect of cultural heritage, with the notion that heritage is not static or fixed, but requires active participation in heritage management. Thus, the project is important in a contemporary context for facilitating 'caring for Country' in a region where access to land is limited, and provides for a more socially-relevant cultural heritage management project, as articulated by Byrne *et al.*:

> The fact that a local community, Indigenous or non-Indigenous, may live in a landscape which is scattered with places where physical traces of past occupation are present, does not in itself create an identity association between those traces and the community. The association comes about through certain activities. These may include the work carried out to protect the traces from erosion or vandalism, the taking of visitors out to see the traces as part of a local cultural tourism venture. It may consist of a community member talking to a class at the local school about the traces, or it may consist simply of the reminiscences about the place which appear in an autobiography written by a community member. (Byrne *et al.* 2003:66–67)

As a landscape management project, the on-ground work entailed incorporation of traditional ecological knowledge with on-ground management and restoration of land. Not only does this collaboration assist in meeting NRM aims in the region (recognised as a high priority for biodiversity, land and water conservation), the work on this land serves as a model for community-driven land management ('caring for Country') that has environmental, social and cultural heritage outcomes. The regional NRM body has partnerships with several major land management and conservation groups that provides assistance to the AHRGAC for on-going management, and directly and indirectly assists in building the capacity of the local community to monitor and manage the land under contemporary land management structures, as Little (2007:2) states:

> A socially useful heritage can stimulate and empower both local community members and visitors to make historically informed judgments about heritage and the ways that we use it in the present.

The clear heritage management framework endorsed as part of this assessment was to continue to foster the existing ongoing partnership and support for future monitoring and management of the property. The original relationships were expanded as specific project outcomes were delivered (on-ground conservation of the wetland ecology and cultural heritage assessments) in efforts to secure long-term resources for community management and development of a structured management framework that includes integration with other key land management authorities.

This project articulated the clear role of archaeology as a useful medium for all negotiations, the instrument for on-ground actions and community engagement (training/employment), linking the divide between natural and cultural heritage management, and developing a landscape-level research programme that investigates the inter-connections between oral history and the archaeological record – itself a means for community empowerment as the lead agent in sustainable property management via research and land care under customary guidance ('caring for Country'). The project is a practical example of what we understand collaborative Indigenous archaeology to be:

> Within collaborative Indigenous archaeology, this perspective requires that archaeologists consider Indigenous perspectives at many times other than during the final interpretation or at the moment of doing 'public outreach' to a descendant community. These perspectives should be acknowledged and often embedded at all stages of the archaeological process, from project formulation to field methods, from excavation recovery to laboratory analysis, from interpretation to writing. In particular, these

incorporations should be fundamental elements of archaeological field schools that focus on Indigenous pasts, for in these complex intersections of teaching and research lies real potential to change the discipline. (Silliman 2008:3–4)

Conclusion

> Our Elders are heavily involved in all things from start to finish. We have had some major projects, some minor projects; it has really created this sense of ownership, of wanting to be involved, of loving the place and wanting to do the right thing to do by it. It has given them back a real sense of pride, I think. (Aunty Vernice Gillies, pers. comm., 2011)

With increasing recognition of the overlap between natural and cultural heritage management, this project demonstrates the potential to implement landscape-scale projects that result in integrated holistic outcomes at a community level. The approach advocated here was to establish a cultural heritage management framework that allows ongoing study of the local archaeology and ecology while land care activities are undertaken at a regular basis, including monitoring of the area's natural and cultural values. The programme continues to provide many community days, ongoing research, formal training projects, that all contribute to a greater understanding of the natural and cultural landscape and community wellbeing. It is argued here that it was the combination of skills provided by both the Traditional Owners and the archaeological team that fostered integration of community objectives with heritage agency and NRM targets; facilitated the engagement process; coordinated on-ground actions; and was able to deliver documents and reports outlining the dual management actions and technical detail specific to the natural and cultural values.

At the same time, it was only through the structured, developed and community mechanisms of control and ownership that allowed the ongoing, successful development of this project and provided a platform, not just for archaeological investigation, but for archaeology to make a very real contribution to the present and the future through living heritage and community NRM. It demonstrates the identified potential and value of this type of collaborative approach for any archaeology project that many researchers have increasingly been advocating and implementing, as for example, Marshall (2002:218):

> The kind of collaborative research fostered by community archaeology will be crucial if archaeology is to have a future. It is the only way Indigenous heritage, descendant communities and other local interest groups will be able to own the pasts archaeologists are employed to create.

Where the Traditional Owners were once frustrated and resentful of the lack of engagement and involvement in previous management regimes, the AHRGAC now approach all agencies and organisations that may have an interest in assisting their efforts for on-ground conservation and management of both natural and cultural heritage values associated with this area. It directly challenges existing structures and issues that cripple many joint-management negotiations and plans across Australia, as Ross *et al.* note:

> In the name of conservation and the protection of biological diversity, Indigenous peoples are being systematically excluded from lands and resources they occupied, utilised, and indeed protected for generations. Labelled poachers and trespassers in newly created parks, wildlife sanctuaries and other protected and often fenced areas, part of the problem, rather than a solution, Indigenous peoples find it increasingly hard if not impossible to participate meaningfully in the sustainable management of their own ancestral territories. (Ross *et al.* 2010:29)

This project has led to the development of a structure whereby land management agencies and organisations seek to become involved in this community owned and managed area. In the protection of archaeological resources and heritage values associated with this cultural landscape, management processes here focused on the dual conservation of the natural and cultural features. In many ways, this simply means integrating the actions required to maintain the ecological and biodiversity values of the local system with the integrated cultural heritage landscape. However, under this structure of understanding heritage as social activity, it provided direct opportunities for community-driven rehabilitation and management. The methods by which the process was to be implemented resulted in a long-term community-based heritage management programme. The activities served to protect the cultural values associated with these places, in recognising that caring for Country is an integral component of identity and cultural vibrancy. In general, the conservation works associated with these properties are aimed at sustainable land use, erosion and salinity control, biodiversity protection and waterways protection. These actions are also those that are required to ensure the protection and conservation of the region's non-renewable cultural resources and reconnect aspects of a fragmented, threatened cultural heritage landscape. This paper has argued that to effectively achieve these integrated outcomes, an understanding of heritage as social activity is required, and the integration of CHM and NRM methods beneath a community-owned/managed structure is perhaps the only effective platform to achieve practical outcomes in this regard. There is great potential for landscape-level archaeological projects to contribute much toward active heritage management, conservation, and sustainable outcomes when developed in partnership with Indigenous communities, and especially when NRM methods and processes are integrated into each stage of a project.

Acknowledgements

This project and paper was developed under the direction and guidance of the Menang Elders and wider community who have welcomed us into their Country, Land and Heritage. The Albany Heritage Reference Group Aboriginal Corporation provides much of the direction and energy to the project, along with a number of other projects in the Albany region. The project was served greatly by a number of younger community members and project officers, especially Jerry Narkle, Ryan Humphries, and Iszaac Webb, along with a number of trainees and volunteers from the Albany community. South Coast Natural Resource Management Inc. (especially project officer Shandell Cummins) continues to provide support to the projects, as does the team at the Department of Indigenous Affairs (Southern Region). The Centre of Excellence in Natural Resources Management, the team at the Department of Water (Albany), Great Southern TafeWA, and the Many Peaks Primary School, have all provided great technical and community support to various components of this project. There are many different facets to this project led by the Traditional Owners that are not covered here, as this paper has only focused on several elements related to the role of archaeology and the integration of natural and cultural heritage management. We thank the reviewers Annie Ross and Mick Morrison, as well as the editors of *Terra Australis*, who helped a great deal towards getting this paper into shape.

References

Anaya, J. 2010. Report by the Special Rapporteur on the situation of human rights and fundamental freedoms of Indigenous people: Addendum – Situation of Indigenous peoples in Australia. UN General Assembly, Human Rights Council, Fifteenth Session, Agenda Item 3.

Aplin, G. 2002. *Heritage: Identification, conservation, and management.* Oxford University Press, Melbourne.

Bradley, J. 2008. When a stone tool is a dingo: Country and relatedness in Australian Aboriginal notions of landscape. In: David, B. and Thomas, J. (eds), *Handbook of landscape archaeology*, pp. 633–637. Left Coast Press, Walnut Creek.

Brown, S. 2008. Mute or mutable: Archaeological significance, research and cultural heritage management in Australia. *Australian Archaeology* 67:19–30.

Byrne, D. 2008. Counter-mapping in the archaeological landscape. In: David, B. and Thomas, J. (eds), *Handbook of landscape archaeology*, pp. 609–616. Left Coast Press, Walnut Creek.

Byrne, D., Brayshaw, H. and Ireland, T. 2003. *Social significance: A discussion paper*. Research Unit, Cultural Heritage Division, New South Wales National Parks and Wildlife Service, Hurstville.

Byrne, D. and Nugent, M. 2004. *Mapping attachment: A spatial approach to Aboriginal post-contact heritage*. Department of Environment and Conservation, Hurstville.

Colley, S. 2007. University-based archaeology teaching and learning and professionalism in Australia. *World Archaeology* 36:2:189–202.

Dortch, C.E. 1984. *Devil's Lair: a study in prehistory*. Western Australian Museum, Perth.

Dortch, C.E. 2002. Modelling past Aboriginal hunter-gatherer socio-economic and territorial organisation in Western Australia's lower South-west. *Archaeology in Oceania* 37:1–21.

Environment Australia. 2001. *A directory of important wetlands in Australia* (3rd edition). Environment Australia, Canberra.

Green, L.F., Green, D.R., David, R. and Góes Neves, E. 2003. Indigenous knowledge and archaeological science: The challenges of public archaeology in the Reserve Uaçá. *Journal of Social Archaeology* 3(3):366–398.

Guilfoyle, D.R., Bennell, B., Webb, W., Gillies, V. and Strickland, J. 2009. Integrating natural resource management and Indigenous cultural heritage: A model and case study from south-Western Australia. *Heritage Management* 2(2):149–176.

Guilfoyle, D.R., Webb, W., Webb, T. and Mitchell, M. 2011. A structure and process for 'working beyond the site' in a commercial context: A case study from Dunsborough, southwest Western Australia. *Australian Archaeology* 73:25–32.

Hemming, S. and Rigney, D. 2010. Decentring the new protectors: Transforming Aboriginal heritage in South Australia. *International Journal of Heritage Studies* 16:1–2,90–106.

Janicke, G. 2010. *Lake Pleasant View ecological assessment*. Report No. CENRM 109. Centre of Excellence in Natural Resource Management, University of Western Australia, Perth.

Little, B.J. 2007. Archaeology and civic engagement. In: Little, B.J. and Shackel, P.A.(eds), *Archaeology as a tool of civic engagement*, pp. 1-22. Altamira Press, Walnut Creek.

Marshall, Y. 2002. What is community archaeology? *World Archaeology* 34(2):211–19.

Morse, K. 2009. Emerging from the abyss: Archaeology in the Pilbara Region of Western Australia. *Archaeology in Oceania* 44: supplement.

Moser, S., Glazier, D., Phillips, J.E., Nemr, L.N., Mousa, M.S. and Aiesh, R. 2002. Transforming archaeology through practice: Strategies for collaborative archaeology and the Community Archaeology Project at Quseir, Egypt. *World Archaeology* 34(2): 220–248.

The Open Mind Research Group. 2006. *State of Indigenous cultural heritage – A survey of Indigenous organisations.* A report for the Department of Environment and Heritage, Canberra.

Prangnell, J., Ross, A. and Coghill, B. 2010. Power relations and community involvement in landscape-based cultural heritage management practice: an Australian case study. *International Journal of Heritage Studies* 16 (1):140–155.

Ross, A., Sherman, K.P., Snodgrass, J.G., Delcore, H.D. and Sherman, R. 2010. *Indigenous peoples and the collaborative stewardship of nature: Knowledge binds and institutional conflicts.* Left Coast Press, Walnut Creek.

Silliman, S.W. 2008. Collaborative Indigenous archaeology: Troweling at the edges, eyeing the center. In: Silliman, S.W. (ed.), *Collaborating at the trowel's edge: Teaching and learning in Indigenous archaeology,* pp.1–21. University of Arizona Press, Tuscon.

Sullivan, S. 2008. More unconsidered trifles? Aboriginal and archaeological heritage values: Integration and disjuncture in cultural heritage management practice. *Australian Archaeology* 67:107–115.

Windle, J. and Rolfe, J. 2003. Valuing Aboriginal cultural heritage sites in central Queensland. *Australian Archaeology* 56:35–41.

8

Traim tasol ... Cultural heritage management in Papua New Guinea

Tim Denham, La Trobe University, Melbourne, Australia

In this chapter, I discuss a range of issues associated with cultural heritage management practice in Papua New Guinea today. I build my discussion around three different types of cultural resource management project that I have undertaken. These increase in complexity, scale and scope. First, I describe a community heritage project among the Kalam of the Simbai Valley, Madang Province. Second, I raise several issues associated with the World Heritage nomination of the Kuk Early Agricultural Site. Third, I make some generalised observations on the expanding commercial cultural heritage management sector. The first two sections are highly specific, whereas the third section is more polemical. Each gives a flavour of the variety of my experiences, reflects the range of practices occurring within the country today, and is intended to be illustrative rather than comprehensive.

I have undertaken fieldwork in Papua New Guinea over a period of 20 years (1990–2010). My first two-month field visit in 1990 among the Kalam at Tsendiap and Tsarep in the Lower Jimi Valley, Western Highlands Province was incidental to a thesis on cultural geography (Denham 1996). Since then, my professional interests in Papua New Guinea have been either academically-oriented archaeological research or cultural heritage management. Archaeological research has focussed on traditional forms of plant use, including early agriculture, in the highland interior during the Holocene (Denham *et al.* 2003). Over the last six years my focus has broadened to include the history of occupation generally, as well as a range of different cultural heritage management projects within the country.

A community project among the Kalam

A community-based ethnoarchaeological field project was initiated among the Kalam in the Simbai Valley of Madang Province in 2007 (Bedingfield and Denham 2008). The project arose out of discussions between Dr Ian Saem Majnep and myself, and was deliberately focussed on the Simbai Valley, as opposed to the Upper Kaironk Valley. The latter had already been subject to considerable multi-disciplinary research (e.g. Bulmer 1977; Majnep and Bulmer 1977, 2007). Sadly, Dr Majnep died shortly before our fieldwork began in 2007, although we were able to pay our respects to his family and *matmat* (burial ground) at Waiak in the Upper Kaironk Valley.

The intention of the research was to link archaeological investigations of rockshelters and open sites to the well-documented oral traditions and ethnoscience of Kalam communities (see Bulmer 1991 for a list of relevant references). Mr Kari Heri, a field officer at the Papua New Guinea Museum and Art Gallery (hereafter 'PNG National Museum'), accompanied the team during both field seasons. Unfortunately, the research project was suspended during the second field season in 2008, which was the first season of excavation, due to an accident (Bedingfield 2009). For various reasons, the project has not yet been restarted.

While at Simbai in 2007 we stayed at the Kalam Guest House at Ñukunt, adjacent to the Kalam Cultural Museum. The Kalam Cultural Museum is a community initiative designed to foster a sense of cultural awareness among community members at Simbai and was part of initiatives to generate tourism in the area, such as the Kalam Culture Festival. The Museum housed a collection of ancient and recent items of cultural interest. Archaeological artefacts donated by community members included stone club heads, stone mortars and pestles, stone bowls, stone files, stone axe-adzes and bone points. These items had been collected adventitiously by community members while gardening, building houses, digging ditches and so on, or had been kept in homesteads or men's houses in the past. Other donated items of recent material culture included spears, *bilas* ('body decorations'), headdresses, various types of animal trap (some of which functioned) and *tapa* ('bark cloth') beaters.

We were requested by the museum curators, Ernest Simgi and Ishmael Yei, to inventory all items of any antiquity. The reasons were threefold: to assist the curators with an inventory, to provide visitors with information, and for safekeeping. Following discussions with community members it became clear that local people had become increasingly worried that items housed in the Museum might be stolen and sold.

We designed and filled a recording sheet with simple descriptions and diagnostic photographs for each artefact (Kalam Cultural Museum *et al.* 2008). We were aided by Dr Pam Swadling (formerly Head of Archaeology, PNG National Museum) who assisted with the classification and description of stone mortars, pestles and bowls. The subsequent inventory was lodged with local and national authorities at the Kalam Cultural Museum and at the PNG National Museum, respectively. Copies are also retained by authors in Australia. The illustrated artefact inventory will ensure that if any artefacts are stolen and are attempted to be sold on the international antiquities market, that they can be identified and returned to their rightful owners.

The recording of artefacts at the Kalam Cultural Museum indicates how field teams can meaningfully contribute to community-initiated cultural heritage management projects while undertaking their own research. This work can be undertaken incidentally while in the field and represents a small reciprocation on the part of the field team with their hosts (see Muke 2000). Similar types of local community initiative seem to be becoming more common across Papua New Guinea today. These initiatives are sometimes designed to attract tourists, but more often they are a means for communities to re-identify, teach and safeguard their own material culture for future generations.

Nominating the Kuk Early Agricultural Site

The archaeological site at Kuk bears witness to multiple phases of wetland manipulation and drainage for cultivation extending from at least 7000–6400 years ago to the present, and potentially to 10,000 years ago (Golson 1977; Denham *et al.* 2003, 2004; Bayliss-Smith 2007). As such, it is important globally for understanding the emergence of agriculture and its contribution to the transformation of human society during the Holocene. As well as its significance to the academic community, the agricultural history of Papua New Guinea is taught in schools across the country, is a source of national pride, and is a potential focus of national identity (cf. Mangi 1994). As such, Kuk is an excellent example of knowledge transfer.

Papua New Guinea became a State Party to the *World Heritage Convention* in 1997. Since that date, the country has worked towards the establishment of a World Heritage site. The first tentative steps for the nomination of Kuk had already begun by 1997 (Golson and Swadling 1998), the year I began my PhD research on the site (Denham 2003), and continued in various forms for the next ten years (Muke *et al.* 2007).

In 2006, John Muke and I were invited to participate in the National World Heritage Action Planning Workshop in Port Moresby. At this workshop, a Tentative List of intended World Heritage sites for the country was formulated, an institutional framework and strategy were developed for the nomination and management of World Heritage sites within the country, and steps were taken to ensure the completion of the nomination of the Kuk Early Agricultural Site. At the workshop, we were then asked by the Department of Environment and Conservation, which is responsible for World Heritage in the country, to complete the nomination process. Together with the assistance of various collaborators and contributions, we were able to submit the nomination in early 2007 (DEC 2007; Muke *et al.* 2007). In 2008, the Kuk Early Agricultural Site was formally accepted onto UNESCO's World Heritage List at the 32nd Session of The World Heritage Committee in Québec City, Canada. I was fortunate enough to be present at that meeting as the sole representative for Papua New Guinea's nomination.

Kuk started as an archaeological site, but it has become Papua New Guinea's first and only World Heritage site. The tortuous history of the Kuk nomination process up until 2007 is the subject of several publications (Ketan 1998; Strathern and Stewart 1998; Denham 1999; Ketan and Muke 2001; Muke *et al.* 2007). It will not be recounted in detail here. The complexity of the nomination process is instructive and a few key aspects are summarised.

First, existing legislation can be used to manage World Heritage sites within the country; namely, there is no need to formulate and enact new legislation to manage World Heritage. For Kuk, enabling legislation includes the *National Cultural Property (Preservation) Act* and associated *Regulations (1965)* (considered below) and the *Organic Law on Provincial and Local Level Governments (1995/1997)*. The latter empowers local communities to generate laws that are nationally binding to protect their own cultural and natural resources. The intention is to get the landholders at Kuk to voluntarily create an *Organic Law* based upon the traditional land use management plan for the site.

Second, the Kuk nomination was successful despite an ongoing land ownership dispute. This is unusual in a global context because ordinarily land disputes undermine successful cultural heritage management projects. In June 1968, the Kawelka landowners sold the land at Kuk to the Australian Colonial Administration of the Territory of Papua New Guinea (Ketan 1998:18; Muke 1998:20–21). Kuk formed part of the endowment of government land inherited by Papua New Guinea at Independence in 1975. The land was used to establish a tea research station, which subsequently became an agricultural research station. In common with many other rural institutions, the research station was mothballed in 1990 and was gradually reoccupied by traditional Kawelka owners during the 1990s (Muke 1998). At the time of nomination in 1997, legal ownership rested with the national government (i.e. Kuk was alienated land) yet traditional ownership and use rights had been reasserted by the Kawelka. Having sold the land, the Kawelka saw it empty and re-occupied it, thereby effectively becoming squatters on their own traditional land. The draft management plan needed to reconcile both claims to ownership: the State's (legal) and the Kawelka's (traditional). At the same time, any solution needed to avoid setting an awkward legal precedent that could potentially legitimate the re-occupation of alienated land elsewhere in the country. The draft management plan effectively formalises and legitimises the *status quo*: the

Kawelka acknowledge the government's legal claim, and the government acknowledges the rights of the Kawelka to occupy and use the land in traditional ways that accord with management guidelines.

Third and following, the traditional land use management plan for Kuk links the distant past to current cultivation practices. The ongoing management of Kuk effectively requires continued occupation and cultivation by Kawelka because they provide a connection between archaeological and contemporary practices through which the site gets it significance as an organically-evolved cultural landscape. Indeed, from this perspective, ongoing cultivation is not to be viewed negatively, rather it adds layers of meaning and significance to the site.

Fourth, the proposed long-term management plan for Kuk was designed to be sustainable within the unpredictable fiscal and political environments of modern-day Papua New Guinea. The long-term management of the site would not be feasible if it relied upon large annual budgets and political stability. The draft management plan was based around low annual inputs of money and was adapted to operate in an environment of recurrent political instability. The management plan, as originally envisaged, incorporated a culturally appropriate sense of reciprocity: provincial authorities were to provide or upgrade existing services, e.g. road grading and clean water, while Kawelka were to accede to management guidelines in terms of prohibited activities, e.g. deep drainage or digging, planting of trees and mechanised cultivation. More grandiose schemes that proposed a visitor's centre and guesthouse may not be sustainable within the current social and economic climate of Western Highlands Province; there is an insufficient number of tourists to make such enterprises self-sustaining.

Since the successful nomination of the site in 2008, there have been some advances within Papua New Guinea in terms of the management of Kuk and World Heritage generally (Denham 2012). For Kuk these advances include the development of provincial and national committees regulating World Heritage, the development of educational materials, dissemination of literature for the general public inside and outside the country (Denham 2008a and 2008b, respectively), and advances in the preparation of a traditional land use management plan. As with many things, however, continued progress in the management of Kuk ebbs and flows; it needs ongoing commitments at the local, provincial and national levels. Foremost it requires constant communication and engagement with Kawelka landholders at Kuk.

The commercial sector

Swadling (1983 in Mandui 2006:380) noted almost thirty years ago, that most forms of modern development in Papua New Guinea erode the 'unwritten' record of the past. Over the last seven years, I have been involved with several contract, or commercial, cultural heritage management projects within Papua New Guinea. Cultural heritage management is undertaken as part of the Environmental Impact Assessment (EIA) process, and most of the larger projects are associated with mining and petroleum exploitation. Cultural heritage forms part of the complex social, economic and environmental impacts of resource extraction projects (Filer and Macintyre 2006; Bainton 2010). Resource exploitation projects offer challenges and opportunities for cultural heritage management within Papua New Guinea (Bainton et al. 2011). These projects are usually subject to complex confidentiality and intellectual property agreements, which can make reference to specific aspects of an individual development problematic.

Academic archaeologists occasionally work on commercial projects in Papua New Guinea, and elsewhere, in order to gain access to sites for research purposes. Archaeological and cultural heritage consultancies now dwarf in terms of value and volume research projects in these fields within the country. The financial and logistical support offered by commercial projects far exceeds that which can be mobilised on research grants, and significant academic contributions can derive from such involvement (e.g. McNiven *et al.* 2011; Spriggs 2012).

The discussion consists of general observations on the practice of cultural heritage management (CHM) within the commercial sector in Papua New Guinea. It is based on my own experiences. This presentation is necessarily general in order to de-identify any individual agency, group or company. Consequently, it is possible to raise several thorny issues, comprising:

1. the respective roles of national and overseas contractors;
2. the role and functioning of the regulatory authority; and,
3. where to now?

These comments are primarily directed towards people outside the country, rather than to those within Papua New Guinea.

The respective roles of national and overseas consultants

Consultant practitioners within the commercial CHM sector are largely either national Papua New Guineans or people based in Australia. National practitioners can be characterised as mostly belonging to two ill-defined groups: those who have worked within archaeology or forms of cultural heritage within the country for many years, even decades; and, those who are newly trained in modern archaeological field methods. The former group includes long-term advocates of cultural heritage practice within the country, even though some have not had regular employment. Some of this group were trained in archaeological field methods decades ago and institutional support for this group over the last few decades has been highly variable. The latter group are younger and relatively inexperienced; although many have benefitted from closer associations with recent cultural resource management and research projects. Additionally, there are people employed in the industry who have received no formal training in archaeology or cultural heritage; their skills and competence vary greatly. Members of any group may work on projects ultimately run by consultant companies owned by Papua New Guineans or overseas.

In recent years, some of the older national practitioners have found it difficult to find regular or permanent work. There are many reasons for this, but it reflects a tension. On the one hand it makes sense to educate a new generation of archaeologists and cultural heritage practitioners in the latest techniques; this is common practice in any country, especially when the industry is rapidly expanding. On the other hand, care needs to be taken to accommodate and include professionals who have often struggled in adverse conditions for decades, and who may in part have helped generate the current cultural heritage management climate within the country.

Most cultural heritage management projects I have been involved with require participation by national practitioners. Some far-sighted resource development companies also require Indigenous training and capacity building to be incorporated into long-term project goals. The lead-in time and planned operating time of many resource sector developments are measured in decades rather than years. Consequently, there is enormous scope for the training of local people, namely individuals selected from the communities impacted by the development, to be educated (at national high school and at university) by the resource company and trained in cultural heritage management. Thereby, members of communities can be trained and employed to manage their own cultural heritage, mediate between community and company, and liaise with provincial and national authorities.

It is fair to say that most overseas contractors and practitioners in cultural resource management had no or limited previous experience working in Papua New Guinea or any other developing nation. At one level, this should not be a problem, as experience and skills are transferable. However, there are some issues that need addressing.

I have heard it said that undertaking cultural heritage management among local communities in Papua New Guinea is the same as working among Aboriginal groups in Australia. Although the techniques of archaeological and cultural heritage practice may be comparable, such statements show an ignorance of the cultural, historical and social contexts within which the work takes place. The history of local communities in Papua New Guinea is different to that among Aboriginal groups in Australia. Their respective histories of colonialism (namely, the relationship between local communities and agents of European/Australian colonial rule) and internal colonialism (namely, the relationship between communities and national entities following formation of independent nation states) are different. Significantly, communities in Papua New Guinea have, for the most part, not been alienated from their land and maintain a degree of sovereignty over the land and its use. Consequently, many groups in Papua New Guinea had maintained a range of traditional cultural practices and their language until relatively recently. Although there may be points of similarity, there are clear differences between Australia and Papua New Guinea, most pertinently relating to culture, archaeology and heritage, as well as to regulatory environment. Ideally, overseas consultants working in the commercial sector in Papua New Guinea should undergo some form of rigorous cultural awareness training before arriving in-country. Such training will hopefully avoid the neo-colonial tendencies that creep into some expatriate practice.

In 2006, Herman Mandui (2006:381) asked "Who will take responsibility for management of Papua New Guinea's cultural heritage?" Ultimately, Papua New Guineans need to be enabled to take full responsibility for the management of their own cultural resources. Through many ongoing initiatives, such as multi-institutional collaboration (e.g. McNiven *et al.* 2011), it is hoped that the current reliance on overseas consultants – especially at managerial and supervisory levels – will give way to an increasing reliance on national practitioners over the next decade. However, such a transition has been a recurrent theme in the recent history of cultural heritage management within the country (Craig 1996).

Role and functioning of the regulatory authority

The Papua New Guinea National Museum and Art Gallery (PNG National Museum) is the regulator of archaeological research and most cultural heritage management (World Heritage issues being a notable exception) within the country. The PNG National Museum issues permits to undertake archaeological and cultural heritage fieldwork, as well as to export materials for analysis; monitors ongoing projects, often through the presence of a field officer; and, maintains the material and report archives for projects conducted within the country.

The *National Museum and Art Gallery Act 1992* made the PNG National Museum responsible for preservation of cultural heritage, including the management, documentation and preservation of archaeological sites and relics within the country. *The Act* states that the Museum is to 'maintain the national register of traditional and archaeological sites, locate and record prehistoric sites and monuments, and carry out the salvage of archaeological excavations as required by the *National Cultural Property (Preservation) Act 1965* and the *Environmental Planning Act 1978*' (i.e. the National File of Traditional and Prehistoric Sites).

The management of cultural property is vested in the Trustees of the Papua New Guinea National Museum and Art Gallery, under the *National Cultural Property (Preservation) Act 1965*, which is the primary legislation relevant to immovable cultural resources, namely archaeological and cultural sites. Provisions of *the Act* pertain to 'any property, movable or immovable, of particular importance to the cultural heritage of the country', including 'any object, natural or artificial, used for, or made or adapted for use for, any purpose connected with the traditional cultural life of any of the peoples of the country, past or present'. Other relevant legislation includes the *Conservation Areas Act 1978* and the *Cemeteries Act 1955*. The *Environmental Planning Act 1978* was repealed and superseded by the *Environment Act 2000*, which unlike its predecessor makes no reference to cultural aspects of human communities.

Against an often challenging operational backdrop, the PNG National Museum is required to regulate multiple, multi-phased cultural heritage management projects, as well as all archaeological research within the country. Some projects are massive by any international standard, running into many millions of Australian dollars, and occur in remote regions spread across the country. Yet, there seems to be a major and disproportionate mismatch between the funding allocated to cultural heritage management projects within Papua New Guinea today, and the resources allocated to the regulation of these activities. Not only is the Archaeology Section at the PNG National Museum under-resourced for such a task, its highly committed staff have not been trained to work in this type of regulatory environment. Until about ten years ago, a greater part of their role was to regulate archaeological research within the country; cultural heritage management was much more limited in scope (Mandui 2006; exceptions include Swadling 1973).

Where to now?

Given the apparent problems and tensions within current cultural heritage management practice in Papua New Guinea, what should happen next? Well, there is no *carte blanche*. It would seem appropriate to develop a strategy that builds upon the range of expertise within the country today, using current professionals, institutions and legislative environment. In this regard, it is necessary to marry short-term goals with long-term change.

First, although a new professional cohort is being trained, this does not mean that those national practitioners who have struggled against the odds and without a regular position or income for decades should be overlooked. These are the Indigenous, national leaders within the field that the next generation will look to for guidance. Their voices and experiences can provide the essential, culturally appropriate insights on how to improve current practice.

Second, there is no easy fix for institutional capacity and regulatory frameworks within Papua New Guinea. Existing institutional structures and practices are set to continue for the foreseeable future. A short-term improvement in the regulation of cultural heritage activities could be assisted, especially drawing on external expertise and resources, through:

1. inter-governmental assistance, both within and outside of country, in the training of staff in cultural heritage regulation. Ideally, this would involve placement within regulatory institutions in other independent Pacific Island nations;
2. revision of standard procedures for the monitoring and reviewing of cultural heritage management projects, including provision for feedback and follow-up to local communities (Mangi 1994), and the training of staff in these procedures;
3. adequate resourcing, including the development of a GIS-based inventory of the National Register, together with training of staff in the maintenance and use of the system; and
4. increased staffing of the Archaeology Section at the PNG National Museum to include a Head, sufficient field officers, National Register officer (GIS/paper archive) and material archive/collections officer.

Cultural heritage legislation within Papua New Guinea is in urgent need of revision. Any legislative reform needs to take into account foreseeable institutional arrangements, any proposed reforms to land ownership laws, as well as practical matters, such as enforcement mechanisms. Legislation needs to be culturally and institutionally appropriate. For example, it would not be appropriate to import or adapt legislation from Australian states with robust and well-resourced regulatory authorities; the institutional context is very different in Papua New Guinea. Any legislation needs to balance the respective rights of individual communities to look after and curate their own heritage, while simultaneously acknowledging the collective right to look after and curate significant heritage at the national level. The sought-for balance between communities and national institutions reflects a unique balance within contemporary Papua New Guinea society.

Third, if codes of practices are developed to regulate archaeological fieldwork and cultural heritage management, it would seem that these need to be appropriate, relevant and sustainable within the Papua New Guinean context. It is not acceptable to have cultural management projects undertaken by untrained practitioners. However, it would be equally inappropriate to advocate codes of practice that are more stringent than those that occur in almost any developed nation. Although well-meaning in trying to raise standards, any such move would almost certainly be counter-productive and unsustainable.

Codes of practice need to be relevant to the regulatory/legislative framework within the country, as well as to cultural, professional and socio-economic contexts. By way of an analogy, have we learnt from debates in economic development over the last 50 years? Advocates of neo-liberal development and modernisation were well-intentioned; they sought to fast-track the economic development and social transformation of 'undeveloped nations' (Chisholm 1982). New technology and new methods were imported from developed nations and superseded traditional practices, often with disastrous consequences (Hettne 1990). Following several waves of critique, it became apparent that development needed to be appropriate to any given context in order to be sustainable (Richards 1985). Similarly, any codes of practice for cultural heritage management need to be appropriate to the modern socio-economic climate and cultures of Papua New Guinea in order to be relevant and sustainable.

Sequential commentators have lamented the perception that cultural heritage management is considered an impediment to economic development in Papua New Guinea (Muke 1998; Mandui 2006), as is often the case elsewhere. Why is this the case? Rather than apportioning blame solely outwards, onto others, should we not take some collective responsibility? Has not a failure of professional leadership, especially in communication to the general public and other spheres of society, also contributed to this perception?

There is great scope for improving commercial cultural heritage management practices within Papua New Guinea. Any such moves will require broad consultation with those voices, national or otherwise, who have experience working in cultural heritage within Papua New Guinea over the long-term. Additional consultation is needed with a wide range of stakeholder groups, including the PNG National Museum, National Cultural Commission, Department of Environment and Conservation and National Research Institute – as well as government agencies and non-governmental organisations associated with resource exploitation and potential land reform. Without an inclusive process, well-intended recommendations are likely to founder.

Turning a corner …

Three different types of cultural heritage management are described above in order to provide an impression of the range of practices occurring across Papua New Guinea at the time of writing. I did not dwell on the specifics of practice, such as field survey, site recording and excavation techniques, these are the subject of field manuals. Similarly, I have not sought to situate cultural heritage management within Papua New Guinea into broader global debates within the field. My observations and discussions are drawn from personal experience and are designed to highlight some of the outstanding issues as I see them at this moment in time.

All cultural heritage within Papua New Guinea belongs to a community. Irrespective of whether the sites or finds connect with current Indigenous knowledge about the past or belong to realms of the past unconnected with contemporary knowledge and practices. All finds and sites are connected to the land (including water-bodies) in some way, and all land has traditional owners. At the same time, these communities, their land and cultural heritage belong to a nation. The balance between local and national interests is not always clear within many spheres of Papua New Guinean society, and future developments across the spectrum of cultural heritage management will need to take into account this dynamic tension.

Another dynamic tension exists in the balance between overseas and national practitioners in all spheres of cultural heritage management, and particularly in the commercial sector. Over the next decade, it is to be hoped that the number of overseas consultants working within the country will diminish; their roles will increasingly be taken up by national practitioners at all professional levels. At the same time, it is hoped that the regulatory environment for cultural heritage within the country will undergo review and that a system tailored and appropriate for the Papua New Guinean context emerges. During this period, the primary role for those based overseas will be to enable this transition in various ways, namely, through mentoring, education and training, provision of equipment and infrastructure, and so on. Care needs to be taken that we do indeed enable change, rather than direct change to conform to frameworks of practice with which we are familiar, namely, those in Australia.

Acknowledgements

I would like to thank the editors of this volume for inviting me to contribute. I would also like to thank several individuals and two reviewers, who all wish to remain anonymous, for comments on drafts of this paper. Their contributions are greatly appreciated. This article was originally written in 2011.

References

Bainton, N.A. 2010. *The Lihir destiny: Cultural responses to mining in Melanesia*. ANU E Press, Canberra.

Bainton, N.A., Ballard, C., Gillespie, K. and Hall, N. 2011. Stepping stones across the Lihir islands: Developing cultural heritage management in the context of a gold-mining operation. *Journal of Cultural Property* 18:81–110.

Bayliss-Smith, T.P. 2007. The meaning of ditches: Interpreting the archaeological record using insights from ethnography. In: Denham, T.P, Iriarte, J. and Vrydaghs, L. (eds), *Rethinking agriculture: Archaeological and ethnoarchaeological perspectives*, pp. 126–148. Left Coast Press, Walnut Creek.

Bedingfield, A. 2009. *Archaeological fieldwork at Simbai, Madang Province, Papua New Guinea, May–June 2008*. Report prepared for the Papua New Guinea Museum and Art Gallery. Monash University, Clayton.

Bedingfield, A. and Denham, T.P. 2008. *Archaeological reconnaissance and community consultation at Simbai, Madang Province, Papua New Guinea, October–November 2007*. Report prepared for the Papua New Guinea Museum and Art Gallery. Monash University, Clayton.

Bulmer, A. 1991. Ralph Bulmer – A bibliography. In: Pawley, A. (ed.), *Man and a half: Essays in Pacific anthropology and ethnobiology in honour of Ralph Bulmer*, pp. 45–54. The Polynesian Society, Auckland.

Bulmer, S. 1977. Between the mountain and the plain: Prehistoric settlement and environment in the Kaironk Valley. In: Winslow, J.H. (ed.), *The Melanesian environment*, pp. 61–73. ANU Press, Canberra.

Chisholm, M. 1982. *Modern world development: A geographical perspective*. Barnes and Noble, New Jersey.

Craig, B. 1996. *Samting bilong tumbuna: The collection, documentation and preservation of the material cultural heritage of Papua New Guinea*. Unpublished PhD thesis, Flinders University of South Australia.

Denham, T.P. 1996. *Understanding the Kalam, understanding ourselves: Reflections on the representation of the Kalam, Bismarck Mountain Range, Papua New Guinea*. Unpublished MS thesis, Pennsylvania State University.

Denham, T.P. 1999. Review of 'Kuk heritage: Issues and debates in Papua New Guinea' by Strathern, A. and Stewart, P.J. (eds). *Archaeology in Oceania* 34:89–90.

Denham, T.P. 2003. *Multi-disciplinary investigation of early and mid-Holocene plant exploitation at Kuk Swamp, Wahgi Valley, Papua New Guinea*. Unpublished PhD thesis, Australian National University.

Denham, T.P. 2008a. Kuk Swamp. *Our Way (PNG Airlines Magazine)* 11:38–41.

Denham, T.P. 2008b. A world cradle of agriculture. *The UNESCO Courier* 6:13–15.

Denham, T.P. 2012. Building institutional and community capacity for World Heritage in Papua New Guinea: The Kuk Early Agricultural Site and beyond. In: Smith, A. (ed.), *World Heritage in a Sea of Islands*, pp. 98-103. UNESCO, Paris.

Denham, T.P., Haberle, S.G., Lentfer, C., Fullagar, R., Field, J., Therin, M., Porch, N. and Winsborough, B. 2003. Origins of agriculture at Kuk Swamp in the Highlands of New Guinea. *Science* 301:189–193.

Denham, T.P., Golson, J. and Hughes, P.G. 2004. Reading early agriculture at Kuk (Phases 1-3), Wahgi Valley, Papua New Guinea: the wetland archaeological features. *Proceedings of the Prehistoric Society* 70:259–98.

Department of Environment and Conservation (DEC) (prepared by Denham, T.P., J. Muke, L. Salas, V. Genorupa and others) 2007. *The Kuk Early Agricultural Site: A cultural landscape*. World Heritage Nomination (successful), Government of Papua New Guinea, Port Moresby.

Filer, C. and Macintyre, M. 2006. Grassroots and deep holes: Community responses to mining in Melanesia. *Contemporary Pacific* 18:215–232.

Golson, J. 1977. No room at the top: agricultural intensification in the New Guinea Highlands. In: Allen, J., Golson, J. and Jones, R. (eds), *Sunda and Sahul: Prehistoric studies in Southeast Asia, Melanesia and Australia*, pp. 601–38. Academic Press, London.

Golson, J. and Swadling, P. 1998. The nomination of Kuk for inclusion on the World Heritage Listing. In: Strathern, A. and Stewart, P.J. (eds), *Kuk Heritage: Issues and debates in Papua New Guinea*, pp. 1–18. Centre for Pacific Studies, Townsville.

Hettne. B. 1990. *Development theory and the three worlds*. Longman, Harlow.

Kalam Cultural Museum., Bedingfield, A., Denham, T.P and Swadling, P. 2008. *Inventory of artefacts, Kalam Cultural Museum, Simbai* [Madang Province, Papua New Guinea]. Monash University, Clayton.

Ketan, J. 1998. *An ethnohistory of Kuk*. National Research Institute, Port Moresby.

Ketan, J. and Muke, J. 2001. *A site management plan for the Kuk World Heritage project in Papua New Guinea*. UNESCO (National Commission PNG) and University of PNG, Port Moresby.

Majnep, I.S. and Bulmer, R.N.H. 1977. *Birds of my Kalam Country*. Auckland University Press, Auckland.

Majnep, I.S. and Bulmer, R.N.H. 2007. *Animals the ancestors hunted*. Crawford House, Belair.

Mandui, H. 2006. What is the future of our past? Papua New Guineans and cultural heritage. In: Lilley, I. (ed.), *Archaeology in oceania: Australia and the Pacific Islands*, pp. 379–382. Blackwell, Oxford.

Mangi, J. 1994. The role of archaeology in nation building. In: Layton, R. (ed.), *Conflict in the archaeology of living traditions*, pp. 217–227. Routledge, New York.

McNiven, I., David, B., Richards,T., Aplin, K., Asmussen, B., Mialanes, J., Leavesley, M., Faulkner, P. and Ulm, S. 2011. New direction in human colonisation of the Pacific: Lapita settlement of South Coast New Guinea. *Australian Archaeology* 72:1–6.

Muke, J. 1998. The death (and re-birth) of Kuk: A progress report on the recent developments at the Kuk prehistoric site. In: Strathern, A. and Stewart, P.J. (eds), *Kuk Heritage: Issues and debates in Papua New Guinea*, pp. 64–86. Centre for Pacific Studies, Townsville.

Muke, J. 2000. Ownership of ideas and things: A case study of the politics of the Kuk prehistoric site. In: Whimp, K. and Busse, M. (eds), *Protection of intellectual, biological and cultural property in Papua New Guinea*, pp. 96–115. Conservation Melanesia Inc., Canberra.

Muke, J., Denham,T.P. and Genorupa,V. 2007. Nominating and managing a World Heritage Site in the highlands of Papua New Guinea. *World Archaeology* 39: 324–338.

Richards, P. 1985. *Indigenous agricultural revolution*. Methuen, London.

Spriggs, M. 2012. Comment. *Australian Archaeology* 75:15–16.

Strathern, A.J. and Stewart, P.J. (eds). 1998. *Kuk Heritage: Issues and debates in Papua New Guinea*. Centre for Pacific Studies, Townsville.

Swadling, P. 1973. *The human settlement of the Arona Valley, Eastern Highland District, Papua New Guinea*. Papua New Guinea Electricity Commission, Port Moresby.

Swadling, P. 1983. *How long have people been in the Ok Tedi Impact Region?* PNG National Museum Record No. 8. National Museum of Papua New Guinea, Boroko.

9

Hierarchies of engagement and understanding: Community engagement during archaeological excavations at Khao Toh Chong rockshelter, Krabi, Thailand

Ben Marwick, Department of Anthropology, University of Washington, USA

Rasmi Shoocongdej, Faculty of Archaeology, Silpakorn University, Thailand

Cholawit Thongcharoenchaikit, Natural History Museum, National Science Museum, Thailand

Boonyarit Chaisuwan, Fine Arts Department, Thailand

Chaowalit Khowkhiew, Faculty of Archaeology, Silpakorn University, Thailand

Suengki Kwak, Department of Anthropology, University of Washington, USA

Introduction

In this chapter we present a case study showing an explicit strategy for local community engagement at an archaeological excavation in southern Thailand. We show how we tailored our approach to engagement to suit different sections of the local community. Our experience and strategies are probably familiar to many archaeologists working in the Southeast Asian region who have independently converged on similar approaches. We review the history of cultural heritage management in Thailand and show that while government policy has focussed resources on tourism at monumental sites, academic work has been most progressive in pioneering local community engagement at archaeological sites. Inspired by this progress, this chapter aims to provide a basic template for public engagement at various scales by explicitly documenting our strategies of local engagement at an excavation we conducted in Peninsular Thailand. We describe a model of understandings of archaeology that we found useful to strategise our engagement with the public. By providing this template we hope to make the process of promoting cultural engagement at archaeological excavations more effective and efficient for future projects.

Background to cultural heritage management in Thailand

Unlike its neighbours, Thailand has never been colonised. This means that the early years of archaeology and cultural heritage conservation in Thailand have taken a different path, especially compared to Vietnam and Cambodia where the École française d'Extrême-Orient of the French government strongly influenced the development of research and conservation of prehistoric sites (Glover 1999; Stark and Griffin 2004). Instead, a series of isolated foreign-led expeditions reporting prehistoric archaeological sites (e.g. Evans 1926; Sarasin 1933; Malleret 1969), combined with interest of the Thai royal family in preserving ancient monuments, defined the history of cultural heritage conservation in Thailand.

Lertrit (2000; 2010) has traced an official concern for protecting historical monuments to the reign of King Chulalongkorn (1868–1910), and during the reign of King Vajiravudh (1910–1925) a section of the Palace's religious affairs office was split off to form a fine arts department concerned especially with Buddhist monuments. In 1925 this department was moved from the palace to the National Museum, under the supervision of the Royal Council. An act of parliament in 1932 established the Fine Arts Department as a section of the Ministry of Religious Affairs, from which it later moved to the Ministry of Education and Ministry of Culture. In 1943, Field Marshal Pibulsonggram (1897–1964), then Prime Minister, set up a new university, the University of Fine Arts or Silpakorn, to train students in art, art history and archaeology to provide staff for the Fine Arts Department. Most of the early fieldwork by the Fine Arts Department of Thailand was concerned with proto-historic and historic ruins, primarily in Phimai, Lopburi and Ayuthya.

The main legal instrument that the Thai Fine Arts Department operates with is the *Act on Ancient Monuments, Antiques, Objects of Art and National Museums, B.E.2504 (1961)*. The act is mostly concerned with the ownership and administration of ancient monuments, antiques, art objects and national museums. The focus on monumental and aesthetic qualities of ancient objects and sites is consistent with the origins of the Department in the royal palace. The early homes of the Fine Arts Department in the Ministries of Religious Affairs, Education and Culture show the close links between heritage management and the official maintenance of national narratives of Thai history and culture. Stark and Bion (2004:118) note that the modern Thai Fine Arts Department is comparable to equivalent agencies in Vietnam, Malaysia and Indonesia as one of a group of 'outstanding examples of heritage management organisations, with well-trained archaeologists and (at least until recently) adequate funding'.

With this bureaucracy in place and the improved accessibility of Thailand to foreigners after the Second World War, larger, more coordinated and more diverse archaeological research and conservation efforts appeared in Thailand (Shoocongdej 2011a). For example, the Thai-Danish Prehistoric Expedition of 1960-1962 resulted in the well-documented excavation of Sai Yok rockshelter (van Heekeren and Knuth 1967) and expeditions by Chester Gorman (1971) and his students from the 1960s onwards.

To summarise, the history of archaeological heritage management in Thailand has been focused on relatively recent prehistory and visually appealing remains, mostly because these sites and artefacts help foster and legitimise Thai national pride (Shoocongdej 2011a). Much of the current scholarly interest in the cultural heritage of Thailand continues this focus with conservation of Buddhist and Khmer monumental sites, prevention of illegal trading of antiques and development of cultural heritage tourism to promote economic growth (Peleggi 1996). A subset of this research is notable for a critique based on Buddhist ideology of contemporary Western conservation strategies (Byrne 1995; Karlström 2005). In the context of cultural engagement, the most politically and economically important element of these current interests is archaeological heritage tourism, which has become a substantial component of public engagement in Thai archaeology because of government policies enacted by the Tourist Authority of Thailand. These government priorities have been criticised as a misuse of archaeological heritage and one of the reactions to this has been a shift of focus by academic archaeologists in Thailand to strengthen local and grassroots community organisations (Lertrit 1997; 2000; Shoocongdej 2011b). In a recent review, Schoocongdej (2011b) describes four long-term archaeological research projects (Sub Champa, Pong Manao, Ban Bo Soak and Ban Rai Rockshelter) run by Thai scholars that include an explicit, multi-component and mutually beneficial engagement with local communities for evaluating the significance of the local archaeological heritage and conserving it. The case study we present here was inspired by the success of these projects and drew on many of their methods. We focus on the specific event of the excavation to show how a fine-grained approach to engaging with site visitors can improve understanding of local prehistory and the process of archaeology.

Excavations at Khao Toh Chong: A case study in community engagement in peninsular Thailand

Our interest in working at Khao Toh Chong was motivated by recent work that describes three viable models of the hominin colonisation of Southeast Asia: a route from south Asia along the coast of Burma; a route directly south from China into northern Thailand and a route from China into northern Vietnam (Marwick 2009). Peninsular Thailand is significant in these models because it is an area where hominins likely travelled through to colonise island Southeast Asia, regardless of which route they entered mainland Southeast Asia (Figure 1). The potential of this region to provide information on the period that we are interested in has been established by the results of excavations at Lang Rongrien rockshelter and the Moh Khiew site cluster, both in Krabi Province.

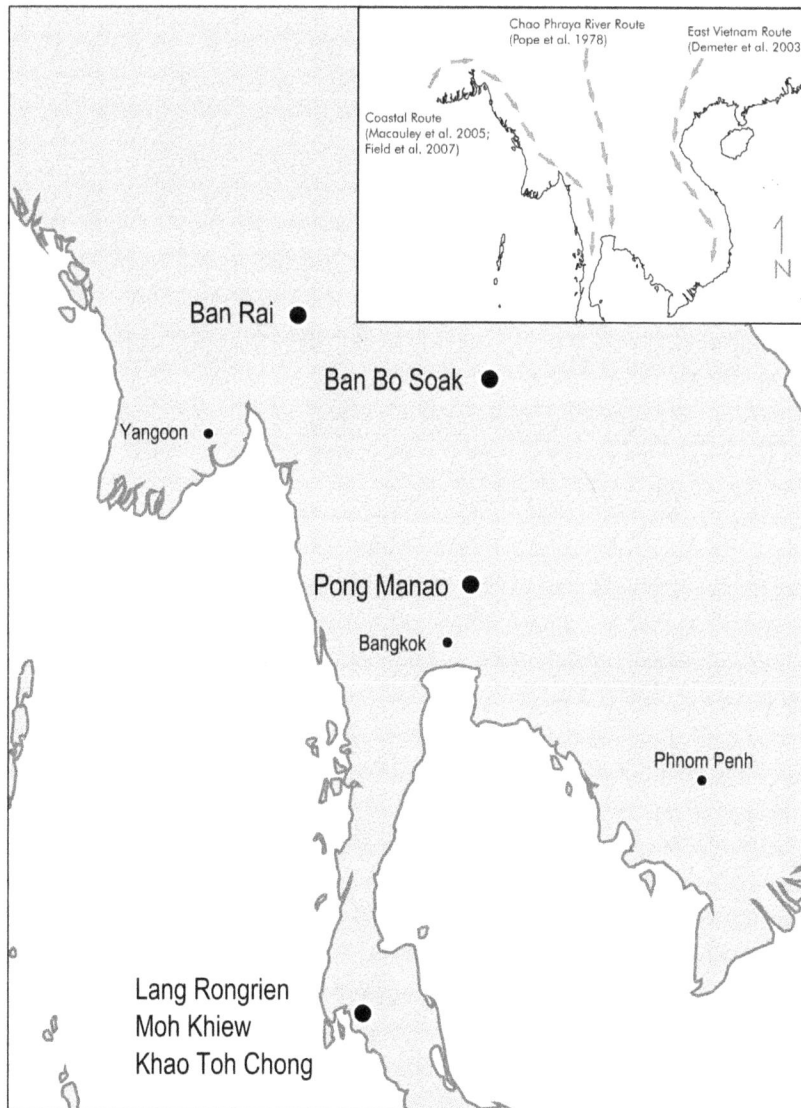

Figure 1. Map of sites discussed in the text.

Source: Produced by authors.

Lang Rongrien is a rockshelter in the Krabi River valley about 12 km east of the coast on the Malay Peninsula in Thailand. Approximately 100 m² of the rockshelter floor – virtually the entire surface – was excavated by Anderson over 1974–1990. Anderson's excavation of Lang Rongrien is one of the pioneering contributions to the establishment of human antiquity in Southeast Asia. For its time, the level of reporting was also remarkable, however the limited detail available on chronology, stratigraphy and the lithic artefact assemblage makes it difficult to use this site to address current research priorities about human evolution, adaptation, colonisation and global diversity. Anderson's excavations did not reach bedrock but terminated in a sterile layer of limestone debris composed of rock fall from the roof of the shelter (Anderson and Mudar 2007:299). This invites the possibility that we might find deeper and older deposits in the area.

Excavations at Lang Rongrien recovered a small (36 pieces) Pleistocene assemblage of flaked stone artefacts dating to 37,265±1000 (PITT-1249). These pieces have not been analysed in detail (i.e. metric and technological data are not available) and their current location is unknown. An older radiocarbon date of 43,000 BP has been claimed by Anderson (1990) for Lang Rongrien, but this date is ambiguous because it is described in published accounts as >43,000 BP with no error range. As this date was at the limit of radiocarbon dating at the time, it is likely that the true age of the deposits is much older. This dating ambiguity and the small Pleistocene assemblage limit this site in its contribution to questions on human evolution and colonisation. However, these details suggest to us that sites of similar antiquity are located in the area.

Other Pleistocene sites in Krabi Province that have been excavated include the Moh Khiew site cluster where excavations have recovered well-preserved human skeletal remains dating to 25,800±600 BP (TK-933Pr) (Matsumura and Pookajorn 2005). Statistical comparisons of cranial and dental measurements of this Moh Khiew skull by Matsumura and Pookajorn suggest that the Moh Khiew specimen is most similar to Australian samples, especially the Late Pleistocene series from Coobool Creek. Excavations by Thai archaeologists began at Moh Khiew in 1990 and are currently ongoing. Many of the publications to date on Moh Khiew have been very brief and we are awaiting publication of the finds at a level of detail that is suitable to engage with questions of human behavior and adaptation. However, the presence of a second confirmed Pleistocene site in Krabi indicates that there is a high probability of new investigations revealing additional sites of similar or greater antiquity.

The site we report on here, Khao Toh Chong (KTC) rockshelter, is located about 10 km south of Lang Rongrien. This site was first recognised as an archaeological resource by two local school teachers who brought it to the attention of the Thai Fine Arts Department. The site is a limestone overhang at the base of a 300 m high karst tower in Thap Prik Village. The rockshelter is about 30 m long with an average of about 10 m from the rear wall to the dripline (Figure 2). The dripline is about 40 m above the ground and a series of large boulders (3-4 m high) at the dripline give excellent protection from the wind and rain as well as trapping sediment in the shelter. The surface of the rockshelter is level fine sediment with no signs of disturbance and about 10 m above the surrounding ground, which is about 60 m above sea level. Similar to Lang Rongrien, KTC has a deep multi-chambered cave system with well-preserved active speleothem formations located about 30 m to the west of the rockshelter. We plan to obtain a speleothem sample from this cave system to analyse as a palaeoclimate archive and complete the other proxies we are analysing.

In June–July 2011 we directed a four-week archaeological field school at KTC that included students from Thailand, Cambodia, Indonesia, Burma, the Philippines, Korea, Vietnam and the USA. The field school excavated two areas of 2 x 2 m to a depth of about 2 m and recorded unusually well-preserved stratigraphic layers and features, including lenses of marine and freshwater shellfish. Analysis of these materials is ongoing, and our initial impressions are that the excavated deposit spans the entire Holocene period and late Pleistocene. There are several distinct changes in the sequence in each of the major categories of evidence we recovered. The stone artefact technology changed from polished adze flakes made from fine-grained sedimentary rock accompanied by ceramics in the upper levels to large flaked cores and flakes made from coarse-grain metamorphic rock in the lower levels. The lower levels also show a change in the faunal assemblage with the appearance of artiodactyl remains (some with cut marks) and a reduction in the remains of small mammals and reptiles that are common in the upper parts of the site. The ceramic assemblage also changes from black sherds in the upper levels to thicker, red sherds with frequent incised decorations in the lower levels and then ceramics disappear altogether when flaked cores and artiodactyl remains appear in the lowest levels. We interpret the lower levels, with their artiodactyl remains, flaked stone artefacts and absence of ceramics and polished stone artefacts, as likely to have been deposited during the late Pleistocene or early Holocene and expect radiocarbon analyses currently underway to confirm this.

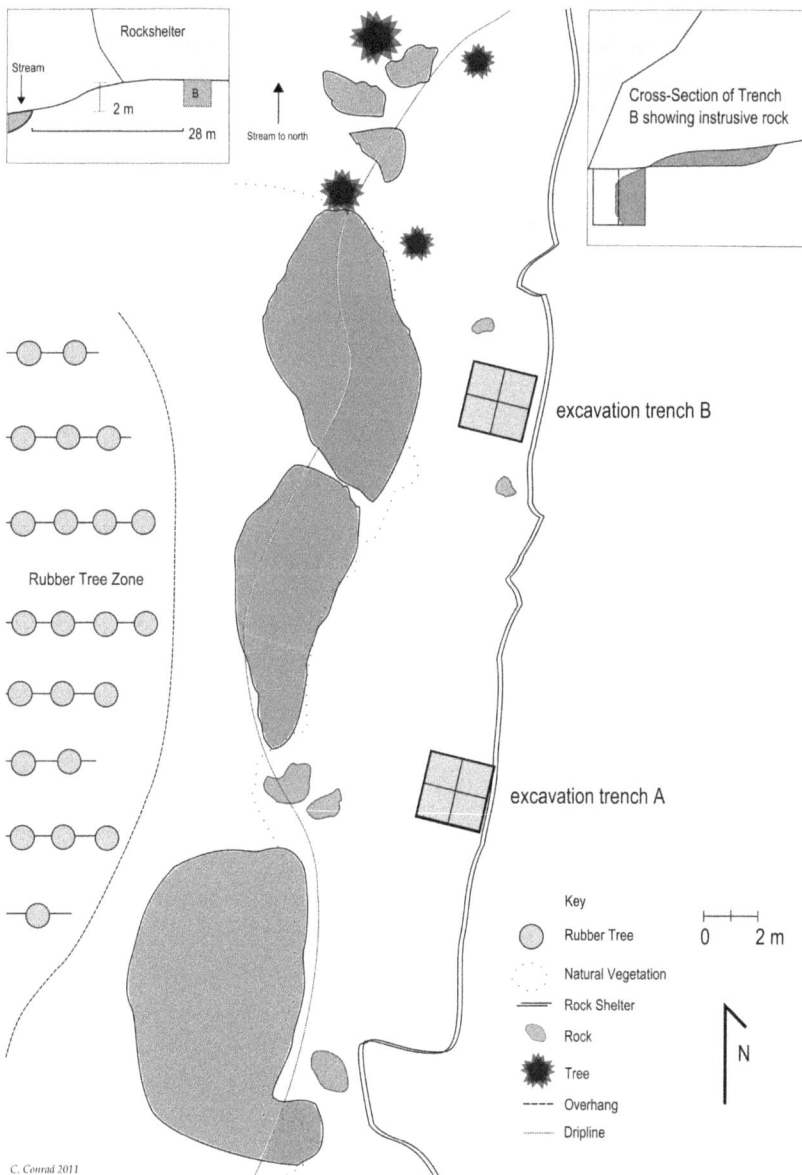

Figure 2. Plan of Khao Toh Chong rockshelter showing areas excavated in 2011.

Source: Drawn by Cyler Conrad.

One of the features of this deposit that makes it unique amongst mainland Southeast Asian sites is that relatively few post-depositional processes have disturbed the site. For example there we did not encounter any burials, there are no animal burrows and there is very limited termite activity at this site. The most striking indication of this is a series of six shell dense layers 0.1-0.2 m thick that vary in the proportion of species from the upper levels to the lower levels. With further analysis we expect the taxonomic variation in these shell layers will yield information about environmental change and human adaptive responses during the Holocene. Probing to a depth of 1 m at the base of our excavations did not locate bedrock, so we expect continued excavations at KTC to yield a long and rich Pleistocene record, comparable to nearby Lang Rongrien and Moh Khiew.

Hierarchies of engagement at Khao Toh Chong

A near-universal feature of archaeological excavations is the crowd of onlookers who are curious to see what is being uncovered. In some cases these crowds are engaged using long-term programs such as ticket sales, guides and permanent barriers to control their movements. However, the resources for such complex visitor management are frequently unavailable. This was the case at KTC where our initial reconnaissance of the site suggested that we could expect a small number of local visitors, but after fieldwork began we received greater attention than expected. The main reason for this increase in attention was the threat posed by the expansion of a limestone quarry operating on the other side of the same karst tower where the KTC rockshelter is located. The quarry is widely regarded by local residents as disruptive because of the heavy vehicles that crowd the narrow local roads and create dangerous driving conditions. The destruction of the karst tower caused by quarrying was remarked by locals to be unwelcome because of its unsightliness. Many were concerned that expansion of the quarry might involve increased heavy vehicle traffic, sound and dust pollution, the loss of agricultural land and disruption to irrigation. As local residents became increasingly aware of the goals of the archaeological work, they realised that it might be relevant to their interests in slowing or stopping the quarry expansion.

The field school excavations at KTC attracted substantial national media attention in Thailand because we revealed that the area near the quarry contained a scientifically significant archaeological record which is threatened by proposed expansion of the quarry operations. This media attention unexpectedly multiplied the amount of local visitors and motivated us to be more strategic in our engagement with visitors with the limited resources available. Like many excavations, our visitors could be classified into four groups: school children, local residents, local government and the national media. While we gave the basic story of what we though prehistoric people were doing to everyone, to ensure maximum impact of our work we crafted slightly different messages for each of these audiences.

With the first group, the school children, we emphasised the value of competence in basic literacy and numeracy (Figure 3). Our message was that if they learn to read and count well then they get to do exciting activities like excavation and receive attention from TV crews. We also highlighted the visceral and physical appeal of the work. By inviting the children to touch ancient things as they were being recovered from the sieves, we encouraged them to vividly imagine a prehistoric person's touch on the same object. These kinds of tactile and immersive experiences seemed to be very engaging for children, perhaps because of the novelty of the idea of a time far before the present when the artefacts were made and used.

The second group were the local adult residents, with whom we most frequently engaged out of the four groups (Figure 4). We showed how the finds in our excavation gave a few insights into the quotidian details of how people used to live in the area. Some residents were impressed to see that their current lifestyle shared some similarities to the prehistoric lifestyle they inferred from the artefacts from KTC. The local residents were the group most interested in the possibility that the significance of our finds would support their opposition to the expansion of the nearby limestone quarry. Unlike the local children, most local adult residents had little interest in any direct physical involvement with the excavation. The local adult visitors were also the group that were most active in practicing local beliefs and superstitions at the site, arranging for Buddhist monks to pray at the excavations and a visit from a local shaman who provided an interpretation of prehistoric use of the site. Byrne (2011) notes that these types of religious activities are common at archaeological sites in most Asian countries and represent an important dimension of local cultural engagement with archaeological sites and objects.

Figure 3. Local children assisting with sieving at the excavation of Khao Toh Chong.

Source: Photo by authors.

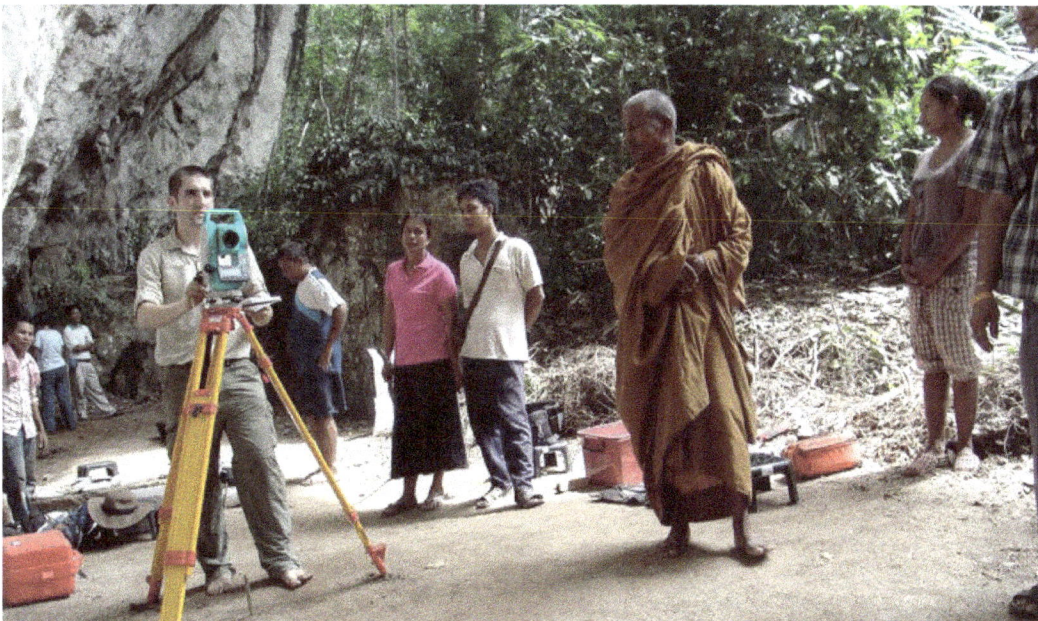

Figure 4. Local adult residents visiting the excavation at Khao Toh Chong with a monk.

Source: Photo by authors.

The third group was local government officers. Our engagement with this group was highly formalised and distinctly different from the school children and local adult visitors. With the local government officers we focussed on communicating the broader social and economic impacts of our project. We described how Thai students and scholars were working closely with ten other nationalities, and how all participants were learning new skills and building relationships that we hoped would improve the capacity of archaeology in Thailand and the home countries of the other participants. We also demonstrated the positive economic impacts of the archaeological fieldwork resulting from hiring locals, buying local food and supplies and making minor infrastructure improvements (like building a bigger footbridge across the fields that lead to the site and installing electrical cables).

Figure 5. Government officials, in this case staff from the Krabi Office of Culture, visiting the excavation at Khao Toh Chong.

Source: Photo by authors.

The fourth group was the national television news media, whose visits were the briefest and least frequent, but most intensive because of the attention they demanded (Figure 6). Our strategy with the media was to engage with them exclusively about the scientific results of the excavation. We knew the video captured during the media visited would be edited to about ten seconds of footage, so we wanted to be sure that viewers could see we were working on basic questions about prehistoric humans as our primary activity. Although the greatest proportion of our public engagement time was spent on educational and advocacy engagement with local children and adults, we believed that for a wider audience it was important to communicate the message that our project was scientific, rigorous and controlled. There were two reasons for this, first is that the local situation of an expanding quarry is of little interest to news-watchers throughout the rest of Thailand. Second, we wanted people to get a rapid understanding of the difference between controlled archaeological excavation and looting. Our strategy was that the most effective way to demonstrate the difference between archaeology and looting in ten seconds was to show how archaeology is characterised by the use of measurement instruments, extensive documentation and slow and systematic excavation procedures.

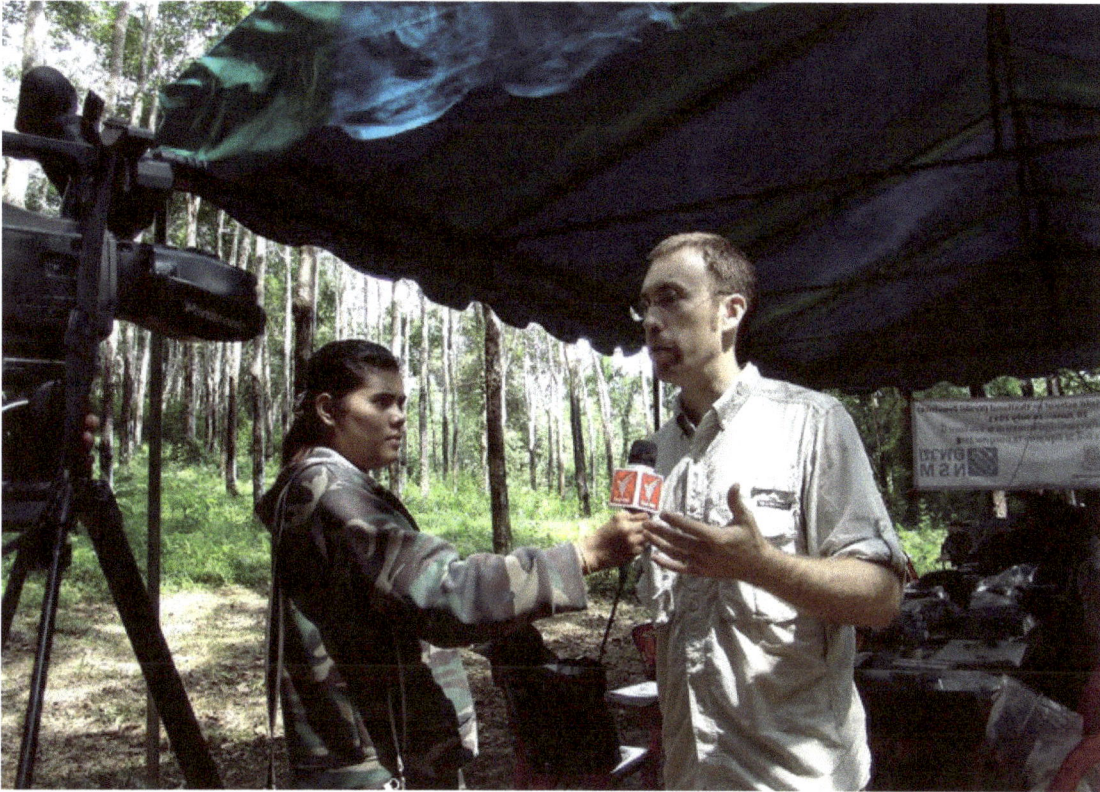

Figure 6. Staff from the Thai Public Broadcasting Service interviewing Ben Marwick at the field lab adjacent to Khao Toh Chong. All interviews were conducted in Thai.

Source: Photo by authors.

Through our engagement with these four groups we shared nearly every aspect of our activities with the public, although in a selective and hierarchical way, depending on the interests, abilities and agendas of each group (and us as archaeologists). There were also details that we had to strategically withhold from all forms of public engagement. In the upper layers of our excavations we found a very small number of valuable ancient objects. These finds were hard to hide from the local residents since they were frequently watching the excavations. The recovery of these objects created a concern that looters would come in the middle of the night to dig up the site and local backyards in search of additional finds to sell as jewellery. In response to this concern, we agreed that the valuable objects were not to be openly discussed at the site or mentioned to any visitors. We also were careful to avoid circulating their photos online, where we had been keeping an illustrated multi-lingual diary of the excavation (see http://afst11.wordpress.com/ and https://www.facebook.com/groups/176618192390525/photos/).

Hierarchies of understanding: Modelling local understandings of what archaeology is about

Implicit in our strategy of public engagement are our expectations and hopes for how the four different groups would understand what archaeology is about. Although we did not formally survey our visitors like Lape and Hert (2011), we suspect there were probably as many different understandings of archaeology as there were individual visitors to the site. To make this diversity manageable, we found it useful to fit a three-dimensional model of archaeological understanding based on Abbott's (2004) updating of Charles Morris' (1946) classic theory of semiotic relations. The three dimensions are syntactic archaeology, semantic archaeology and pragmatic archaeology.

Syntactic archaeology refers to the possibility that by inflicting too much of our technical details on the visitors we convey the understanding of archaeology as a set of abstract arguments about prehistoric human life. The outcome is that we contribute toward an understanding that archaeologists most value the elements of surprise, elegance or counter-intuitiveness in our accounts of prehistoric people. People holding this view of archaeology typically see it as an ivory tower game that cannot justify the use of public resources such as grant funds. Obviously we strived to discourage this type of understanding amongst our visitors. We label this a syntactic archaeology because it – unfortunately – suggests that the public understand archaeology as academic work producing explanations that do not involve any readily tractable meaning.

Semantic anthropology is a more positive understanding of archaeology, where our work is recognised as an attempt to generate meaning by reference to objects with agreed-upon meanings and truth and falsity. People holding this view understand that we can explain aspects of prehistoric life to the point where we can give a sufficient account to solve a problem. For example, when visitors ask 'what did they eat?' we pointed to the shellfish and burnt animal bones, named the species and described their habitats and give our interpretation of how the food was prepared. This way of understanding archaeology views it as a transposing activity where a question about the past is moved into the common-sense world of the immediate where it becomes immediately comprehensible. We translated a phenomenon from one sphere of analysis to another until a final realm is reached at which we are intuitively satisfied. Here we convey the understanding of anthropology as process of answering questions, which we believe was the most frequent type of understanding of what archaeology is amongst our visitors and the one we felt most comfortable promoting.

Pragmatic archaeology is archaeology that results in intervention. That is, we make an explanation of the archaeological record that allows a current issue beyond the realm of the archaeological research to be managed more effectively. Archaeology will of course never cure an epidemic or eradicate poverty, but in our case we might imagine it improving people's quality of life by limiting the expansion of an intrusive industrial operation or giving them an increased sense of belonging, community legitimacy and familiarity with the place they live by showing continuity of lifeways from prehistory into the present. This conveys the understanding of archaeology at intervention, where we produce explanations that allow action to be taken beyond the immediate context of the work. While we did not actively encourage this type of understanding of archaeology amongst our visitors, it was clear to us that many local residents hoped that the archaeological fieldwork would result in an intervention. In general, archaeologists rarely convey this pragmatic understanding to the public because we rarely encounter a narrow neck of causality (Abbott 2004:9), where archaeological work is part of a small number of mechanisms that can be identified and controlled in the scheme of causes of phenomena of broad importance to the public.

Conclusion

In this chapter we have described our strategy for engaging with different sections of the local community that visited our excavations at Khao Toh Chong. We drew on successful models of community engagement from more established archaeological projects elsewhere in Thailand. We tailored our engagement to four different groups that we identified in our visitors: local children, local adults, local government officials and national media. The specific method of engaging with each group was based on our perception of their interests and the way they appeared to understand the purpose of archaeological research. As a final observation, we note that while the development of Southeast Asian archaeology has included the incorporation of Western theories and methodologies into local archaeological practices, a distinctive practice of cultural engagement has emerged independent of Western traditions.

In her review of the uses of archaeology in Southeast Asian countries, Shoocongdej (2011a:724) observes that Southeast Asian archaeologists appear to have a heightened awareness of their professional responsibilities to the communities and societies. The motivation for this distinctive practice comes from the belief that archaeological evidence and control over interpretations of this evidence do not belong to one particular group; instead these belong to everyone to whom to the evidence is relevant (Shoocongdej 2011b). This belief is most likely a result of Western imperial activity in Southeast Asian and Western uses of the discipline of archaeology (Shoocongdej 2007). Our hope is that more explicit documentation of these distinctive local practices, such as we have presented here, will make cultural heritage management more efficient and effective by improving cooperation and communication between stakeholders (Lertcharnrit 2010; Shoocongdej 2011a).

Acknowledgements

Thanks to the students who participated in the field school and made vital contributions to our public engagement activities: Borisut Boriphon, Jess Butler, Praewchompoo Chunhaurai, Cyler Conrad, Kim Sreang Em, Fitriawati, Anna Hopkins, Kate Lim, Supalak Mheetong, Kyaw Minn Htin, Chonchanok Samrit, Thanh Son, Rachel Vander Houwen and Hannah Van Vlack. Thanks to local residents Mr. Suthep Chantara and Mr. Niwat Watthanayommanaporn, who first brought the archaeological potential of this site to our attention. Ben Marwick was supported by a Luce/ACLS Postdoctoral Fellowship (Grant # B9471 01 180401 7301) and an award from the Schwartz Endowment for International Education from the Office of Global Affairs, University of Washington.

References

Abbott, A.D. 2004. *Methods of discovery: Heuristics for the social sciences.* WW Norton and Co., New York.

Byrne, D. 1995. Buddhist stupa and Thai social practice. *World Archaeology* 27(2):266–281.

Byrne, D. 2011. Cultural heritage and its context in Southeast Asian local-popular culture. In: Miksic, J. (ed.), *Rethinking cultural resource management in Southeast Asia preservation, development, and neglect*, pp.3–14. Anthem Press, London.

Evans, I. 1926. Stone implements from Chong. *Journal of the Federated Malay States Museums* 12(2):35–57.

Glover, I. 1999. Letting the past serve the present – some contemporary uses of archaeology in Vietnam. *Antiquity* 73(281):594–602.

Gorman, C. 1971. The Hoabinhian and after: Subsistence patterns in Southeast Asia during the Late Pleistocene and early recent periods. *World Archaeology* 2(3):300–320.

Karlström, A. 2005. Spiritual materiality: Heritage preservation in a Buddhist world? *Journal of Social Archaeology* 5(3):338–355.

Lape, P. 2011. Archaeological practice in Timor Leste: Past, present and future. In: Miksic, J. (ed.), *Rethinking cultural resource management in Southeast Asia preservation, development, and neglect*, pp.67–90. Anthem Press, London.

Lertcharnrit, T. 2010. Archaeological resource management in Thailand. In: Messenger, P.M. and Smith, G.S. (eds), *Cultural Heritage Management: a Global Perspective,* pp.176–187. University Press of Florida, Gainesville.

Lertrit, S. 1997. Who owns the past? A perspective from Chiang Saen, Thailand. *Conservation and Management of Archaeological Sites* 2(2):81–92.

Lertrit, S. 2000. Cultural resource management and archaeology at Chiang Saen, northern Thailand. *Journal of Southeast Asian Studies* 31(1):137–161.

Malleret, L. 1969. Histoire abrégée de l'archéologie Indochinoise jusqu'à 1950. *Asian Perspectives* 12:43–66.

Marwick, B. 2009. Biogeography of middle Pleistocene Hominins in mainland Southeast Asia: A review of current evidence. *Quaternary International* 202:51–58.

Morris, C. 1946. *Signs, language and behavior*. Prentice-Hall, Oxford.

Peleggi, M. 1996. National heritage and global tourism in Thailand. *Annals of Tourism Research* 23(2): 432–448.

Sarasin, F. 1933. Prehistoric researches in Siam. *Journal of the Siam Society* 26(2):171–202.

Shoocongdej, R. 2007. The impact of colonialism and nationalism in the archaeology of Thailand. In: Kohl, P. Kozelsky, M. and Ben-Yehuda, M. (eds), *Selective remembrances: Archaeology in the construction, commemoration, and consecration of national pasts*, pp.379–400. The University of Chicago Press, Chicago.

Shoocongdej, R. 2011a. Contemporary archaeology as a global dialogue: Reflections from Southeast Asia. In: Lozny, L. (ed.), *Comparative archaeologies: A sociological view of the science of the past*, pp. 707–729. Springer, New York.

Shoocongdej, R. 2011b. Public archaeology in Thailand. In: Okamura, K. and Matsuda, A. (eds), *New perspectives in global public archaeology*, pp. 95–111. Springer, New York.

Stark, M.T. and Griffin, B. 2004. Archaeological research and cultural heritage management in Cambodia's Mekong Delta: The search for the 'Cradle of Khmer Civilization.' *Marketing Heritage: Archaeology and Consumption of the Past*, 117–141.

van Heekeren, H.R. and Knuth, E. 1967. *Archaeological excavations in Thailand. Volume 1. Sai Yok. Stone Age settlements in the Kanchanaburi Province*. Munksgaard, Copenhagen.

10

Local heritage and the problem with conservation

Anna Karlström, University of Queensland, Brisbane, Australia

In a recently published book on South African heritage, Lynn Meskell suggests that nature protection and conservation predicts all discussions of the cultural past and that the overlapping discourses of natural and cultural heritages are dominated by the natural, reflected in contemporary biodiversity and conservation politics (2012:4, 100). As these discourses are intertwined and difficult to separate from each other, I believe that current debates within cultural heritage conservation politics can also, and reversely, be used to enrich nature conservation discussions. Therefore, the focus for this article is not exactly a formally Protected Area regulated by law or guarded by nature conservationists, which according to the main objectives of this volume is the general scene for most of the contributions. Instead of exploring the interface between local groups and nature conservation, it explores connections between local groups and heritage conservation,[1] directed by contemporary heritage discourse.[2] This discourse includes traditional Western definitions of heritage that focus on material and monumental forms of tangible heritage, and conservationist ideal aiming at maintaining heritage as an unchanging monument to the past (Smith 2006:6, 29–34), a conservationist ideal that to a large extent is shared with contemporary nature conservation discourse. The relationship between local groups and heritage conservation illustrates the same kind of problem that can occur in the interface between local groups and nature conservation, where local groups represent 'other' views that sometimes challenge the conservationist ideal. I use my example from an area around Vientiane in Laos to highlight the different approaches to and meanings of conservation, i.e. that conservation not always is linked to permanence, but rather to change, and argue for the importance of local groups' involvement in heritage management projects, nature and/or culture. The examples I use here are part of my PhD research, conducted in Laos from 2001 to 2006, and based on long-term involvement and archaeological fieldwork including a mosaic of methodologies, such as excavations, interviews, participant observation and archive studies (Karlström 2009).

---

1   I want to mention something about how I use 'preservation' and 'conservation', since these concepts might appear interchangeably in the text. Within biology there is a preference to talk about nature conservation, but in the heritage field there is no clear, or general, distinction between heritage preservation and heritage conservation. However, the act of maintaining heritage, keeping it from the present for the future, is often called conservation in British English and preservation in American English. I mainly use conservation. When I refer to its technical meaning, i.e. the method or the profession of the conservator, it is always in connection to 'restoration'. In base form I use 'preserve' rather than 'conserve'.

2   Commonly referred to as Authorised Heritage Discourse (AHD), since Laurajane Smith coined the term in her book *Uses of Heritage*, which was published in 2006. I continuously refer to the 'contemporary heritage discourse(s)' throughout this article.

The problem with conservation

There are primarily two perceptions, as I have argued elsewhere (Karlström 2009:193–196), that are decisive for the dominating conservationist ideal. The first is the presumption that a link with the past is necessary if we want to know who we are, to have an identity, and that this link provides certainties in an uncertain world. The other is the presumption that we need physical remains from the past to understand history. If we lose these remains, we will not be able to understand the past and we might as well lose our identities. The fear for this loss is, as I see it, the main reason for and motive behind the desire to preserve, which is a generally accepted starting point within contemporary heritage discourse. On the other hand, in a world where the Buddhist notion of material *impermanence* governs the perception of reality, conservation of material culture and heritage becomes a contradiction in terms (Karlström 2005:347–348). With these contradictions as a point of departure I argue, in this article, that culture heritage is something that we create because we, archaeologists and heritage managers, think that conservationist ideals are universal. To a large extent we also ignore not only local groups' views on what heritage is but also the fact that these views differ from the established views within the contemporary heritage discourse and that this fact results in that no universal ideals within heritage management can ever exist.

I consider the focus on material authenticity and the idea that heritage values are universal and should be preserved for the future, and preferably forever, as the main problems with conservation. However, is it possible to impose a conservationist ideal and frames of reference in contexts and worlds where non-conservationist ideals might prevail? Well, I don't think so. My argument for this answer lies within the following exploration of the concepts *restoration and conservation*, *destruction and decay*, and *consumption, looting and loss*, which all relate to and are dependent/ depend on conservationism: restoration and conservation are tools for keeping heritage from destructive influences. However, the meanings of the concepts differ, depending on the context. Restoration must not necessarily presuppose conservation. In turn, conservation can be a destructive force, and destruction might be needed for the conservation of certain heritage values. By illustrating the complexity between the two extremes conservation and impermanence, and the complexity between the different worlds in which they exist, I argue that we cannot continue to urge a universal frame of reference that recommends conservationism. This fundamentalist ideology of heritage conservation might even be dangerous (cf. Holtorf 2006). The different worlds I am talking about do not represent the dichotomies 'Western vs Eastern', or 'Christian vs Buddhist', or 'We vs the Other' straight off. There are obviously different worlds also in what seems a common world. Thus, the purpose of illustrating this complexity with my examples from Laos is not to show how a proper 'Buddhist heritage management' should be carried out. My purpose is rather to open up for other frames of reference than the one imposing conservation within heritage discourse, and to call for new heritage discourses to be created, including perceptions and values of local groups, might they be Buddhists or not.

A particular restoration-conservation debate, propelled by Viollet-le-Duc and Ruskin in mid nineteenth century Europe, is often brought up as the starting point for the ongoing debate about whether to restore/reconstruct or conserve heritage remains. Today, the term restoration refers to the act of returning something to its authentic or former state, but without adding new material (if additions are allowed they must be distinguishable from the original) and not necessarily aiming at unity in style. That is also how restoration is defined in the Venice Charter, which was approved in 1964. Whereas restoration aims at preserving and revealing historic and aesthetic values, based on respect for original materials, conservation is today dominated by scientific methodology, knowledge and values. Central to the contemporary field of conservation is a belief in scientific inquiry and that there is a fundamental need to preserve the integrity of the physical object. Regardless if reconstruction, restoration or conservation is argued for, the different approaches are all aiming at the same, namely to *maintain authenticity and the feeling of originality*. Wanting everything to be as close to the original as possible, for as long as possible, is a generally accepted starting point that prevails within contemporary heritage discourse and the present conservation ethics. Moreover, the final stage in the restoration and conservation processes is a *complete* thing or building. The consequence is often an unacquainted denial of a thing's life between construction and decay among contemporary heritage specialists. The practices, which occur in between, are explained as religious in nature, or supernatural (Byrne 1995; also this publication), or too subjective to be taken into account in modern scientific heritage practices. However, from a general Lao perspective it is what happens in between, concerning maintenance and restoration that is meaningful. But then, the significance and meaning of the term restoration is different from what was described above in connection to modern scientific heritage conservation practices.

Nevertheless, it is important to notice that the contemporary heritage discourse of course exists in Laos. At the two World Heritage sites, Vat Phou and Luang Prabang, this is officially the prevalent notion, and also among heritage managers working for the government in ministries and museums. In my first survey of Vientiane, I spent some time with officials working for different ministries, companies and organisations that were involved in urban planning, road construction, irrigation, mining etc. I interviewed, discussed and distributed questionnaires, through which I explored the heritage management at different levels. Most answers confirmed an awareness of existing international guidelines and legislations, and referred to the contents of national constitutions and laws concerning environmental impact assessments, which cover investigations of archaeological sites and cultural heritage. This awareness has significantly increased during the last decade, and laws and regulations become increasingly efficient. Still, at this level it is very much a question about priorities and money. Thus, in line with nationalistic ideologies about a glorious past propagated by the Lao government, only the monumental and spectacular remains from past times are prioritised, such as the World Heritage sites and, in Vientiane, for example That Luang, Vat Ho Phra Keo and Vat Sisaket.[3] In other words, different worlds might as well be represented by different groups in society: local or non-local groups and groups located in the periphery or in the centre of economic and political power.

Restoration-conservation

Let us now move on to Vientiane and explore the concept of *restoration*. Often it is similar or equivalent to building a new monument, as an act of making merit.[4] Restoration in this context means something radically different from what is implicit in modern scientific principles of conservation and preservation. It is rather a *restoration of an idea of the prestige of the original*, than of the physical form of the original.

Phonesay – maintaining value through change

In Ban Phonesay, one of the hundred villages of Vientiane city, there is a temple named Vat Phone Say Sethathirat. Commonly it is called just Vat Phonesay. After the capital of Lan Xang[5] was moved from Luang Prabang to Vientiane almost 500 years ago, King Sethathirat had this temple built as a gift to his wife. This was also part of the procedure of stating power and royal kinship, which was marked by the building expansion in Vientiane during this period. The main temple construction, the *sim*, differs from most other in Vientiane. It is a low building, in traditional Lao

---

3   That Luang, Vat Ho Phra Keo and Vat Sisaket are historic temple sites in Vientiane that now are museums, and among the main tourist attractions in the city today.

4   In Buddhism, and particularly Theravada Buddhism, merit accumulates as a result of good deeds, acts or thoughts and is carried over to later in life or to a person's next birth.

5   Lan Xang was the first united Lao kingdom, established in 1356.

style, and almost the only temple in its original shape from this period in Vientiane today, even though it has been restored now and again over the centuries. Close to the *sim* stands a stupa in disrepair (Figure 1). With my untrained eyes and limited knowledge, I characterised its condition as miserable when I first documented it in the initial phase of the survey. I soon understood that the heap of bricks and the Buddha statues, some complete and others at different stages of decay, were not at all neglected but rather looked after with the greatest care. Small pieces of gold leaf are every so often added to the deteriorated corpses of the Buddha statues in an act of merit-making. Minor repairs as well as more extensive construction works are continuously carried out to maintain the stupa and the statues, activities carried out by villagers and monks and important for the everyday use of and religious practice at the temple (interviews with villagers and monks in Ban Phonesay, February 2002). It is also, through this repeated restoration process, that the prestige, or spiritual value, of the object is maintained.

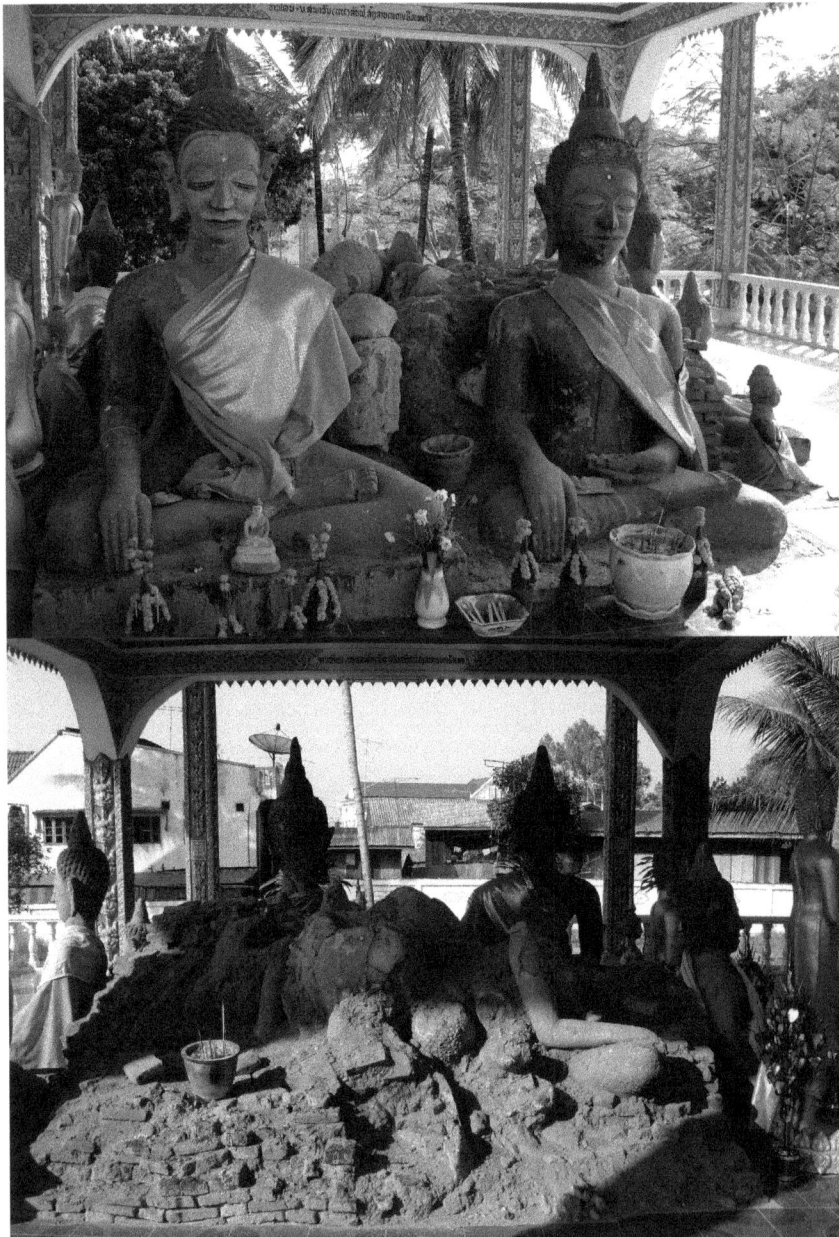

Figure 1. Buddha statues being restored at Vat Phonesay, front and back of the old stupa.

Source: Photo Anna Karlström.

It is important to clarify here, that the reality in Laos is not a strictly doctrinal Buddhism, but rather a mix of animism and Buddhism (see Holt 2009 for a more detailed exploration about how Buddhism is related to Lao conceptions of spirits). Within this popular Buddhism different sorts of objects, such as images, amulets and stupas, that serve as mediums for concentration, aim at bringing to mind the person of the Buddha and inspiring to find the right way. These representations are often referred to as 'reminders'. It is not the physical form and fabric of these reminders that are of importance, but rather that the Buddha's attainment is symbolised by them, and as such, they act as a 'field of merit' (Tambiah 1970:45). True for all these reminders are that they are attributed with power, or *arepower* (cf. Holbraad 2007:189-225). This clearly illustrates the merging of Buddhism and animism, as these obviously Buddhist objects (Buddha images, stupas and amulets) become animated with spiritual power through different kinds of animistic and magical sacralisation rituals. By empowering objects, they become storage places for the spiritual values. And it is the spiritual value that has to be maintained.

It is also important to note the concept of authenticity here, and its different meanings. In contemporary heritage discourse *material authenticity* is one of the foundations and ascribed objects that are true and in their original state. This concept of authenticity privileges mainly unchanged conditions and presupposes a linear time perception, where appreciation and value grow the closer we come to the original state. However, authenticity in popular Buddhism is more about to what extent the object is empowered. It is not dependent on age or the material's originality. Exploring religious practices and beliefs and the production and use of images in Thailand, Denis Byrne writes that it is more relevant to talk about *authenticity established via performance* in a Southeast Asian context than about material authenticity (Byrne 1993). In this way, new images are constantly produced and recreated and must establish their own identities, which give them authenticity. Referring back to what I mentioned above about 'reminders' acting as 'fields of merit', we see that the value of the objects at Vat Phonesay has nothing to do with their form and fabric. It is what they represent and to what extent the object is loaded with significance and power that is important. This is also an example of the different conception of authenticity, which is established here through performance (Karlström in press). The restoration of the ancient Buddha statues at Vat Phonesay is an act of adding and changing the objects' form and fabric and by doing so establishing authenticity and maintaining the place's heritage values.

Ou Mong – maintaining value by building something new

Now almost twelve years ago, one of the oldest temples in Vientiane, Vat Ou Mong, was totally demolished. The *sim* ('ordination hall'), with its interior walls covered with ancient murals, was wiped out because it was to be replaced with a new *sim*. This was done as an act of merit-making that would enhance the beauty and prestige of the temple compound, and the merits of people's lives (Potkin 2001; Karlström 2009:16). Construction workers had already started to build a new *sim* at the temple compound when the demolition of the old *sim* was completed. As the walls grew higher, inscriptions with donors' names and the amounts of contribution were added, and signs were put up on the temple yard telling the same. I cannot say exactly when the construction work started, but it lasted over several years. When I passed the temple six years ago it was not yet completed, earlier this year it was. Even after the installation ceremony, when a temple receives its formal authority, building activities often continue. Following the inherent meaning of the merit-making act, *the completion of a temple is subordinate to the process of its construction*, which also challenges the general idea within contemporary heritage discourse that the final stage in the restoration and conservation processes is a complete thing or building. A temple under construction offers (in this case, the villagers living near the temple) a chance to donate money or provide volunteer labour, which means making merit. Adding, removing and elaborating over time are necessary parts of the merit-making act. With this more or less institutionalised maintenance practice, one can argue that the notion of completion is not relevant in this context. It is the (more or less constant) act of restoration that is important, rather than the result of it, after its completion.

In popular Buddhism, merit-making and other practices involving materiality are superior to the notion of impermanence. Things *are* important and significant. Not as remains from past times, but rather as part of the religious belief and practice. The stupa and the Buddha statues at Vat Phonesay are not defined as historical documents, worthy of preservation because of their ancient origin and material authenticity. At Vat Ou Mong, we also see that the ancient *sim* was not regarded valuable because of its material qualities. The demolition of the old *sim* and the construction of a new one at Vat Ou Mong illustrate another form of restoration practice, slightly different from the maintenance of the stupa and Buddha statues at Vat Phonesay (Figure 2). A restoration practice that has to do with the 'pouring through' of spiritual values. Whereas the Vat Phonesay stupa and Buddha images are reminders and important as instruments through which the significance of Buddha's life reaches people in the present, the Vat Ou Mong *sim* is a shell or a container, a storage place for spiritual values. And whereas the Vat Phonesay reminders need constant restoration through the adding of things to them (such as the gold leaf) for the significance to be maintained, the destruction of the Vat Ou Mong *sim* is necessary for the significance and the spiritual values first to be liberated and then free to enter, or pour through, to the new shell, the new *sim*, which is a result of the villagers' wish to gain merit (pers.comm., villagers in Ban Ou Mong 2000). Building a new *sim* is the most meritorous of acts the villagers can ever carry out, and therefore the restoration of the Vat Ou Mong *sim* is important primarily as a part in the merit-making act. Despite these differences, the essential meaning of the restoration practice is here shared: restoration is *present-oriented* and implies that *things are added* and *constantly change*.

Figure 2. The old *sim* at Vat Ou Mong being demolished in December 2000 and the new *sim* completed.

Sources: Photo/film by Alan Potkin and photo by Anna Karlström.

So, in contrast to the contemporary heritage discourse where restoration means *returning* a structure to its previous state and focusing on *form and fabric and material authenticity*, restoration means, in general in Lao society, *returning* the structure's *prestige and spiritual values*, by *turning* it to something new as a result of *adding or changing* its physical form and fabric. At Vat Ou Mong, the restoration means that the old building is replaced by a new building; the physical *sim* is *turned* to something new. By letting the spiritual values exit the old and enter the new, the prestige of the *sim* is *returned*. Similarly, the restoration of the ancient stupa and the Buddha statues at Vat Phonesay is an example of a process where the structures are constantly *turned* to something new as things are added, but by doing so, they are *returned* to their previous prestige. Returning prestige and spiritual values (not returning to original form and fabric) has to do with the act of merit-making and is an essential part of the religious practice and belief, i.e. what most people interviewed in the Vientiane survey considered their main heritage.

Destruction and decay

I will here continue to explore the concepts of destruction and decay and their relation to conservation, and I argue that they are mutually dependent. Although many of the problematic issues connected to the ideology of heritage conservationism have been discussed over the last century, destruction and decay are still most commonly regarded as threats to and in opposition to conservation. Recently, critical voices against the presumption that destruction and decay are threats to our cultural heritage have been heard and primarily triggered by conflicting values, wars and global terrorism (Meskell 2002; Holtorf 2006; Dolff-Bonekämper 2008; González-Ruibal 2008).[6] In addition to these, there are others who focus on Asia, and bring up examples that well relativise destruction and conservation (cf. Byrne 1995; Johnson 2001; Lahiri 2001; Wijesuriya 2001). I want to illustrate here that the situation in such a seemingly uncontroversial context as Laos also prove to be reason enough for questioning that destruction is only a threat to cultural heritage, and that this in turn challenges the entire perspective of conservationism and its fundamental position within the contemporary heritage discourse.

According to the Buddhist notion of impermanence, decay is inevitable. Decay is a constant reminder of death and therefore, in accordance with the ideas of rebirth, essential for any celebration of life (Robinson and Johnson 1997:34–42; see Shanks and Tilley 1987:116 for similar ideas about materiality within archaeology). It is crucial for rebirth and finally enlightenment, the ultimate goal. Consequently, conservation as the opposite of decay and destruction becomes, in a strictly Buddhist perspective, a contradiction in terms. Even so, the notion of rebirth, and consequently the idea of decay as essential for rebirth, is valid also within popular Buddhism. In the case of Vat Ou Mong, the destruction of the *sim* was not a consequence of a strictly canonical Buddhist practice, but rather a result of the kind of popular Buddhism that is practiced there by the majority. The merging of Buddhism and animism is obvious if we look at the (Buddhist) *sim* as a storage place for spiritual values and power. These spiritual values and power in turn animate the *sim* through different kinds of (animist) sacralisation rituals and the destruction of the old *sim* is necessary for this power to pour through and animate the new *sim*.[7]

---

6   For example the Bamiyan Buddha statues, the Berlin Wall and the Twin Towers have been brought up as examples of how the destruction and loss of monuments and sites rather can *create new meanings* and *produce heritage*. Furthermore, Lynn Meskell speaks about the Bamiyan Buddhas as 'negative heritage', because for those who destroyed them, the Taliban, they represented a site of negative memory, and thus the act of destruction was a political statement (2002:561).

7   The distinction between what is Buddhist and what is animist might be a bit simplified here in this example, but it is just to show that they co-exist. In common and everyday religious practice it is impossible to distinguish Buddhism from animism, they are intertwined and operate within a total field (cf. Tambiah 1970:41).

When it comes to another religious structure, the stupa, the situation is slightly different. At Vat Phonesay, for example, the ancient stupa is the instrument through which the significance of Buddha's life reaches people in the present, and is thus not only an impermanent container for spiritual values. Here, its *decay is part of the restoration act*, in which all stages in the circle of life are represented, and which ends with rebirth and the possibility of final extinction (pers. comm. villagers and monks in Ban Phonesay 2002). Destruction of the stupa at Vat Phonesay is not necessary, because its spiritual power needs not to pour through.

One of the other sites that were investigated within this project, Viengkham, provides yet another idea of how destruction and decay might be related to conservation within popular Buddhism. In connection with the excavations in 2004, there were discussions about moving the village's temple back to the place where the old temple site had been and where we at that time excavated. In contrast to the demolition of the *sim* at Vat Ou Mong, the two decayed temple structures at this old temple site could be left just as they were if a new temple was to be established there. Because, at that particular site, the two mounds had increasingly turned into animistic objects, from having been primarily Buddhist structures. Not only the mounds but also the entire site had become imbued with animistic rather than Buddhist beliefs. Most of the stories about the site evolved around animistic beliefs and practices and had to do with *phii* (spirits). They involved otherworldly and supernatural explanations. *Phii* was the reason behind the strong hesitation from the residents of the village to participate in the excavations initially. And *phii* was also the explanation as to why bad things happen to people who remove objects from the site. This is because *phii* occupy these ruins and are their guardian spirits. Decay had turned the ancient temple foundations into animated objects, and as such, they could remain even though a new temple was to be constructed at the same site. As empowered and magic objects, the ruins help to protect the prestige of the village and its residents (Figure 3). To retain a decayed structure, and just restoring or maintaining its spiritual values through merit-making acts, has nothing to do with its form and fabric. It sustains good relations with ancestral and guardian spirits and gives protection against bad and evil spirits (cf. Tambiah 1970).

Figure 3. The old temple site in Viengkham, still in use during the excavation in February 2004.

Source: Photo Anna Karlström.

In conclusion, these three examples illustrate other approaches to destruction and decay than those dominant within contemporary heritage discourse. At Vat Ou Mong, destruction of the old *sim* was necessary for the spiritual values to be maintained. If the old *sim* had remained, the spiritual values would have been stuck within it and wasted, not free to enter the new *sim*. In Phonesay decay was part of the restoration act as the ancient stupa is the instrument through which the significance of Buddha's life reaches people in the present. And in Viengkham, decay has transformed the temple ruins from being Buddhist monuments to being animistic objects containing guardian/protective spirits where spiritual values are conserved through merit-making acts. Accordingly, destruction and decay are sometimes needed for the appreciation of certain heritage expressions.

Consumption, looting and loss

Consumption is inherently destructive. It involves elimination, the using up of resources and the destruction of material culture itself. A common idea, both within academia and among the general public is that consumption is a threat to society and its spiritual and moral values, that it is a danger to both society and the environment. It is seen in opposition to production, which is associated with creativity and considered the manufacture of value. However, consumption must not only be about buying things or equivalent to modern mass consumption and used as a critique against capitalism. Other approaches have been argued for over the last two decades within, for example, material culture studies. These approaches challenge predominant dichotomies and try to see beyond consumption and destruction as opposed to production and creativity. Instead, they emphasise the relation *between* consumption and creativity, and consider consumption as a way of developing relationships with things (cf. Miller 2008). Such approaches draw attention to the appreciation of consumption and should be applicable also within the heritage discourse. I would argue that this is necessary if we want to continue working with applied heritage management in Buddhist contexts (and of course elsewhere, but I illustrate it here through my examples from a Buddhist context) as Buddhism encourages spending rather than saving, which is reflected in economic systems examined by Melford E. Spiro in the 1960s. His case study was Burmese, but Buddhist notions are also more generally applicable. He introduces his paper in *American Anthropologist* as follows:

> The Buddhist world view, and especially its notions of rebirth and karma, provide a cognitive orientation within which religious spending is a much sounder and much more profitable investment than economic saving ... (1966:1163)

By spending, and directing the surplus towards merit-making, the spender becomes a consumer. Consequently, in this case and as we can see through the examples above in the previous paragraph, consuming heritage becomes the prerequisite for maintaining its value.

Plundering and looting are needless to say huge problems within the field of archaeology and heritage conservation. It is an issue that has been discussed over the last two decades, often in connection to debates on the ethics of archaeology (cf. Zimmerman, Vitelli and Hollowell 2003; Scarre and Scarre 2006), with focus on subsistence digging (when people dig up artefacts to sell and use the money to support a subsistence lifestyle), the commercial use of artefacts and the illicit nature of looting (cf. Brodie, Doole and Renfrew 2001), and on collecting and archaeological excavation as destructive activities. Different kinds of looting represent different problems and require different approaches. Following a discussion of the diverse moral claims that surrounded subsistence digging among members of the World Archaeology Congress (WAC) in 2003, Julie Hollowell points out that there is a wide variety of positions taken in either justifying or critiquing subsistence digging (2006:73–93). These issues will not be discussed further here.

What I want to bring up is rather the double function of looting that exists in many contexts (Laos among others); a fact that has almost never been raised in recent debates about looting and the commercial use of artefacts. On one hand, the plundering of an object is illegal. Plundering damages the archaeological record, there is a loss of information about the past and a loss of a heritage that is owned by all of humanity. These arguments are the ones most often heard in the debates, so I will therefore not repeat them. But on the other hand, plundering might be part of the local religious practice and belief, and therefore necessary if we want the traditions and cultures that are based on these beliefs and practices (i.e. certain cultural heritage) to be maintained, and that is what I want to bring up here.

In a Buddhist context, abandonment, decay and impoverishment are continuously balanced against the process of maintenance and restoration as we have seen already. Suddenly a religious structure is considered worn out and in no use for merit-making. What remains then are the sacred objects, objects animated with power through rituals impregnated by Buddhist as well as animist ideas, but free for anyone to plunder. Plunderers are often pious Buddhists, seeking sacred objects to use in their everyday religious life. The plundering could be regarded as a release of the objects, which allows them and their spiritual values to 'pour forth' into the greater world and be of further use, because the object itself holds more value than the fact that it was buried under or placed inside a religious structure (Karlström 2009:208–210). Plunderers would then rather be looked upon as relievers than looters. This is also the case when it comes to the structures themselves, which are considered looted as people remove parts from them. Such cases are the stupas, which are believed to contain fragments of the relics from the historical Buddha's cremation. These are relics that were distributed throughout Southeast Asia along with Buddha's teachings some hundred years after his death. Denis Byrne explains that the 'radiant power of a relic transmits itself to the physical fabric of the stupa encasing it'. Therefore, he says, it is common when a stupa deteriorates that fragments of it are 'taken away to be encased within new stupas, the empowered fabric of the old stupas thus seeding new ones' (2007:159).[8] Similar occurrences also take place in other parts of the world that are not necessarily Buddhist. In David Matsuda's example from Latin America, he demonstrates how the local people he worked together with regards the unearthed (looted) artefacts as gifts from the ancestors: a 'seed' given by real or mythological patrons to be 'harvested', or excavated, by later generations (1998:87). Another example of a similar situation is Julie Hollowell's experience of working in Alaska. She describes that:

> On St Lawrence Island, digging for artefacts is part of every Islander's heritage, an activity that can usually strengthen one's connections with the past. Artefacts are regarded as gifts left by the ancestors that, if they allow themselves to be found, are meant for use in today's world (2006:88).

Within archaeology and heritage discourse today, there is the prevalent view that looting results in a loss of heritage. However, heritage can sometimes be created thanks to loss. One example is the destruction of the Bamiyan Buddha statues in Afghanistan, which caused strong reactions across the world. This was defined as a crime against culture, an unacceptable destruction of cultural heritage. To destroy such symbolically loaded structures, which before the destruction were identified by international organisations as valuable cultural heritage, is certainly a strong action demonstrating a wish for another social order. However, one can also argue that UNESCO's defence against destruction was an equally strong action (Turtinen 2006:43ff). In fact, the destruction in itself was the main argument for appointing the site as World Heritage and for inscribing it on UNESCO's List of World Heritage in Danger. If heritage is about

---

8   This also explains why it is believed that thousands of stupas in this part of the world now contain such fragments.

remembering the past and contributes to people's identities, then destruction and consumption of the archaeological record – the loss of heritage – help us remember even better and might strengthen our identities even more. With these examples, it should be obvious that we need a far more broad-minded discussion about lost and/or gained heritage because, today, heritage is being created more rapidly than it is being lost.

Conservation as change

If we acknowledge conservation as change it helps us to open up for and better understand the altering meanings of restoration-conservation, destruction and decay, and consumption, looting and loss. Owe Ronström finishes his book on heritage politics at a World Heritage site in Sweden by arguing that conservation is change (2007:292):

> Preserve or change? What is really what? To preserve something only because it is old is a fairly new idea, and the product of such acts of conservation is always something entirely new. There has always been change, the one thing in the world that does not change. Therefore, the problem is conservation, and not change. And after all – is there any more thorough change than the act of conservation?[9]

If we now recall the story about the demolition of the *sim* at Vat Ou Mong, we might ask ourselves which approach to heritage conservation would here be the most appropriate. Is it possible to bridge the obvious gap between a sophisticated conservationist sensibility where the Lao cultural patrimony should be preserved, and the perceptions and priorities of the local community where the old *sim* is laid in ruins, the villagers make merit and by doing so take part in and hand over the intangible and ever-changing heritage of a Buddhist community? If it is possible, *how* can we then bridge this divide, to best meet as many demands as possible? Or is it desirable to even try? Can the *sim* be included in the heritage conservation process at all?

These are all difficult questions with no simple or straight answers. I think general alternative strategies for dealing with a heritage that is constantly changing are not easily found. We might be better off trying to debate conservation ethics in a somewhat more respectful way, where a situated, particular and non-essentialist approach is argued. What is needed is imagination and sensitivity, to put heritage conservation into practice in a constructive and intelligent way, so that the people involved recognise their rights in justifying the same values, as they consider important and sacred. A baseline for this approach must be to acknowledge the different worlds we have to deal with and accept that the frames of reference within the contemporary heritage discourse cannot and should not be used unswervingly for other realities, other worlds. If we depart from the things themselves and treat things as meanings, rather than immediately assuming that they signify, represent or stand for something, it might be easier to recognise the systems wherein things get their significance, including our own (cf. Strathern 1990). Following this 'meta' perspective we will then be able to acknowledge that there are different worlds, rather than worldviews. There are different realities, rather than different appearances of reality. Henare, Holbraad and Wastell, following Latour (2002; in Henare *et al.* 2007:11), conclude this by stating that:

> For if cultures render different appearances of reality, it follows that one of them is special and better than all the others, namely the one that best *reflects* reality. And since science – the search for representations that reflect reality as transparently and faithfully as possible – happens to be a modern Western project, that special culture is, well, ours.

---

9   My translation from Swedish: 'Bevara eller förändra? Vad är egentligen vad? Att bevara något enbart för att det är gammalt är en tämligen ny idé, och resultatet av sådant bevarande blir alltid något alldeles nytt. Och äldst av allt är förändringen, det enda I världen som inte förändras. Därför är det också bevarandet och inte förändringen som är problemet. Och egentligen – finns det någon mer genomgripande förändring än bevarande?'.

After acknowledging the existence of different worlds, it would have been easy just to keep to our own and separate it from the other. In the case of Vat Ou Mong, a scenario like that may have resulted in our thinking 'let them destroy their cultural heritage'. But if scholarship, political commitment and sensitivity are one and the same (which I hope we all strive for) we have to engage in our different worlds and realities, and look at the differences and similarities to better understand other frames of reference. Even though there are several ways of approaching the ethics of heritage conservation in different parts of the world, the universalist position taken by the contemporary heritage discourse is dominating. UNESCO and ICOMOS represent this discourse. The World Heritage concept initially challenged the national view of cultural heritage. Now it 'has accordingly been challenged in the name of local and indigenous interests, and pressing questions have been raised about its meaning and ethical status' (Omland 2006:242). Nevertheless, the concept rests on the fundamental idea that a heritage can be held in common. Even though the contemporary heritage discourse aspires to pluralism, it is not comfortable with the immaterial and spiritual. It is something that is seen as irrational and therefore regarded by the West as pre-modern (Byrne 1993), but still authentic. This fragile, exotic Other runs the risk of disintegration when coming in contact with the West. When UNESCO initiated the Convention for the Safeguarding of Intangible Culture Heritage, to meet the local and indigenous interests, these 'endangered authenticities' (Clifford 1988:5) were expected to adjust themselves to and accept the contemporary heritage discourse. One of the main purposes of the Convention is to safeguard the intangible heritage. Safeguarding here means 'measures aimed at ensuring the viability of the intangible cultural heritage, including the identification, documentation, research, *preservation*, *protection*, promotion, enhancement, transmission, particularly through formal and non-formal education, as well as the revitalisation of the various aspects of such heritage' (UNESCO 2003, my emphasis). Even if these endangered authenticities and intangible heritage are taken into consideration and conservation strategies are formulated in consultation with local groups, the fundamental aim and necessity of conservation is, as I have already argued, still unquestioned. The problem here is our privileged position in the Western world to direct and decide the framework. We are interested in alternative histories, but not in alternative heritages (cf. Byrne 1991 and Omland 2006), and definitely not in alternative frameworks and different worlds. Because, We want the Other to remain exotic, to confirm our own identities and stating difference and power, placing the Western culture at the top of the civilisation process. Thus, the intangible heritage as defined and preserved by UNESCO is irrelevant in other contexts than the Western.

Conclusion

To conclude and answer my own questions above, I would argue that as long as the heritage conservation process only follows the framework of the contemporary heritage discourse and UNESCO's universalist ideology, the *sim* at Vat Ou Mong should not be included. Within the heritage management framework advocating conservation, the *sim* in question is considered archaeological, a valuable piece of heritage. In the case of merit-making, the *sim* is not at all conceived or constructed in this way. The *sim* might well be included in a heritage management process if another frame of reference is used, a frame of reference found within the Lao context. Then it would most likely be included owing to other reasons than the ones stated above, by UNESCO. It would also be managed in a different way. In one respect, the *sim* is already included in a heritage management process, but one that is directed by another heritage discourse where destruction and decay does not necessarily contradict construction and conservation, and where it is the prestige and the spiritual values that are restored rather than the physical form and fabric.

I assume that if we aim at returning something to its previous physical state and material authenticity by not adding new materials, it makes us believe that we do not change it. We simply put things back into their original state. But, by showing that restoration in Laos is present-oriented, that things are added and *do* change, I hope to have illustrated that conservation in the contemporary heritage discourse is basically about the same thing. It does not matter if we return something to its previous state or if we turn something into a new state, returning and turning are both active processes resulting in change and that something new and more valuable, material or immaterial, is created. *Things are not conserved because they are valuable; it is rather so that they become valuable because they are conserved.* We want to preserve the past, but instead we create our own imagination of the past (Edson 2004:339). We want to preserve to prevent loss, but instead we create something new. Heritage is created, destroyed and recreated, and therefore it has been argued that the past *is* a renewable resource (cf. Holtorf 2001). We must acknowledge the past and the things that remain from the past as renewable resources, as something changeable. Otherwise, it means that we assume what people in the future will appreciate and that they will value things in exactly the same way as we do today. And that, in turn, means a kind of 'future-imperialism', a colonisation of future perceptions.

To me, heritage is therefore always both product and process. It may be something monumental or intangible, or even lost. It may be a conserved structure, where the integrity of the physical object, its form and fabric is preserved: a heritage that is created to maintain authenticity and the feeling for originality. It may be a decayed structure, where new things are added and worn out parts are removed: a heritage that is created to maintain an idea of the prestige of the original. It may be a structure that has been deliberately destructed and lost: a heritage that is created to maintain spiritual values. It may be a new structure: a heritage that is created to maintain the possibility for rebirth. What is common is that heritage is created from our different needs, needs from within both the contemporary heritage discourse and local groups.

References

Brodie, N., Doole, J. and Renfrew, C. (eds) 2001. *Trade in illicit antiquities: The destruction of the world's archaeological heritage*. McDonald Institute for Archaeological Research, Cambridge.

Byrne, D. 1991. Western hegemony in archaeological heritage management. *History and Anthropology* 5:269–276.

Byrne, D. 1993. *The past of others: Archaeological heritage management in Thailand and Australia*. PhD thesis, The Australian National University.

Byrne, D. 1995. Buddhist stupa and Thai social practice. *World Archaeology* 27(2):266–281.

Byrne, D. 2007. *Surface collection: Archaeological travels in Southeast Asia*. AltaMira Press, Lanham.

Clifford, J. 1988. *The predicament of culture. Twentieth-century ethnography, literature, and art*. Harvard University Press, Cambridge.

Dolff-Bonekämper, G. 2008. Sites of memory and sites of discord: Historic monuments as a medium for discussing conflict in Europe. In: Fairclough, G., Harrison, R., Schofield, J. and Jameson, Jnr. J.H. (eds), *The heritage reader*, pp. 134–138. Routledge, London.

Edson, G. 2004. Heritage: Pride or passion, product or service? *International Journal of Heritage Studies* 10(4):333–348.

González-Ruibal, A. 2008. Time to destroy: An archaeology of supermodernity. *Current Anthropology* 49(2):247–279.

Henare, A., Holbraad, M. and Wastell, S. (eds) 2007. *Thinking through things: Theorising artifacts ethnographically*. Routledge, London.

Holbraad, M. 2007. The power of powder: multiplicity and motion in the divinatory cosmology of Cuban Ifá (or mana, again). In: Henare, A., Holbraad, M. and Wastell, S. (eds), *Thinking through things: Theorising artifacts ethnographically*, pp. 189–225. Routledge, London.

Hollowell, J. 2006. Moral arguments on subsistence digging. In: Scarre, C. and Scarre, G. (eds), *The ethics of archaeology: Philosophical perspectives on archaeological practice*, pp. 69–93. Cambridge University Press, Cambridge.

Holt, J.C. 2009. *Spirits of the place: Buddhism and Lao religious culture*. University of Hawaii Press, Honolulu.

Holtorf, C. 2001. Is the past a non-renewable resource?. In: Layton, R., Stone, P. and Thomas, J. (eds), *The destruction and conservation of cultural property*, pp. 286–297. Routledge, London.

Holtorf, C. 2006. Can less be more? Heritage in the age of terrorism. *Public Archaeology* 5:101–109.

Johnson, M. 2001. Renovating Hue (Vietnam): authenticating destruction, reconstructing authenticity. In: Layton, R., Stone, P. and Thomas, J. (eds), *The destruction and conservation of cultural property*, pp. 75–92. Routledge, London.

Karlström, A. 2005. Spiritual materiality: Heritage preservation in a Buddhist World? *Journal of Social Archaeology* 5(3):338–355.

Karlström, A. 2009. *Preserving impermanence: The creation of heritage in Vientiane, Laos*. PhD thesis, Uppsala University.

Karlström, A. In press. Authenticity: rhetorics of preservation and the experience of the original. In: Samuels, K. and Rico, T. (eds), *The rhetoric of heritage*.

Lahiri, N. 2001. Destruction or conservation? Some aspects of monument policy in British India (1899–1905). In: Layton, R., Stone, P. and Thomas, J. (eds), *The destruction and conservation of cultural property*, pp. 264–275. Routledge, London.

Matsuda, D. 1998. The ethics of archaeology, subsistence digging, and artifact looting in Latin America: Point, muted counterpoint. *International Journal of Cultural Property* 7(1):87–97.

Meskell, L. 2002. Negative heritage and past mastering in archaeology. *Anthropological Quarterly* 75:557–574.

Meskell, L. 2012. *The nature of heritage: the new South Africa*. Wiley-Blackwell, Hoboken.

Miller, D. 2008, So, what's wrong with consumerism? *RSA* Summer: 44–47.

Omland, A. 2006. The ethics of the World Heritage concept. In: Scarre, C. and Scarre, G. (eds), *The ethics of archaeology: Philosophical perspectives on archaeological practice*, pp 242–259. Cambridge University Press, Cambridge.

Potkin, A. 2001. *Multimedia and digital imaging in environmental and cultural conservation, Vat Ou Mong peri-demolition archive*. Cultivate Understanding Multimedia, Vientiane.

Robinson, R. and Johnson, W.L. 1997. *The Buddhist religion: An historical introduction*. Wadsworth, London.

Ronström, O. 2007. *Kulturarvspolitik: Visby – från sliten småstad till medeltidsikon*. Carlsson, Stockholm.

Scarre, C. and Scarre, G. (eds) 2006. *The ethics of archaeology: Philosophical perspectives on archaeological practice*. Cambridge University Press, Cambridge.

Shanks, M. and Tilley, C. 1987. *Social theory and archaeology*. Polity Press, London.

Smith, L. 2006. *Uses of heritage*. Routledge, London.

Spiro, M.E. 1966. Buddhism and economic action in Burma. *American Anthropologist* 68(5):1163–1173.

Strathern, M. 1990. Artefacts of history: events and the interpretation of images. In: Siikala, J. (ed.), *Culture and History in the Pacific*. Finnish Anthropological Society, Helsinki.

Tambiah, S.J. 1970. *Buddhism and the spirit cults in north-east Thailand*. Cambridge University Press, Cambridge.

Turtinen, J. 2006. *Världsarvets villkor. Intressen, förhandlingar och bruk i internationell politik*. PhD thesis, Stockholm University.

UNESCO. 2003. *Convention for the safeguarding of the intangible cultural heritage*. UNESCO, Paris.

Wijesuriya, G. 2001. 'Pious vandals': restoration or destruction in Sri Lanka? In: Layton, R., Stone, P. and Thomas, J. (eds), *The destruction and conservation of cultural property*, pp. 256–263. Routledge, London.

Zimmerman, L.J., Vitelli, K.D. and Hollowell-Zimmer, J. (eds) 2003. *Ethical issues in archaeology*. AltaMira Press, Walnut Creek.

11

The WCPA's Natural Sacred Sites Taskforce: A critique of conservation biology's view of popular religion

Denis Byrne, Office of Environment and Heritage, NSW, Australia,

University of Technology, Sydney, Australia

The environmental anthropologist, Anna Tsing (2005:12), refers to modern-day nature conservation as a form of 'globally circulating knowledge'. In this chapter I focus on the way local religious systems have attracted the interest of conservation biologists who have come to see that in many parts of the world – and their attention is particularly on the developing world – religious beliefs and practices turn out to have 'conservation outcomes' (e.g. Verschuuren 2007; Mallarach 2008; Wild and McLeod 2008; Verschuuren and Wild 2010). These take the form of sites and landscapes which, primarily because of their religious significance to particular cultural groups, have retained 'high conservation value'. There are, for instance, sacred mountains like Adam's Peak in Sri Lanka (Wickramasinghe 2003), the *lulic* (sacred) forests of East Timor (McWilliam 2003), and the sacred groves of India (Boraiah *et al.* 2003), all of them being situations where forest environments have been preserved (though usually not in a primary state).

Increasingly, knowledge about these places and the religious beliefs and practices that have sustained them flows along international circuits, particularly under the auspices of the International Union for Conservation of Nature (IUCN), with the direction of flow mainly being from the south to the north and from the east to the west. There tends to be a reverse flow of conservation management and protected area management expertise.

Tsing takes as the topic of her book, *Friction: An ethnography of global connection* (2005), the tensions that commonly develop between commercial interests, conservationists, and the Indigenous groups who inhabit environments whose resources and biodiversity have attracted global interest. The relations between these groups and interests often occur in the form of collaborations but, she observes: 'Collaboration is not a simple sharing of information' (2005:13). She goes on to note that: 'In standardizing global knowledge … truths that are incompatible are suppressed. Globally circulating knowledge creates new gaps even as it grows through the frictions of encounter' (2005:13). Looking to the relatively recent engagement of conservationists with what have in nature conservation circles come to be called 'sacred natural sites',[1] I suggest that the knowledge now circulating globally about this form of the 'sacred' tends to be characterised

---

1   The term 'sacred natural sites' appears to have been coined by participants in the IUCN's Task Force on Cultural and Spiritual Values, established in the 1990s, and now known as the Cultural and Spiritual Values Specialist Group. See http://iucn.org/about/union/commissions/wcpa/wcpa_what/wcpa_governance/wcpa_cultural/ (Accessed: 06/2012).

by a certain elision or gap. Reading through the case-study literature on 'sacred natural sites' it becomes clear that most of these places are locally believed to embody or be occupied by spirit beings or deities and to be invested with supernatural powers – for instance, the miraculous power to cure illness or bring rain. The term 'numinous', indicating the presence of a divinity, is appropriate for them (Levy *et al.* 2006:11–14). Local people's relationships with these places are characterised by magical practices directed at gaining access to their efficacy. It is also evident in the case-study literature, though, that its authors, who are predominantly conservation biologists, shy away from the idea of the magical-supernatural, as though there were something disreputable about it. My own impression, gained when it was requested that I avoid the term 'supernatural' in a chapter I was writing for one of these volumes (Byrne 2010), was of a certain squeamishness at the B-grade-movie flavour that the term seemed to the editors to convey.

In the volume in question, *The Importance of Sacred Natural Sites for Biodiversity Conservation* (UNESCO 2003), most of the contributions, which are in case-study form, begin with a fairly cursory description of the religious significance of certain areas (mostly forests) and then move to describe the biodiversity conservation measures which conservation advisors have implemented there. There is very little discussion of just how religious belief and practice has acted to conserve biodiversity. Indeed, the conservation measures put in place appear to be directed at conserving biodiversity for the sake of biodiversity rather than for its local religious value (a notable exception in that volume is Wickramasinghe's (2003) chapter on Adam's Peak, Sri Lanka). The volume, overall, has an instrumentalist feel to it: environmental protection seems to be regarded as the inherent function and endpoint of religion.

Nature conservation in the non-Western world

The points I make about the 'sacred natural sites' literature, then, are firstly that there is an avoidance of the magical-supernatural and, secondly, that in the context of the 'sacred natural sites' program, religion is equated with conservation. I will return to the latter point later, turning now to what I see as an inclination to domesticate popular religion by filtering out the magical. To begin with, I suggest this represents a protestantisation of the religious systems conservationists encounter in the non-Western world, which is to say that conservation biology imposes on them a worldview that developed out of the sixteenth century Protestant Reformation. The Reformation introduced a relatively narrow view of the ways in which God was manifest. Whereas in medieval Christianity God was a living presence in the landscape, manifest in the miraculous efficacy flowing from saintly people, sacred relics and sacred places, in the Protestant view, particularly the Calvinist view, religion was to be a matter between one's soul and a god who dwelt in heaven (Eire 1986). This effectively removed God from nature as an active, causal force, opening the natural world up to understanding through learned inquiry. It is in this sense that one speaks of the 'disenchantment' of the post-Reformation landscape.

The Reformation also produced what Charles Taylor (2007:300) refers to as the 'buffered' European self, a self with a mind and body more bounded and insulated from the cosmos than the medieval European self. In terms most relevant to us this meant a self that was less vulnerable to malign and potentially malign spirits – ranging from Satan to nature spirits – whose presence in a plethora of numinous objects and places meant that ordinary people in medieval times lived in a climate of anxiety which is difficult for we moderns now to imagine. This buffering was reinforced by Renaissance humanism, the scientific revolution and the Enlightenment, all of which, not incidentally, comprised the cultural-intellectual environment within which the field of natural history emerged, natural history in turn being the generic field in which conservation biology has its antecedents. In summary, the Reformation was a key step in the emphatic rejection of a European conception of nature which has important generic resonances with the religious beliefs and practices which have produced and sustained the 'sacred natural sites' found in the non-Western world.

I need to make clear here that I am not suggesting the religious systems of traditional societies in the non-Western world are directly comparable to those of pre-Reformation Christianity, nor, of course, that they are survivors of a lineally prior form of human religion. My point, rather, is that the Reformation ushered in a religious worldview that, in its rejection of the supernatural, marked it off from medieval Christianity and also put it at odds not just with popular religion in the non-West (e.g. the animist traditions of folk religion in Southeast Asia, as well as popular Buddhism and popular Islam in that part of the world) but with popular Catholicism as it continues to flourish in the Mediterranean and many parts of northern Europe as well as in Central and South America, Africa, and the Philippines. These traditions of folk religion all have in common a belief that the divine is manifest as 'real presence' in the environment and that the numinous quality of 'sacred' landforms, trees, animals, and artefacts confers agency upon them. To say that Protestantism (and Counter-Reformation Catholicism) is at odds with this religious view is no exaggeration: through the nineteenth century and continuing today, a huge amount of Christian missionary effort in the non-West has been directed at eradicating it. The nature conservation movement is not engaged in this missionary project but its inclination to de-emphasise the supernatural, often to the point of effacing it, amounts I suggest to a protestantisation of non-Western popular religion. The same point can be made about cultural heritage practice in Asia (Byrne 2009).

It may seem to represent a curious back-flip that Western nature conservationists and conservation biologists would now embrace 'sacred natural sites', with the implication this seems to carry that the secular-rationalist underpinnings of Western conservationism are being set aside. I argue, though, that these underpinnings are very evident in the approach conservationists have taken to the 'sacred natural sites' program. As noted earlier, there is a tendency to 'domesticate' the magical-supernatural realm that contextualises these sites; a tendency to portray people's relations with the sites in terms of a reverence for nature which is compatible with the post-Reformation worldview. This aligns with the kind of 'distant' reverence for nature we see emerging in Protestant and Counter-Reformation Catholic Europe, a view of nature as sacred (insofar as it was created by God) but not numinous. Put more simply, nature is taken to be a work of God rather than a realm of gods. This is a long way from the popular religiosity of the 'sacred natural' in Southeast Asian cultures in which the tree, the rock formation, or the cave is an independent divine agent, animated with supernatural force. The instrumental-rationalist aspect of the post-Reformation Western view gives humans a monopoly on agency: humans act on nature, either to exploit or conserve it. This view concedes that nature has effects on humans, as when a hurricane or drought destroys our crops, but the notion that nature – in the form for instance of a tree, a rock, or a forest – could launch a spiritual assault on us, or could miraculously intercede to help us, is alien to it.

It is worth remembering that Western nature conservation, as a discourse and field of practice, is not a recent arrival in the non-Western world; it has quite a long history there. Mary Louise Pratt (1992:31) observes that during the European Age of Discovery, spanning the fifteenth to the seventeenth century, 'One by one the planet's life forms were to be drawn out of the tangled threads of their life surroundings and rewoven into European-based patterns of global unity and order'. Western nature conservation consolidated its position in what we now refer to as the developing world as part of the West's imperial-colonial expansion following the Age of Discovery (e.g. Grove 1995; Dunlap 1999; Beinart and Hughes 2007). Much of the work of Western botanists and zoologists in the colonies was directed at helping develop 'under-used' landscapes and at more efficiently exploiting natural resources (e.g. Kathirithamby-Wells 2005). But the unrestrained harvesting of natural resources often proved to be unsustainable, resulting in incidents of environmental collapse and a general environmental degradation that by the second half of the nineteenth century began to seriously alarm colonial authorities (Grove 1995; Griffiths and Robin 1997; Adams 2003:29–33). Something similar was occurring in the settler colonies: in Australia, for instance, a series of droughts led to the ruination of farmers and to severe soil loss (Flannery 1994).

As environmental historians have shown, modern nature conservation thus developed to a large extent out of colonialism's effort to ameliorate the environmental damage it itself had inflicted (e.g. Grove 1995). Nature conservation was thus integral to the project of colonialism and its position in the colonial and the post-colonial state has in many respects been one of mutuality, though the field may perceive itself as being opposed to state projects and elite interests.

Getting into bed with Asian modernity

Across Asia, a characteristic feature of nationalist movements and of the independent post-colonial nation states they gave rise to has been their antipathy towards 'superstition' (Byrne 2012). The face of Western civilisation that was presented to Asia by Western missionaries, traders, engineers and other experts during the late nineteenth and early twentieth centuries was one that emphasised rationality and science. Most intellectuals in Asia willingly embraced this view as they embarked on their own programs of national reform and modernisation. In China, Japan and Siam, intellectuals realised that only by matching the West's science and technology would they be able to prevent their societies being completely overrun. In Burma, the East Indies and the other Western colonies in Asia, Indigenous nationalists saw Western-style rationalism as a key to being able eventually to shake off their colonial masters. For most modernising reformers and nationalists in Asia, popular religion, steeped in 'superstition', was seen as an impediment to this kind of 'progress'.

In China, popular religion encompasses cults to local deities and to natural divine forces (e.g. Dean 1993; Feuchtwang 2001; Chau 2006). In Theravada Buddhist Thailand and Cambodia, it comprises animist cults, local cults to ancestral spirits, and the unorthodox popular practice of Buddhism where the emphasis is upon magical efficacy for this-worldly benefits rather than with transcendence (e.g. Tambiah 1970; Wijeyewardene 1986; Jackson 1999). The anti-superstition campaigns launched by reform-minded modernists in China were directed against village-based cults to local deities and to a lesser extent against popular Buddhism. In the first four decades of the twentieth century, as the anti-superstition campaigns unfolded, 'probably more than half of the million Chinese temples that existed in 1898 were emptied of all religious equipment and activity' (Goossaert 2006:308). This of course was *prior* to the 1949 Communist Revolution and its suppression of superstition and takes no account of the destruction of temples wrought by the Cultural Revolution (1966–76).

Rather than weaning people off belief in the supernatural, however, modernity was experienced by many in China as a disaster that only the supernatural could mitigate. Detailed ethnographies and histories that have become available in the last decade or so (e.g. Jing 1996; Mueggler 2001; Flower 2004; Yang 2004; Chau 2006) paint a picture of local people struggling to preserve not just temples and statues of gods but a 'moral landscape' (Flower 2004) in which every stream, crossroads, field, tree, and shrine was integrated into an ecology of sacred cause and effect, a landscape occupied by gods, ghosts, living people and the spirits of their ancestors. Flower (2004), for instance, shows how catastrophes like the Great Leap Forward famine of 1959–61 were seen locally as being partly a consequence of the damage inflicted on the divine landscape.

The establishment of post-colonial communist states in Vietnam, Cambodia and Laos saw the implementation of anti-superstition campaigns similar to those in China. The Democratic Republic of Vietnam (1954–76) in North Vietnam energetically set out to eradicate popular participation in life cycle rituals, the giving of offerings to spirits, and the practice of paying for the services of spirit mediums, all of which were considered to waste the precious resources of the poor (P. Taylor 2007:29). In Cambodia, the Khmer Rouge suppressed all religion (Kiernan 1996) while in Laos the communist Lao People's Revolutionary Party, coming to power in 1975, disbanded the two main Buddhist clerical orders and aimed to eliminate superstition, along with other social evils. This policy continued through the 1990s in Laos though it is considered, by one of the key commentators, to have been 'one of the least successful of the regime's campaigns' (Evans 1998:68).

Siam, like China, was never colonised in a primary sense but, as well as being exposed to Christian missionary influence that was virulently opposed to 'superstition', the royal government and the elite were deeply conscious of European standards of civilisation and were highly sensitive to Western criticism of themselves as being in any way uncivilised (e.g. Thongchai 2000). Elsewhere (Byrne 2009) I have reviewed in some detail the manner in which the modern nation state in Siam (and in Thailand, after the country's name was changed in 1939) has sought to undermine belief in the magical-supernatural which has been common both to popular Buddhism and the localised cults of folk religion that many or even most Thais have at least some involvement in.

I am not, of course, equating the activities of nature conservationists in Asia with the anti-superstition campaigns reviewed above. What I do want to do is point to a certain alignment between the way modernity has played out in Asia over the last century or so and the modernist character of Western nature conservation. Both nature conservation and heritage conservation tend to be seen as an obvious good and for that reason, I suggest, most practitioners in those fields are not inclined to spend much time examining the larger political and historical context of their work. These can lead these fields to become complicit – unintentionally and probably unconsciously – in authoritarian programs of the modern state that in Asia and elsewhere have taken such a heavy toll on local cultures over the modern period. It is sobering to consider, as James Scott has done so eloquently in his book *Seeing Like a State* (1999), that all the efforts to suppress popular religion that have been mentioned in this section were inflicted on local people 'for their own good'. They were intended to make the world a better place; they were seen by the state and elite interests, in other words, as an obvious good.

(Un)level playing fields

From the preceding section it will be clear that the modern state in Asia has adopted a view of nature which has much in common with the post-Reformation European view. It would be mistaken, however, to think that Asia has simply imported a Western secular-rationalist construction of nature or that it has bowed to the West's ambitions to universalise its own worldview. The modern state in Asia has opposed superstition firstly because superstition has been seen as an obstacle to the kind of progress that would allow it economical and political equity with the West and, secondly and perhaps most significantly, because the magical-supernatural has been seen to constitute a threat to the state's ability to project its authority evenly across its territorial domain.

In Henri Lefebvre's (1991) account of modern spatiality, capitalism encouraged the spread of an 'abstract' kind of space which was uniform, repetitive, quantifiable, predictable and manageable by the centralised institutions of the modern state. Modernity endeavours to mould the spaces it dominates (i.e. peripheral spaces) and it seeks, often by violent means, to 'reduce the obstacles and resistance it encounters there' (Lefebvre 1991:49). The obstacles it encounters are those forms of 'absolute' space comprised of places brought into being by individuals, families and communities by their bodies, imaginations and actions. It is not difficult to see how the landscapes of popular religion in Asia, populated as they are by numerous local sites of the divine, each one of which might be considered to be a node of supernatural power, inevitably imposes limitations on the modern state's ambition to dominate space. Perhaps the best theorised account of modern spatiality in Asia has emerged from Mayfair Yang's (2004) study of Wenzhou in Zhejiang Province, China. She finds that in common with capitalism, state socialism in China energetically set about corroding place-based local cultures. She examines how the Maoist state and its successors set about dismantling three categories of local religious space: the ritual space of lineages, the space of the tomb, and the space of local deity cults. Maoist centralisation of production and collectivisation of agriculture entailed the repression of superstitious practices and saw joint property of lineages confiscated, lineage halls and temples closed (and turned over to other uses such as schools) or demolished, and traditional burial discouraged or forbidden in favour of cremation. Yang also traces the re-emergence since the 1980s of each of these categories of formerly banned public ritual space in response to local resistance to, and state relaxation of, the social-spatial regime.

One of the prime instruments of the kind of spatial levelling that the modern state strives for is the map. In recent years, environmental anthropologists working in parts of Asia occupied by minority 'tribal' groups have devoted considerable attention to the way colonial governments, and then post-colonial governments, sought to extend their centralised authority across the territory of such groups partly via the technology of mapping. On colonial era maps, the forest habitats of these groups in places like Borneo were typically classified as 'wasteland' or 'barren' land and, as such, were appropriated by the state as unoccupied natural resource zones (Roseman 2003; Tsing 2003; Sowerwine 2004). The field systems and fixed villages of the lowland agricultural areas, by contrast, typically did find their way onto colonial maps and the people of these areas were recognised as holding traditional title to their fields and plantations.

The post-colonial governments of Malaysia, Indonesia and Vietnam, rather than rejecting these maps, further elaborated them. The classification of tribal habitats as primary forest simplified the process of treating them as state resources that could be allocated as logging concessions and mining leases to national and international companies without reference to the people who inhabited them. Conscious of the repercussions of being left off the map, concerned anthropologists in the 1990s began working with local groups to produce maps that gave visibility to local constructions of terrain. The term 'counter-mapping' has been adopted for what might be described as this tactical deployment of cultural mapping (e.g. Peluso 1995; Brosius *et al.* 2005). These counter-maps show forests where former swiddens are regarded as storied historical sites, where individual trees are often known by their own names, and where ridges and gullies are intricately inscribed with the territories of spirits and deities.

The wilderness trope in nature conservation discourse, or what one commentator has termed 'distant-nature conservationist mindset' (Campbell 2005:285), has inclined conservationists to see tropical forests and other 'high conservation value' environments more as habitats of biodiversity than habitats of culture. Directing their attention to tropical zone forests in places like Indonesia, conservationists readily fell in with existing land classification systems that classified forests as existing in a 'natural/primary' state or as having been in this state before being 'vandalised' by shifting agriculturalists. The colonial mapping regimes thus found new subscribers among those who longed for unacculturated natural landscapes that were available to be saved. In this regard, as Ben Campbell (2005:301) provocatively asserts, uncritical conservationism 'belongs to a colonial genealogy of perceiving foreign lands as *terra nullius*'. While the counter-mapping approach was developed mainly in response to the marginalisation of shifting agriculturalists, the push for increased agricultural productivity by post-colonial states and their Western advisors has often entailed a rationalisation of agriculture in which the role of religious belief and ritual in regulating agriculture locally has been downplayed or suppressed. Stephen Lansing (1991), for example, famously documented the way, in Bali, both Dutch colonial agricultural managers and Green Revolution reformers badly misread the key importance of ritual cycles and water temple networks in sustainably regulating irrigations flows through rice farms, leading to a range of problems including loss of soil fertility, crop disease outbreaks, and pollution of rivers and coral habitats.

The counter-mapping approach outlined above would seem to have obvious implications for the field of cultural heritage management. Elsewhere (Byrne 2008:612–613) I have drawn attention to the way Indigenous groups and their culturally constructed landscapes were left off colonial and post-colonial maps in New South Wales after Aboriginal people were displaced from their former country. From the mid-twentieth century the Aboriginal dimension of these landscapes was mapped as archaeological, which is to say that archaeological surveys populated the landscape with sites which, up until the 1980s, were described in a way that made no reference to contemporary Aboriginal people. Even after the 1980s there has been a tendency for places of spiritual significance to contemporary Aboriginal people not to be recorded in NSW. This is suggestive of a form of spatial levelling of the kind described by Lefebvre (1991) in which the local space of a cultural minority is reimagined in terms that make sense to the highly centralised modern state, in this case via a centralised heritage inventory.

It is interesting to reflect that the IUCN's 'sacred natural sites' taskforce would very likely perceive itself to be directed against precisely the kind of uncritical conservationism I have been describing. And very clearly it does represent a deliberate move by a section of the conservation biology field to counter the more people-unfriendly approaches still adhered to by many of their colleagues. Dan Brockington's (2002) term 'fortress conservation' neatly captures the essence of these approaches. And yet the tendency for the 'sacred natural sites' movement to downplay or even efface the magical supernatural demonstrates, I think, how easily well-intentioned initiatives can find themselves slipping into step with state programs that are distinctly unfriendly to local groups.

Getting beyond culture–nature duality

One of the hallmarks of Western modernity has been a dualistic or binary view of culture and nature. Nature is seen as categorically distinct from humanity and has been conceptualised as a resource for humanity. In Valerie Plumwood's (2002) critique of the Western relationship with nature she pointed to the way Christianity has taken the spirituality of nature as something to be overcome or, alternatively, has perceived it 'in the largely instrumental terms of leading us to a higher, non-earthly place' (2002:219). She views culture-nature dualism as not just an assertion of difference but a definition of one *against* the other. Debbie Rose (2005) has developed her own critique of Western modernity's culture-nature dualism from her experience in working as an anthropologist with Aboriginal groups in northern Australia whose relationships with nature are characterised by continuity and stand in marked contrast to the relationship of discontinuity that characterises Western relationships with nature. In the Aboriginal system, 'sentience and agency is not a solely human prerogative' (Rose 2005:302). The focus in this chapter is on the way the supernatural force attributed to certain non-humans (including certain sentient beings and artefacts) brings them into dialogue with us. Rose (2005:300), exploring the meaning of the Aboriginal phrase 'country tells you', makes clear that an essentially similar human responsiveness to agency in nature takes the form of reading a myriad of signs in the natural environment.

Changes in the colour and smell of particular plants, for instance, indicate that it is time to start burning the bush. She writes that, 'Within the communicative matrix of country, people respond to the patterns of connection and benefit' (Rose 2005:300).

Lesley Head (2007:838) has recently noted that the critique of nature-culture dualism has scarcely registered in the field of archaeology where humans are still treated as if they were external to the natural system. In Australia, she observes, archaeologists now recognise that Aboriginal people regard plants and animals as well as waterholes and hills as sentient entities. She notes that we do not, however, regard them that way ourselves. We continue with the rationalist-positivist view of non-human species as constituting a kind of setting or background for our own lives or as resources for our own projects (Head 2007:838).

When Bruno Latour (2004:232) states that: 'Traditional societies do not live in harmony with nature; they are unacquainted with it', what he is gesturing towards is a situation in which nature in these societies is not perceived to be a separate category. These societies are unacquainted with the construct of nature as it exists in the West. 'Non-Western cultures *have never been interested in nature*' (2004:43, emphasis in the original) he thus maintains: 'they have never adopted it as a category; they have never found a use for it'. So, rather than asserting that these cultures have no relationship with nature, Latour is saying their relationship is a non-hierarchical one in which nature is simply not conceived of in instrumental terms. This is part of a broader position on Latour's part in which he maintains that in order for us to endure on this planet in the future, we will need to transcend nature-culture duality and position ourselves as actors living in democratic concert with non-human actors.

I maintain that the 'sacred natural sites' movement still exists substantially within the frame of nature-culture duality. In this frame local people who, for instance, worship sacred groves, are posited as revering nature in a manner that 'naturally' leads to nature conservation. This conception, however, seems unable to account for those situations in which so-called traditional cultures have been engaged in what we would see as acts of environmental destruction. Or rather, it seems able to account for such a situation only by assuming the religious systems of these cultures have collapsed or that the people involved have been coerced by external forces. This positions locals as merely passive in the face of change rather than crediting them as possible agents of change.

Yet if we think of the deforestation of large parts of northeast Thailand during the 1960s and 70s it is hard to see local people as other than active agents in this, which is not to say that external forces were not also at play. In northeast Thailand, forest area declined from 42% in 1961 to 14% in 1985 (Hunsaker 1996:3) in the face of commercial logging but also as a result of agricultural expansion driven largely by population growth. If we attend to the literature on spirit cults in Thailand (e.g. Tambiah 1970; Darlington 2003; Kamala 2003) it seems reasonable to think that many people living in the northeast at that time would have been concerned about the spiritual consequences of this assault on what in many cases were sacred trees and forests. There would, for instance, surely have been concern about forest spirits taking retribution by causing crop failure, illness among people and livestock, or other disasters.

But we also know that the practice of popular religion provided a means for propitiating, placating, and compensating forest spirits. Spirit cults in Thailand are characterised by an attribution to spirits of appetites, foibles and reactions very similar to those experienced by humans. Although spirits are considered capable of acting unilaterally against humans, this commonality creates a basis for engaging and negotiating with them. Spirit mediums provide a line of communication between people and spirits which facilitates this. The potentiality to placate spirits whose rights have been infringed gives people space to manoeuvre in terms of their own economic strategies.

It is thus possible to envisage a situation in which 'sacred' nature is not untouchable; a situation in which religious ritual enables people to pursue economic ambitions that entail altering nature and enables them to offset resultant spiritual retribution. Like all of us, conservation biologists involved in the 'sacred natural sites' movement need to be wary of an idealised essentialist framing of Indigenous and other groups in the developing world that fails to account for the enormous pressure these groups are under as they seek to improve their economic position.

This kind of negotiation or 'offsetting' is also seen at a commercial level. There is, for instance, the case of Burmese timber merchants involved in logging in Siam in the late nineteenth century who built some of the well-known wooden Buddhist temples in northern Thailand in Lampang, Prae, Chiang Mai and elsewhere as a means of making merit to offset retribution by forest spirits (Zaw 2002). On the other hand, there appears to be an understanding that forest spirits are themselves able to adapt to environmental change. Turton (1978:124) shows how 'non-specific forest spirits' become specific when forest areas are cleared for houses or cultivation. They become tutelary spirits of the fields that occupy the space of the former forest. Also, people can deflect or resist retribution by spirits using spiritual means. Invulnerability tattoos can function in this way (Turton 1991) as can supernaturally empowered Buddhist amulets (Tambiah 1984).

Thai Buddhism, for its part, seems to have had an ambivalent attitude to forests. On the one hand, the forest monk movement (Tambiah 1984) valorises forests as providing an environment conducive to the practice of meditation. But on the other hand, in the 1970s many so-called 'development monks' encouraged forest clearance as part and parcel of the push to 'civilise' minority groups dwelling in upland forest areas and to combat communism (Darlington 1998). Kamala (1998:466–468) cites the case of a wandering monk who settled in an area near Nong Khai from 1858 to 1962, helping local villagers to overcome their fear of spirits and clear the forests for rice fields.

Conclusions: Conservation and materialism

Earlier, I referred to Christian missionising in Asia as targeting belief in the supernatural. There is a loose parallel, I suggest, between this missionising and the way nature conservationists and cultural heritage practitioners have gone to the region bearing the good word of the Western conservation ethic. Convinced of the universal goodness of our brand of conservation we easily find ourselves making assumptions about the meanings that nature and old things have for locals.

The most obvious assumption we make may be that the religious attention we see people devoting to their sacred groves or old temples equates with a desire on their part to conserve these things. What this misses or ignores is the interactional form of this attention which I would describe as an engagement between human and non-human actors. There is agency on both sides of such engagements. I have been concerned above with the spiritual force animating the kinds of non-human actors which come under the category 'sacred natural sites'. But outside the realm of nature – in the case, for instance, of an old temple or stupa in Thailand – we find that people may conceive of the sacred integrity of a site as not dependent on maintaining the present physical state of the built structure. In the case of an old, dilapidated stupa, people are more likely to honour its miraculous force by completely encasing it in a completely new masonry shell covered with glittering new 'bathroom' tiles than they are by stabilising or restoring its existing fabric (Byrne 1995). In the case of a sacred grove, the spirit or deity embodying it may well be cadastral in nature (Mus 1975:11), meaning it is topographically anchored to (or embodied in) a particular site and will remain *in situ* following the clearance of the forest.

Western conservationists, though they may not themselves believe in the supernatural force of the old temple building or the sacred grove, nevertheless will often acknowledge local belief in this force and will try to make allowance for it in their management prescriptions. However, even where this is the case (and most often in places like Thailand allowance is not made) the conservationist's focus still tends to be solidly on the tangible materiality of the building or grove. Despite the quite genuine respect that I believe members of the 'sacred natural sites' movement have for Indigenous religion, at the end of the day what they are primarily interested in is the 'natural' rather than the 'sacred'. This, I believe, has far more to do with our own materialism than it does with the Indigenous valuation we think we are observing.

The fields of nature conservation and cultural heritage conservation can be seen as grounded in the particular kinds of materialism associated with modernity. We might note that the aspect of popular religion in Asia that I have been addressing – that to do with belief in the supernatural agency of non-human beings and objects – was known in early Western modern history as fetishism. Hal Foster (1993:255) observes that the Protestant Reformation embodied a denial:

> … that any material object has a special capacity of spiritual mediation. Yet in this religious point is hidden an economic agenda: to denounce as primitive and infantile the refusal to assess value rationally, to trade objects in a system of equivalence, in short, to submit to capitalist exchange.

The 'sacred natural sites' movement, on the face of it, seems to reject the Reformation's stand against the notion of divine immanence in nature, relics, and other holy objects and thus seems to contest the Reformation's assertion that sacramental force is exclusive to God in heaven. But as Foster points out, the Reformation's assertion that divine, miraculous force is absent from the earthly realm has an economic agenda – and here we have the linkage between Protestantism and capitalism that Weber famously (2002) theorised in the first decade of the twentieth century: that the value of things was determined by their exchange value. Extending on Marx's analysis, Foster (1993:255) sees this move as one in which 'commodity fetishism partly replaced religious fetishism, or at least compensated for its partial loss'. What I am leading to here is the proposition that 'conservation', as we know it, is an expression of modernity's commodity fetishism. Heritage practitioners do not merely ignore or marginalise the supernatural in their conservation work on old temples, they substitute the fetishism of the supernatural for the fetishism of the commodity. The latter expresses itself in terms of the amassing of archaeological and art historical descriptive data on old places and the 'banking' of it in heritage inventories. On the nature conservation side of the table, commodity fetishism takes the form of the quantification of biodiversity.

Breaking down the culture-nature divide involves far more than some kind of integration of the fields of cultural and natural heritage conservation. It involves a critical deconstruction of our modern historical experience. In many ways it also entails a project of decolonisation. Noting that both the colonisation of nature and the West's colonisation of the 'outer' world involved categorising non-humans and non-Europeans as radically Other (Plumwood 2002:53), what is called for in both cases is a decolonisation that entails surrendering notions of innate superiority, substituting democratic relations for hierarchical ones, and replacing instrumentalism with effective communication.

References

Adams, W.M. 2003. Nature and the colonial mind. In: Adams, W. and Mulligan, M. (eds), *Decolonizing nature: Strategies for conservation in a post-colonial era*, pp. 1–15. Earthscan, London.

Beinart, W. and Hughes, L. 2007. *Environment and empire*. Oxford University Press, Oxford.

Boraiah, K.T, Vasudeva, R., Bhagwat, S.A. and Kushalappa, C.G. 2003. Do informally managed sacred groves have higher richness and regeneration of medicinal plants than state-managed forest reserves?

Current Science 84(6):804–808.

Brockington, D. 2002. *Fortress conservation: The preservation of the Mkomazi Game Reserve, Tanzania*. Indiana University Press, Bloomington.

Brosius, P.J., Lowenhaupt Tsing, A. and Zerne, C. (eds) 2005. *Communities and conservation*. AltaMira Press, Walnut Creek.

Byrne, D. 1995. Buddhist *stupa* and Thai social practice. *World Archaeology* 27(2):266–81.

Byrne, D. 2008. Counter-mapping in the archaeological landscape. In: David, B. and Thomas, J. (eds), *Handbook of landscape archaeology*, pp. 609–616. Left Coast Press, Walnut Creek.

Byrne, D. 2009. Archaeology and the fortress of rationality. In: Meskell, L. (ed.), *Cosmopolitan archaeologies*, pp. 68–88. Duke University Press, Durham.

Byrne, D. 2010. The reinchanted earth: numinous sacred sites. In: Verschuuren, B. and Wild, R. (eds), *Earth: Nature, culture and the sacred*, pp. 53–61. Earthscan, London.

Byrne, D. 2012. Anti-superstition: campaigns against popular religion and its heritage in Asia. In: Daly, P. and Winter, T. (eds), *Routledge handbook of heritage in Asia*, pp. 295–310. Routledge, London.

Campbell, B. 2005. Changing protection policies and ethnographies of environmental engagement.*Conservation and Society* 3(2):280–322.

Chau, A.Y. 2006. *Miraculous response: Doing popular religion in contemporary China*. Stanford University Press, Stanford.

Darlington, S.M. 1998. Ordination of a tree: the Buddhist ecology movement in Thailand. *Ethnology* 37(1):1–15.

Darlington, S. M. 2003. Practical spirituality and community forests. In: Greenough, P. and Tsing, A.L. (eds), *Nature in the global south*, pp. 347–366. Duke University Press, Durham.

Dean, K. 1993. *Taoist ritual and popular cults of southeast China*. Princeton University Press, Princeton.

Dunlap, T.R. 1999. *Nature and the English diaspora*. Cambridge University Press, Cambridge.

Eire, C.M. 1986. *War against the idols*. Cambridge University Press, Cambridge.

Evans, G. 1998. *The politics of ritual and remembrance: Laos since 1975*. Silkworm Books, Chiang Mai.

Feuchtwang, S. 2001. *Popular religion in China: The imperial metaphor*. Routledge, London.

Flannery, T. 1994. *The future eaters: An ecological history of the Australian lands and people*. Reed Books, Sydney.

Flower, J.M. 2004. A road is made: Roads, temples, and historical memory in Ya'an County, Sichuan. *Journal of Asian Studies* 63(3):649–685.

Foster, H. 1993. The art of fetishism: Notes on Dutch still life. In: Apter, E. and Pietz, W. (eds), *Fetishism as cultural discourse*, pp. 251–265. Cornell University Press, Ithaca.

Goossaert, V. 2006. 1898: The beginning of the end for Chinese religion? *Journal of Asian Studies* 65(2):307–337.

Griffiths, T, and Robin, L. (eds). 1997. *Ecology and empire: Environmental history of settler societies*. Keele University Press, Edinburgh.

Grove, R.H. 1995. *Green imperialism: Colonial expansion, tropical island edens and the origins of environmentalism*. Cambridge University Press, Cambridge.

Head, L. 2007. Cultural ecology: The problematic human and the terms of engagement. *Progress in*

Human Geography 31:837–846.

Hunsaker, B. 1996. The political economy of Thai deforestation. In: Hunsaker, B., Mayer, T., Griffiths, B. and Daley, R. (eds), *Loggers, monks, students, and entrepreneurs: Four essays on Thailand*, pp. 1–31. Center for Southeast Asian Studies, DeKalb.

Jackson, P. 1999. The enchanted spirit of Thai capitalism: The cult of LuangPhorKhoon and the post-modernization of Thai Buddhism. *South East Asian Research* 7(1):5–60.

Jing, J. 1996. *The temple of memories: History, power, and morality in a Chinese village*. Stanford University Press, Stanford.

Kamala, T. 1998. *The wandering forest monks in Thailand, 1900–1992: Ajan Mun's lineage*. Unpublished PhD thesis, Cornell University.

Kamala, T. 2003. *The Buddha in the jungle*. Silkworm Books, Chiang Mai.

Kathirithamby-Wells, J. 2005. *Nature and nation: Forests and development in peninsular Malaysia*. NIAS Press, Copenhagen.

Kiernan, B. 1996. *The Pol Pot Regime: Race, power, and genocide in Cambodia under the Khmer Rouge, 1975–79*. Yale University Press, New Haven.

Lansing, S.J. 1991. *Priests and programmers: Technologies of power in the engineered landscape of Bali*. Princeton University Press, Princeton.

Latour, B. 2004. *Politics of nature: How to bring the sciences into democracy*. Harvard University Press, Cambridge.

Lefebvre, H. 1991. *The production of space* (Translated by D. Nicholson-Smith). Blackwell, Oxford (Original French edition,1974).

Levy, R.I., Mageo, J. and Howard, A. 2006. Gods, spirits, and history. In: Mageo, J.M. and Howard, A. (eds), *Spirits in culture, history, and mind*, pp 1–10. Routledge, New York.

McWilliam, A. 2003. New beginnings in East Timor forest management *Journal of Southeast Asian Studies* 34(2):307–327.

Mallarach, J. (ed). 2008. *Protected landscapes and cultural and spiritual values*. World Commission of Protected Areas, Heidelberg.

Mueggler, E. 2001. *The age of wild ghosts: Memory, violence, and place in southwest China*. University of California Press, Berkley.

Mus, P. 1975. *India seen from the east: Indian and Indigenous cults in Champa* (Translated by Ian Mabbett). Monash Papers on Southeast Asia 3, Centre of Southeast Asian Studies, Monash University, Melbourne (Original French edition,1933).

Peluso, N.L. 1995. Whose woods are these? Counter-mapping forest territories in Kalimantan, Indonesia. *Antipode* 27(4):383–406.

Plumwood, V. 2002. *Environmental culture: the ecological crisis of reason*, Routledge, London.

Pratt, M. 1992. *Imperial eyes: Travel writing and transculturation*. Routledge, London.

Rose, D. 2005. An Indigenous philosophical ecology: Situating the human. *The Australian Journal of Anthropology* 16(3):294–305.

Roseman, M. 2003. Singers of the landscape: Song, history, and property rights in the Malaysian rainforest. In: Zerner, C. (ed), *Culture and the question of rights*, pp.111–41. Duke University Press, Durham.

Scott, J.C. 1999. *Seeing like a state: How certain schemes to improve the human condition have failed*. Yale University Press, New Haven.

Sowerwine, J.C. 2004. Territorialisation and the politics of highland landscapes in Vietnam: Negotiating property relations in policy, meaning and practice. *Conservation and Society* 2(1):97–136.

Tambiah, S.J. 1970. *Buddhism and the spirit cults in north-east Thailand*. Cambridge University Press, Cambridge.

Tambiah, S.J. 1984. *Buddhist saints of the forest and the cult of amulets*. Cambridge University Press, Cambridge.

Taylor, C. 2007. *A secular age*. Harvard University Press, Cambridge.

Taylor, P. 2007. Modernity and re-enchantment in post-revolutionary Vietnam. In: P. Taylor (ed.), *Modernity and re-enchantment: Religion in post-revolutionary Vietnam*, pp. 1–56. Institute of Southeast Asian Studies, Singapore.

Taylor, J. 2008. *Buddhism and postmodern imaginings in Thailand: The religiosity of urban space*. Ashgate, Farnham.

Thongchai, W. 2000. The quest for "*Siwilai*": a geographic discourse of civilizational thinking in the late nineteenth and early twentieth-century. *Journal of Asian Studies* 59(3): 528–549.

Tsing, A.L. 2003. Cultivating the wild: honey-hunting and forest management in southeast Kalimantan. In: Zerner, C. (ed.) *Culture and the question of rights*, pp. 24–55. Duke University Press, Durham.

Tsing, A.L. 2005. *Friction: An ethnography of global connection*. Princeton University Press, Princeton.

Tuan, Y. 1977. *Space and place: The perspective of experience*. University of Minnesota Press, Minneapolis.

Turton, A. 1978. Architectural and political space in Thailand. In: Milner, G. (ed), *Natural symbols in South East Asia*, pp. 113–132. School of Oriental and African Studies, University of London, London.

Turton, A. 1991. Invulnerability and local knowledge. In: Chitakasem, M. and A. Turton, A. (eds), *Thai constructions of knowledge*, pp. 155–82. University of London, London.

UNESCO. 2003. The Importance of Sacred Natural Sites for Biodiversity Conservation. Proceedings of the international workshop held in Kunming and Xishuangbanna Biosphere Reserve, People's Republic of China, 17–20 February 2002.

Weber, M. 2002. *The Protestant ethic and the spirit of capitalism* (Translated by P. Baehr and G.C Wells). Penguin Books, Harmondsworth (Original German edition, 1904 and 1905).

Wild, R. and C. McCleod (eds). 2008. *Sacred natural sites: Guidelines for protected area managers*. IUCN, Gland.

Verschuuren, B. 2007. *Believing is seeing: Integrating cultural and spiritual values in conservation management*. Foundation for Sustainable Development and IUCN, Gland.

Verschuuren, B. and R. Wild (eds). 2010. *Earth: Nature, culture and the sacred*. Earthscan, London.

Wickramasinghe, A. 2003. Adam's Peak sacred mountain forest. In: *The importance of sacred natural sites for biodiversity conservation: Proceedings of the international workshop held in Kunming and Xishuangbanna Biosphere Reserve, People's Republic of China, 17–20 February 2003*, pp. 109–118. UNESCO, Paris.

Wijeyewardene, G. 1986. *Place and emotion in northern Thai ritual behaviour*. Pandora, Bangkok.

Wild, R. and McLeod, C. (eds). 2008. *Sacred natural sites: Guidelines for protected area managers*. Task force on the cultural and spiritual values of protected areas, IUCN, Gland.

Yang, M. 2004. Spatial struggles: postcolonial complex, state disenchantment and popular reappropriation of space in rural south-east China. *Journal of Asian Studies* 63(3):719–755.

Zaw, A. 2002. A tale of two temples. *The Irrawaddy* 10(9):1.

12

Sacred places in Ussu and Cerekang, South Sulawesi, Indonesia: Their history, ecology and pre-Islamic relation with the Bugis kingdom of Luwuq

David Bulbeck, The Australian National University, Canberra, Australia

Introduction

This contribution describes and analyses the sacred places in the environs of the 'land where the gods descended' (Reid 1990)—the twin villages of Ussu and Cerekang in East Luwu, South Sulawesi. Ussu and Cerekang were the focus of several months of fieldwork for the 'Origin of Complex Society in South Sulawesi' (OXIS) project undertaken by anthropologists, archaeologists and historians (Darmawan *et al* 1999; Bulbeck and Caldwell 2000; Fadillah and Sumantri 2000). OXIS project members documented the setting and vegetation of the Ussu and Cerekang sacred places, archaeological sites and landscape use in their vicinity, and local understanding of how these places related to stories in the cycle of epic Bugis poetry known as *La Galigo*. The information additionally allows an empirical assessment of the widespread belief (Andaya 1981:17–19; Pelras 1996, 2006) that the sacred places of Ussu and Cerekang encapsulate the origins of the Bugis kingdom of Luwuq[1] during the 'Age of *La Galigo*' at the dawn of South Sulawesi's history.

Braam Morris (1889) was the first to publish the theory that the *La Galigo* stories reflect the beginnings of Luwuq as South Sulawesi's original kingdom approximately a thousand years ago. The theory married two older widespread concepts – the Bugis tradition of a primordial Age of *La Galigo*, and Luwuq's reputation as the oldest South Sulawesi kingdom – concepts respectively established by the late seventeenth and the eighteenth century (Macknight 1993:26; Caldwell 1998). The theory is, however, absent from South Sulawesi's early historical texts which stem from works composed after the circa fourteenth century development of the Bugis script. For instance, the 'King List of Luwuq', represented in 18 known Bugis texts, commences with the archetypal rulers Simpurusia, Anakaji and Wé Mattengngaémpong (Caldwell 1998), who Braam Morris treated as Luwuq's earliest rulers post the Age of *La Galigo*. Detailed claims on early Luwuq abound but in the guise of oral traditions. For instance, stories summarised by Darmawan *et al* (1999:vi; Darmawan 2000) place Luwuq's tenth to thirteenth century origins at Ussu, prior

---

1   Please note this article's use of English transcriptions of Bugis pronunciations for terms that refer to Bugis personages and political entities, and *bahasa Indonesia* spellings for place names. For instance, Luwuq finishes with a glottal stop (represented by a 'q') when the Bugis kingdom is referred to, but Luwu is the official Indonesian spelling for the three present-day regencies (Luwu, north Luwu and east Luwu) that fell within Luwuq's domain of authority at the time it was incorporated within the Dutch colonial administration.

to shifting to Mancapai (southeast Sulawesi) during the fourteenth century reign of Anakaji, and to Noling (south of Palopo city) and Malangke in the fifteenth and sixteenth centuries respectively (see Figure 1).

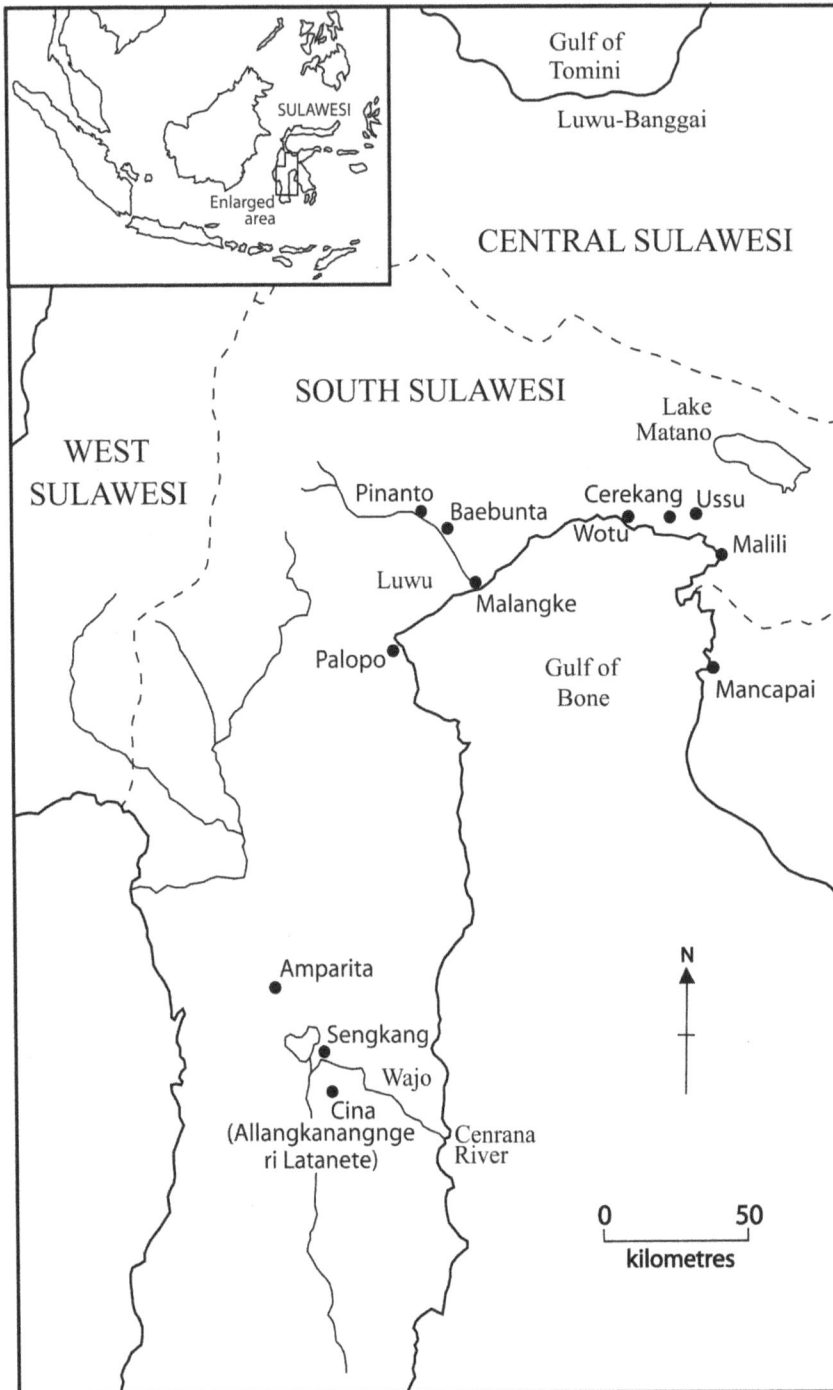

Figure 1: Location of Ussu, Cerekang, and other Sulawesi places mentioned in the text.

Source: Bulbeck and Caldwell 2000: Maps 2 and 8; field notes 15/11/1997.

The present-day inhabitants of Ussu and Cerekang are Muslims, as are most Bugis following the conversion of South Sulawesi's ruling families to Islam in the early seventeenth century. However, the mystical beliefs of the Tossuq ('Ussu people', Pelras 1996:59) based in Cerekang differ sharply from the tenets of orthodox Islam. Tossuq adepts believe that their sacred places are the abode of their deified ancestors, and if they die in Cerekang they too will be deified and rejoin their ancestors in the world of *La Galigo* (Daeng M. pers. comm. 1998). The *La Galigo* heroes continue to administer Luwuq from its hidden centre in *Manurung* ('descended from the sky') in Cerekang, which pairs with Ussu—ancient Luwuq's visible centre (Caldwell 1993). These beliefs of the Tossuq mystics are contested in Wotu, which lies a short distance to the west (Figure 1). According to Wotu tradition, the Cerekang inhabitants moved there from Wotu, bringing with them stories that reflect the true occurrence of the *La Galigo* events in Wotu (Darmawan *et al.* 1999:25–27). Be that as it may, there is a qualitative difference between Wotu, with just two *La Galigo*-associated sites, both accessible to the general public (Darmawan *et al* 1999:32–33; Fadillah 2000), and the rich landscape in Ussu and Cerekang of *La Galigo*-associated places closed to unauthorised entry.

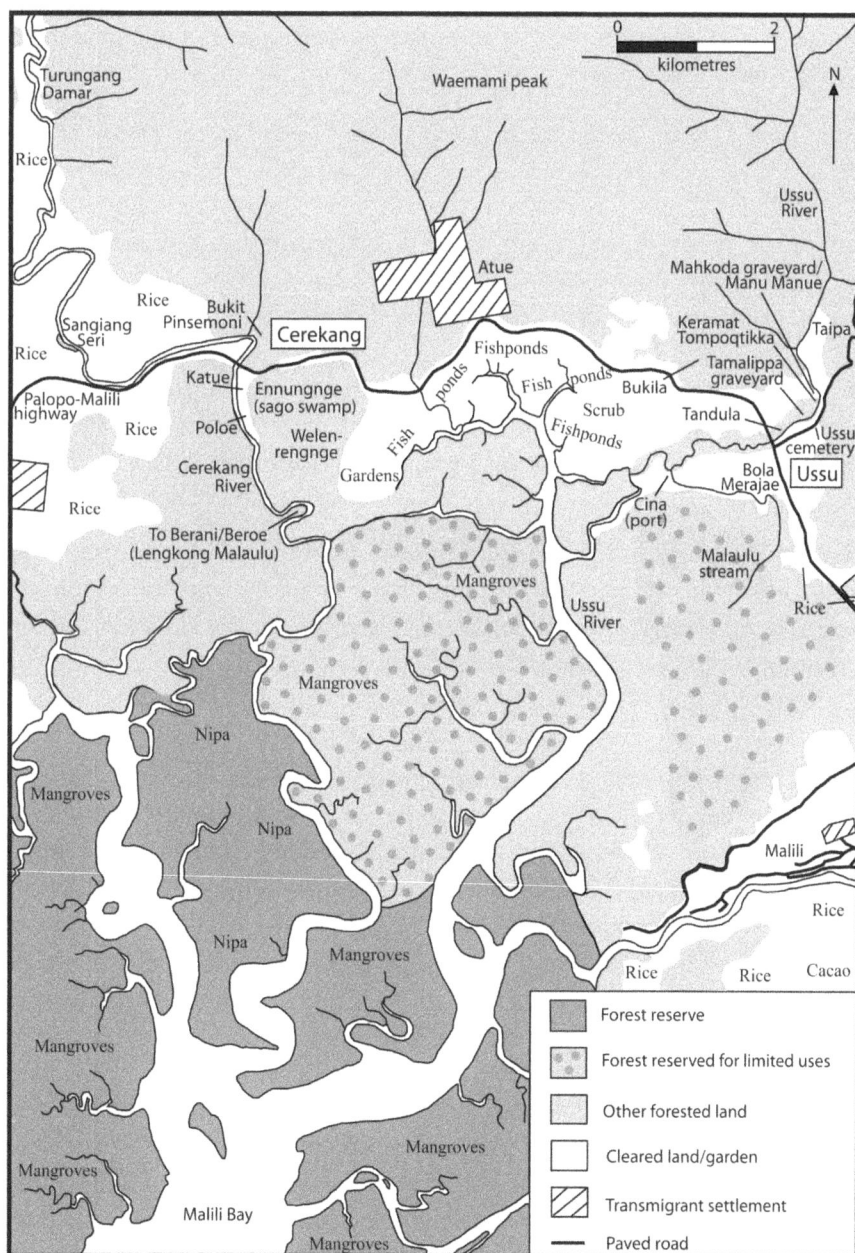

Figure 2: Archaeological sites and sacred places in Ussu and Cerekang in the context of local land use.

Source: Adapted from Bakosurtanal 1992; Gunawan 2005: Gambar 2; OXIS field notes 1997 and 1998.

The heritage significance of the Ussu and Cerekang sacred places has been recognised by South Sulawesi's historical conservation authority, Suaka Peninggalan Sejarah dan Purbakala Sulawesi Selatan, which has appointed a Cerekang resident to act as official custodian of these places. As will become clear in due course, from the point of view of historical heritage conservation, the main value of the Ussu and Cerekang sacred places would be their enclosure of late pre-Islamic and transitional Islamic sites in untouchable condition—untouchable even to archaeologists. The Ussu and Cerekang sacred places have additional heritage value as the material representation of the Tossuq understanding of their ancestry, an example of the widespread belief in Southeast Asia of the continuing power of the past transported into the present (Byrne 2011). In addition, according to Darmawan and Dirawan (2003), prohibitions against the exploitation of these places' natural resources exemplify the role of traditional beliefs and knowledge in conserving local natural heritage and ensuring sustainable development. The Cerekang residents rely on their forested landscape for a wide variety of natural resources, including sago, fruits and forest medicines, timber and nipa palm thatch, and fish and prawns from the tidal stretches of the Cerekang River (Gunawan 2005).

Sacred sites at Cerekang

Cerekang village lies east of the bridge on the Palopo-Malili highway over the upper tidal reaches of the Cerekang River. The Cerekang drains south into the Malili Bay at the head of the Gulf of Bone, running past belts of mangroves and nipa palms recognised by the East Luwu government as forest reserve (Figure 2). A government census of Manurung county (Cerekang) in 2003 counted 567 families, of which 430 had a primary economic reliance on gardening and 104 on fishing, while 110 included family members who were skilled in nipa thatch work. Cerekang accounts for a significant proportion of South Sulawesi's nipa thatch production (Gunawan 2005).

Cerekang village was established in its present location in the 1930s when the Dutch colonial government built a highway through Luwu between Palopo and Malili. Prior to the 1930s Cerekang was situated further downstream, in the vicinity of the Poloe site with its evidence of seventeenth to nineteenth century habitation (Bulbeck and Caldwell 2000:41). The Cerekang sacred places are strung along both sides of the Cerekang River.

Bukit Pinsemoni

Bukit Pinsemoni is a low ridge with two peaks to the north of the Palopo-Malili highway immediately east of where it crosses the Cerekang River (Figure 2). According to Tossuq tradition this is the place where Batara Guru, the world's first human, descended from the skies, and it is also where the gods have their palace. Cerekang adepts additionally believe that an invisible mosque on the hillside marks the Waemami spring whose pure water derives from Mecca, although this claim is disputed in Ussu (field notes 13/6/1998; Darmawan *et al* 1999:36,38,64–65). Details of this oral tradition recorded by Pelras (1996:58–59) are slightly different: Batara Guru had descended at the Waemami stream before building his palace that used to stand on the Pinsemoni hilltop (the Waemami stream referred to here is different from the Waemami stream, approximately 2 km to the east, where a transmigrant settlement called Atue has been established, and which passes through fishponds where it enters the Ussu River —field notes 1/8/1998, 3/8/1998). Batara Guru features in the origin story in the *La Galigo* cycle, whose other stories commemorate the acts of the five generations of his descendants. According to the *La Galigo* stories, when the god of the upper world, Daeng Patotoé, heard that the middle world was completely empty, he sent down his eldest son, Batara Guru, to populate it (Koolhof 2003).

Bukit Pinsemoni is barred from entry except to adepts of the highest degree, and the trees in the surrounding sacred forest may not be felled (Pelras 1996:59). According to informants in Ussu, this is to prevent looting of the antiques likely to lie there or any other disturbance to this place believed to be the ancient centre of Luwuq (Darmawan *et al* 1999:64–65). Notwithstanding the place's secrecy, the present writer was allowed to traverse Bukit Pinsemoni on 2/1/1995. My impression that the forest there was secondary was confirmed by Geoff Hope's opinion (viewed from Cerekang) on 10/2/1999. I had been advised to expect to see megaliths at the hilltop's ritual centre, which is named Wé Mattengngaémpong (the same name as the third, female archetypal ruler in the 'King List of Luwuq'), but saw only three rocks, which appeared natural but were of a comfortable height for sitting on. These rocks lay at a clearing free of leaf litter, where, according to Anthon Andi Pangerang (pers. comm. 1995), an iron cylinder set in a tree should also have been visible.

The leaf-free clearing is indicative of continuing human attention, which is not unreasonable given that Tossuq adepts of high degree are permitted entry. Reports of artefacts suggest occupation in the past, and the secondary nature of the forest points to human utilisation in days gone by. As Caldwell (1994) noted, Bukit Pinsemoni would have had advantages as the palace centre of a trading kingdom. It occupies a naturally defensive position at a point along the Cerekang River with an anchorage depth of five to seven metres and navigable access downstream to the sea. Whether Bukit Pinsemoni played any role in Luwuq's history, and what that role may have been, cannot be ascertained owing to the extreme sanctity that surrounds the hill. While this state of affairs is unfortunate from the point of view of historical documentation, it may well be ideal from the perspective of heritage conservation, depending on the exact nature of what lies on Bukit Pinsemoni.

Ennungnge

Ennungnge is a sago swamp and buffer zone near the east bank of the Cerekang River immediately south of the Palopo-Malili highway (Figure 2). According to Geoff Hope (pers. comm. 1999), the buffer zone contains weedy secondary regrowth that could have sprung up as recently as the 1930s, when the Cerekang residents moved to their present location. In Cerekang, Ennungnge is known as the first garden, the place where Luwu's two main subsistence crops, rice and sago, were first cultivated. Reportedly, cultivation ceased there because the area dried up, and nowadays cultivation (or entry of any kind) is forbidden. The area dried up because it is the place where the sun rises, which accounts for its additional name of *Tompoqtikka* ('land of the rising sun') (Darmawan *et al* 1999:37–38). *Tompoqtikka* features in the *La Galigo* stories as a contemporary of Luwuq, for instance, as the country where a son of Batara Guru, Batara Lattuq, obtained his bride (Koolhof 2003).

The OXIS archaeologists carried out excavation and survey along the levee of the Cerekang River abutting Ennungnge. The main excavation, at Katue, revealed evidence of a village involved in iron smelting and maritime trade during the first millennium CE (Bulbeck 2010). The upper deposits at Katue, along with surface finds in the vicinity, reflect the use of this area for gardening and light habitation spanning the seventeenth and nineteenth centuries, as confirmed by local accounts (Bulbeck 2003). The other excavated site was Poloe which, as mentioned above, revealed traces of seventeenth to nineteenth century habitation.

The archaeological evidence points to a hiatus in occupation in the area of Ennungnge between the eleventh and sixteenth centuries. Probably, Ennungnge continued to be cultivated during the second occupation phase, between the seventeenth and nineteenth centuries, prior to the relocation of the residents to present-day Cerekang in the 1930s (Bulbeck and Caldwell 2000:41).

Beroe (Lengkong Malaulu)

Beroe lies inside a sharp bend of the Cerekang River on its west bank (Figure 2). It is reportedly a former village near an old graveyard, in Indonesian called *To Berani* ('the brave'; Figure 3), where La Massagoni, the war leader of Sawérigading, lies buried (field notes 15/7/1998). Sawérigading, a grandson of Batara Guru, is the Bugis archetypal culture hero and the main character in the *La Galigo* stories. Darmawan *et al* heard different accounts of the same place, whose name they recorded as *Lengkong Malaulu* ('bend at Malaulu'). According to these stories, *Lengkong Malaulu* hosts a bustling village beneath the river, and it was here that Sawérigading slipped into the Underworld to be with Batara Lattuq. The numerous crocodiles that can be seen around here are residents of the underwater village, who transform back into humans when they submerge. Cerekang people who pass by *Lengkong Malaulu* are forbidden to tarry here, damage the vegetation or pluck the leaves (Darmawan *et al* 1999: 39–40). As noted by Gunawan (2005), the Cerekang residents believe that crocodiles are of human descent, and stage an annual crocodile worship ceremony.

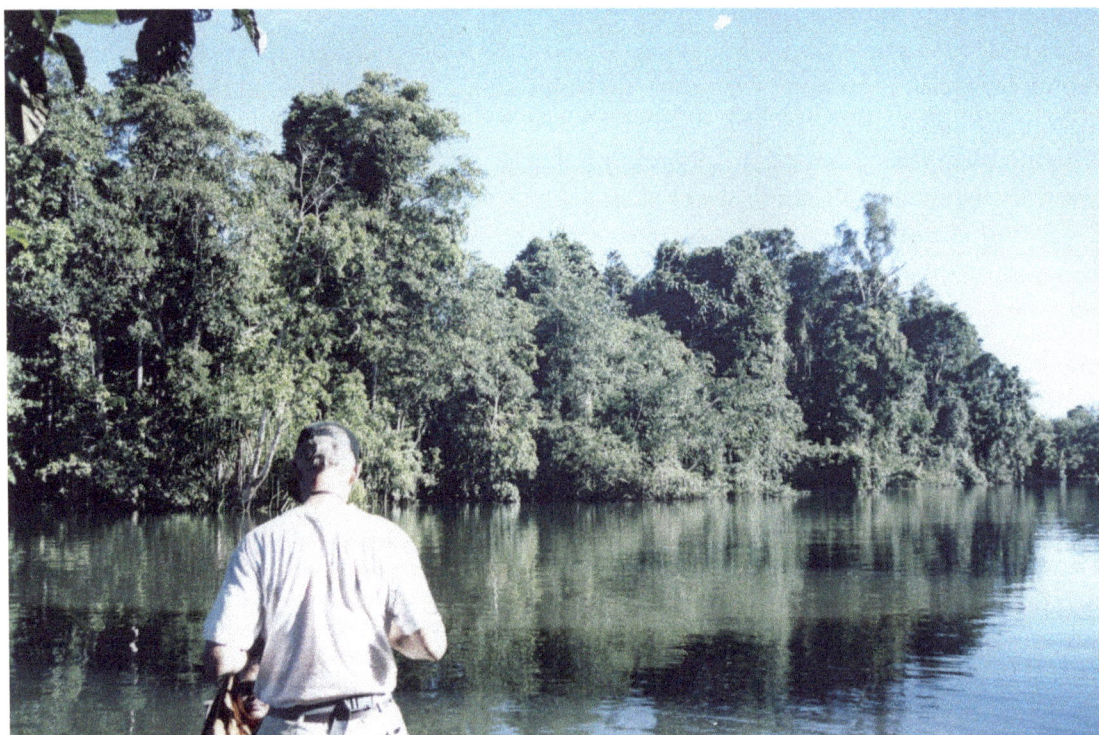

Figure 3: To Berani forest, photographed from the Cerekang River on 17 June 1998.

Source: David Bulbeck.

Additional information that the author recorded on 15/7/1998 was that the To Berani graveyard had been reportedly looted for antiques. On 31/7/1998 Daeng M. accompanied me to the foreshore of Beroe, which is an area of mangroves and nipa palms flooded at high tide. The sacred area behind it, where collection of timber is forbidden, appeared to me to contain secondary forest with a dispersed canopy. My question on what sort of burials lay here elicited the response that the Cerekang people were now required by Islamic custom to inter the deceased in wooden coffins and so they no longer use *balubu* ('large ceramic jars'). This response evokes the pre-Islamic Bugis practice of burying the cremated remains of the deceased in large ceramic jars (Druce *et al* 2005) but need not imply great antiquity; the main find at Poloe was a seventeenth to eighteenth century *balubu* placed on the ground above a nineteenth century occupation deposit (Bulbeck and Caldwell 2000:41). In summary, Beroe may have been a former village but there is no reason to assume habitation prior to the seventeenth to nineteenth centuries.

Welenrengnge (Mangkutta)

Welenrengnge (Mangkutta) is a low ridge vegetated with tall, creeper-cloaked trees around half a kilometre northeast of Beroe (Figure 2; Figure 4). Cerekang people believe this to be the place where the giant Wélenréng tree, which stood on the peak of Bukit Kemmengnge in *Positana* ('centre of the land'), was cut down. When Batara Lattuq and his people tried to fell the tree, first they were attacked by birds, land animals and children of the gods which were defending the tree, and then they found that their iron axe could not cut through the trunk. Wé Tenriabéng (see below) explained that the tree could be felled only with a gold axe wielded by an aristocrat. When this was done, the falling tree sliced through *Bulu Poloe* ('broken mountain') on the north side of Malili, and the splinters were scattered to the sea. The Mangkutta people rode the splinters and thus became the Bajau sea people who avoid staying on land (Darmawan *et al* 1999:40–41). The South Sulawesi Bajau, for their part, profess origin stories that resemble the Cerekang story in their elements but differ in how the elements link together (see Liebner 1998). Some of the Bajau stories are closer than the Cerekang story to the famous *La Galigo* episode in which Sawérigading felled the sacred Wélenréng tree with the help of his twin sister Wé Tenriabéng. In this episode, the tree then transformed into the fleet of ships which brought Sawérigading to Cina in Wajo (Figure 1) where he married Wé Cudaiq, the princess of Cina (Fachruddin 1999; Koolhof 2003:21).

Figure 4: Welenrengnge forested ridge, photographed from Mangkulili on 17 November 1997.

Source: David Bulbeck.

In 1997, OXIS archaeologists were escorted to Welenrengnge, identified as the place where Sawérigading had felled the sacred Wélenréng tree. We were shown two locations along the southern margin, Mangkulili 1 and 2, where local farmers had reported coming upon earthenware sherdage during their gardening activities (Bulbeck and Prasetyo 1998). Because our surface survey did not encounter any definite habitation debris, and because the sites were never revisited for excavation, the implications of the sites for early habitation in the vicinity of Mangkutta, let alone on the ridge itself, are unknown.

Tomba (Sangiang Seri)

Sangiang Seri is located within an expanse of rain-fed rice fields on the west bank of the Cerekang River where it traverses north of the Palopo-Malili highway (Figure 2; Figure 5). In *La Galigo* (Koolhof 2003:16), the rice goddess Sangiang Seri is the granddaughter of Batara Guru, who turned into rice plants where she was buried following her death seven days after she was born. Another *La Galigo* story, designed to convince its audience of the respect that rice deserves, relates how all the crops (including Sangiang Seri) turned into inedible substances after the ruling couple of Tompoqtikka threw out the rice they had cooked for a feast that no-one attended (Koolhof 2003:17).

Figure 5: Sangiang Seri gallery forest, photographed on 17 November 1997.

Source: David Bulbeck.

The Sangiang Seri site was reportedly looted in 1974 or 1975. When the residents discovered what was happening, a mob armed with bush knives chased the looters out of town. There are vague reports of *balubu* and plates having been looted, but what the residents were actually able to see was the traces left by the looters – holes to a depth of 80 cm. When OXIS archaeologists visited the site on 16/11/1997, the looted area of around 4 m diameter had the appearance of raised earth features, some of which had looter's holes, as well as signs of general disturbance. Archaeological excavation of the site was ruled out because of its sacred nature. However, Sangiang Seri is the most likely example of a circa fifteenth to sixteenth century cemetery burial ground encountered by the OXIS team in the Cerekang area.

Turungang Damar ('place where dammar is brought down')

Darmawan *et al* (1999:38) recorded *Turungang Damar* as a sacred site because it was the place where Daeng Patotoé brought down the first dammar tree seeds, whence they were transported across Luwu. However, as noted by Caldwell (1993) and Bulbeck and Prasetyo (1998), it is also a small port on the upper Cerekang River used for loading dammar on to small prahus for cartage downriver (Figure 2). The OXIS team was given the go-ahead for whatever activities we wanted to undertake there, and so a team conducted a survey and excavated two test pits in 1998. In fact the field workers found a place to stay in a hut built as the first stage of a planned transmigrant settlement for *Turungang Damar*.

Excavation revealed a shallow habitation deposit with abundant charcoal, dammar and pottery, as well as stone artefacts and modern glass. Surface survey recovered imported ceramic sherds that may date only to the seventeenth century and later. A charcoal sample from the base of the habitation deposit dates at two sigma to between the fifteenth and seventeenth centuries (Bulbeck and Caldwell 2000:41). Accordingly, the use of *Turungang Damar* as a transhipment point for dammar collected from the hinterland can be securely dated back to the seventeenth century but not necessarily any earlier.

Sacred sites at Ussu

The village of Ussu lies on the Palopo-Malili highway approximately halfway between Cerekang and Malili. The stream that courses past the village, known as the Ussu River, runs through forest and clearings before joining the main branch of the Ussu River which runs past protected mangrove forest into Malili Bay (Figure 2). Ussu is locally reputed to have been an ancient port, and there is archaeological evidence for continuous habitation at the village of Ussu since at least the sixteenth century (Bulbeck and Caldwell 2000:42), in contrast to the twentieth century relocation of the Cerekang inhabitants to their current location. In addition, the Ussu inhabitants are more orthodox in their practice of Islam than the Cerekang Tossuq, and a mosque and Islamic graves can be found in the vicinity of the village, whereas these visible markers of devotion to Islam are conspicuously absent from Cerekang. According to Pelras (1996:59), the damascene iron used in making *pamor luwuq* krisses during Majapahit and later times is specifically referred to across Luwu as *bessi Tossuq* ('iron of the Ussu people'), and according to information told to OXIS the Ussu kris used to be handmade by the gods but not at any particular location (field notes 17/11/1997).

Tamalippa

Tamalippa extends for several kilometres along the Ussu River northeast of Ussu (Figure 2), although not all this area is sacred. Its other name is Tompoqtikka which, as noted above, was supposedly a contemporary of early Luwuq whose name means 'land of the rising sun'. Ussu residents believe that this was the place where Batara Guru brought rice down from the skies to feed them, and also where sago first appeared prior to becoming the most important foodstuff for the Luwu populace. Because of its holy connotations, Ussu people choose to be buried within the vicinity of Tamalippa. Rowdy behaviour is forbidden, as is collection of timber or leaves, and the water in the streams is applied to the head and face when wishing a departing person a safe journey. The clumps of bamboo along the border of Tamalippa may be used only for making burial containers for the deceased (Darmawan *et al* 1999:61–62).

OXIS archaeologists surveyed several burial sites within or abutting Tamalippa. One site is the Ussu Islamic cemetery, which lies downslope from a forested plateau. Survey on 2/1/1995 noted over 100 graves of modern appearance and Chinese Qing ceramics postdating the seventeenth century. Our attention was drawn back to the site after we viewed six fourteenth to seventeenth century Chinese and Vietnamese wares reportedly encountered when a hole was dug for a grave, and heard that the scrub immediately north of the cemetery was often looted for antiques (field notes 8/6/1998, 10/6/1998). We excavated two test pits in this scrub area but encountered only sparse, modern remains. Closer to the Ussu River is the Tamalippa Islamic graveyard, inspected on 2/1/1995 before the OXIS team members realised it was sacred-secret. From the visible remnants the ten (or more) graves appeared to date back to the eighteenth and nineteenth centuries, and we did not record signs of looting. A third Islamic burial spot holds the commemorated graves of Opu Nenena Cimpaq, who was the Ussu headman at the time of the Dutch colonial administration (i.e. early twentieth century), and his companion. There are no restrictions on visiting the graves as such but to reach it visitors are required to remove their shoes when walking through Tandula, which is a sacred part of Tamalippa (field notes 17/11/1997). A fourth grave site, higher up the Ussu River, is the abandoned Mahkoda Islamic graveyard (*mahkoda* may refer to a ship's captain, which is *nakhoda* in Indonesian). The remnants were still visible of approximately ten graves and these appeared to be of nineteenth century antiquity (field notes 5/8/1998).

The Mahkoda graveyard abuts Manu Manue, an area recently cleared for a garden by a man who had built a field house there. Survey and three test pits into the shallow habitation deposit yielded earthenware pottery, four stone artefacts, a pearl, and basal charcoal with a radiocarbon date at two sigma between the seventeenth and twentieth centuries (Bulbeck and Caldwell 2000:42). On the other side of the Ussu River, on the border of Tamalippa, lies Taipa, another area known as an old village. Survey and excavation proved it to be a small twentieth century settlement, perhaps a single household, with a child's grave (Islamic) on its periphery (field notes 9/6/1998). Across the river from the Tamalippa graveyard is the sacred location of Keramat Tompoqtikka where the Luwu rulers were reportedly installed in order to establish their authority in Ussu. OXIS team members were permitted to make a brief survey here and recorded ironstone lumps, earthenware pottery and a Qing Chinese sherd (field notes 13/6/1998). As a further indication of the recency of the disuse of Tamalippa as a habitation cum gardening area, Geoff Hope (pers. comm. 1999) noted that the secondary forest in the sacred area (viewed near the Ussu Islamic cemetery) need be no older than 50 years, and that Tamalippa's large trees could have been left standing when the surrounding forest was cleared.

Accordingly, Tamalippa's present land use (or rather, sparsity of land use) may date from the 1930s when the Palopo-Malili highway was built. Transport access may have encouraged a concentration of the residents in Ussu which, as noted previously, has remained occupied since at least the sixteenth century. The small dispersed graveyards in the vicinity, which include an eighteenth century example in Ussu itself (field notes 13/6/1998), were abandoned as the Ussu Islamic cemetery became a central burial place, apparently involving the reuse of a circa fourteenth to seventeenth century pre-Islamic burial ground. Whether any places in Tamalippa were sacred-secret prior to the 1930s would be a matter of speculation.

Bulu Bajo (Bukila)

Bulu Bajo ('Bajau Peak') is a steep peak on the north margin of the Palopo-Malili highway. The trees in the forest there are forbidden to be logged. The story given above for Welenrengnge is repeated by the Ussu people in connection with this place, to the point where the falling tree cut through Bulu Poloe. However, the story ending that results in the origins of the Bajau is different. In this Ussu version of the story, when the Wélenréng tree fell, numerous eggs from the tree fell at Bulu Bajo, which created a flood that inundated the peak and washed the inhabitants away to become the Bajau sea people (Darmawan *et al.* 1999:62–63). This event aligns the Ussu story with the majority of South Sulawesi Bajau origin myths, which trace the dispersal of the Bajau to the flood caused by the eggs falling from the Wélenréng tree (Liebner 1998:123). Reportedly, Bajau people still regularly visit the site (field notes 5/8/1998), as corroborated by the frequent reference to Ussu or Cerekang in Bajau origin stories (Liebner 1998).

The author was escorted along a ridge through secondary forest to the peak on 5/8/1998. On the peak was a cemented conglomerate structure which (according to information from the village head) appears to have been built by the Dutch colonial administration. Two earthenware sherds were observed on the path leading up to the ridge. The site inspection is consistent with the oral accounts of low levels of visitation to this place until the present.

Bola Merajae

Bola Merajae ('great house') is a forested expanse that falls within the precincts of Malaulu, which is the village between Ussu and Malili. Geoff Hope (pers. comm. 1999) advised that the forest would be at least a century old although there may have been some more recent felling of its largest trees. Depending on how large Bola Merajae is, it may impinge on the adjacent forest area reserved by the East Luwu government for limited uses (Figure 2). Bola Merajae is sacred to the Cerekang people as the palace of Wé Tenriabéng, Sawérigading's twin sister, and her beautiful ladies-in-waiting. The palace had a hundred halls and attractive decorations all around it. The palace's perimeter was protected from all invaders by troops bearing krisses, spears and swords made of *bessi Tossuq* decorated with a supernatural *pamor* of exceptional potency. In addition, according to informants in Ussu, during the time of the Darul Islam insurgency in the Luwu region (1952–1965), a group of outsiders entered Bola Merajae and encountered lots of earthenware sherds and Chinese antiques, indicating that this used to be a population centre (Darmawan *et al* 1999:60–61).

Figure 6: Bola Merajae sacred forest where the Malaulu stream exits.

Source: David Bulbeck.

OXIS members were not granted permission to enter the site but were allowed to excavate a site on the Malaulu stream where it leaves the forest reserve (Figure 6). The two test pits yielded five radiocarbon dates and an earthenware pottery sequence spanning the period between approximately 2,000 and 300 years ago. Approximately half of the earthenware in the middle of the sequence, which would date to around the fourteenth and fifteenth centuries, consisted of a peculiar kind of ware, with frequent textile impressions internally, labelled 'soft pottery' (Bulbeck 2003). Similar pottery, apparently made locally rather than brought from Bola Merajae, makes up around ten percent of the fifteenth to sixteenth century earthenware in Malangke and Pinanto to the west; soft pottery sherds also occur as a rare component in fifteenth to seventeenth century contexts at Katue and Poloe, near Malili and as far north as Lake Matano. Although the Bola Merajae soft pottery hardly qualifies as a palatial production, it would appear that this area was the source of the potters whose distinctive wares enjoyed considerable popularity in late pre-Islamic Luwu (Bulbeck 2009).

On 9/8/1998 a local landowner accompanied the author along the Malaulu stream to where it enters the Ussu River. The landowner stated this place was the former port of Cina, which evokes the Cina in Wajo renowned in the *La Galigo* stories, although in the landowner's opinion the Cina referred to here was Hong Kong. (This identification of 'Cina' with Chinese homelands, rather than Cina in Wajo, is also common in those places in Central Sulawesi where immigrant Bugis have introduced Sawérigading stories—see Nourse 1998; Basri and Siojang 2003). The Islamic graves and pottery (earthenware and imported) seen during the survey all appeared to be of nineteenth to twentieth century age, as did the brass crockery and iron spear which the landowner showed me as goods he had inherited from his forebears.

Discussion

The sacred sites of Ussu and Cerekang have a significance that extends well beyond these two villages. As noted above, when a new ruler of Luwuq came to power, Tamalippa was on the list of places where the ruler's installation was to be re-enacted as part of the legitimisation of rulership over all the territories of Luwuq. In addition, the holy water from Pinsemoni is sold commercially by the bottle for its purifying qualities, and was used in 1998 to soothe an ethnic conflict that had erupted in the Baebunta area (Darmawan *et al* 1999:38–39). Ussu and Cerekang have considerable importance in several of the origin myths of the Bajau dispersed across South Sulawesi (Liebner 1998). Cerekang's reputation as the place where the gods descended and Sawérigading performed his legendary feats extends not just across Luwu but throughout South Sulawesi's Bugis lands (Andaya 1981:17).

The mystique of Ussu and Cerekang derives from the widespread belief in the great antiquity of the events recounted in the *La Galigo* stories. This is not a belief ascribed to by the Tossuq adepts themselves, since they do not embrace a linear concept of time that would make them place the *La Galigo* events before or after anything else along a chronological scale. To be sure, there were places where pivotal events in the past first happened, but these are also the places where the *La Galigo* gods – the ancestors of the Tossuq – still haunt the landscape and watch over affairs in Luwu. Tossuq adepts are aware that there was a Dutch colonial period and the present-day Cerekang village was relocated during this period, but for them the Cerekang residents of that time now dwell with the other gods in Cerekang's sacred places. I do not know if anyone has ever asked a Tossuq mystic whether the location of the Ussu and Cerekang sacred sites is a product of the resettlement of the population following the construction of the Palopo–Malili highway, but I suspect that a question of this nature would be either answered in the negative or simply not understood.

This is, however, an important consideration for interested parties who do have a linear concept of time, particularly scholars (following Braam Morris 1889) who intellectualise the Ussu and Cerekang traditions into a log of claims on Luwuq's origins at the dawn of South Sulawesi history. As noted above, the cessation of gardening in Ennungnge and the light utilisation of Tamalippa can be dated to around the time when the Palopo-Malili highway was built, as may also be the case with the abandonment of Beroe. A longer time may have elapsed since gardening or tree felling ceased at Bukit Pinsemoni, Bola Merajae, Welenrengnge or Bukila, but there is nothing to indicate old forest cover at any of these places. Of course, protective reservation of these places in recent centuries would be compatible with a scenario in which Ussu and/or Cerekang had hosted the birthplace of Luwuq. However, the archaeological evidence would be incompatible with such a scenario (Figure 7). The OXIS archaeologists looked hard for pre-1600 sites in Ussu and Cerekang, but could find only a weak signal for the fourteenth to sixteenth centuries (by general Luwu standards), and nothing at all for 1000–1300, the so-called Age of *La Galigo*.

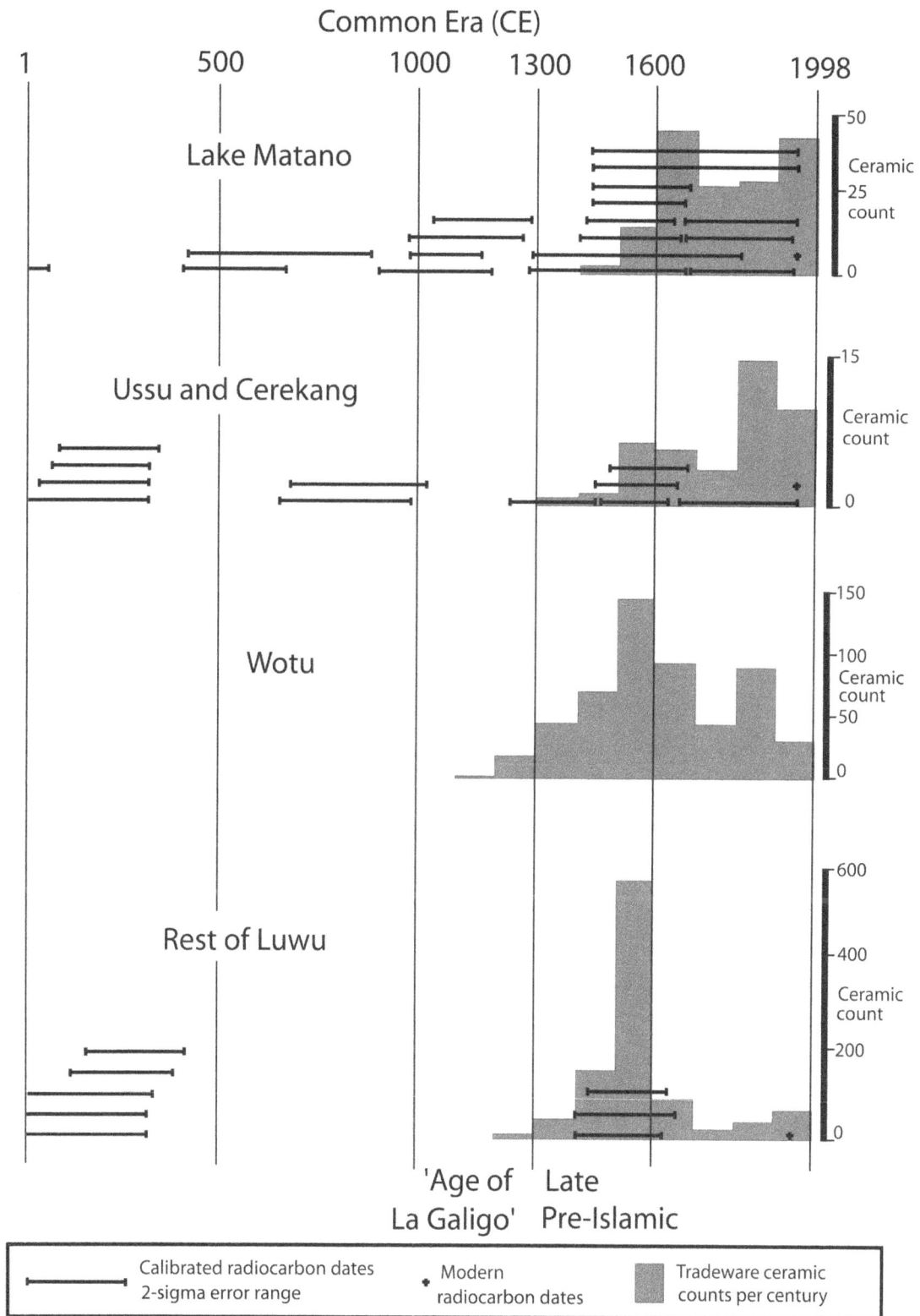

Figure 7: Chronological data from Luwu sites.

Source: Taken from Bulbeck and Caldwell 2000, except 'Swatow' wares dated to 1600–1700 rather than 1550–1700.

Luwu's best archaeological evidence corresponding to the Age of *La Galigo* comes from Lake Matano (Figure 1), where the radiocarbon dates document continuous occupation from around 500 CE onwards (Figure 7). These dates reflect an evolving technology in roasting and smelting the local iron ore to produce the famed *pamor luwuq*. However, the lack of archaeological evidence at coastal Luwu sites between 1000 and 1200 CE suggests that the processed iron was traded northward, through Luwu-Banggai, during this period (Bulbeck 2010:158). Ceramic evidence from Tambu-Tambu in Wotu (no radiocarbon dates are available) points to the establishment of Wotu as a coastal trading station by the thirteenth century (Bulbeck and Caldwell 2000:52), but Wotu was a non-Bugis competitor rather than a predecessor of early Luwu (Fadillah 2000). Ceramic imports to Wotu continued to rise throughout the fourteenth to sixteenth centuries, although not as markedly as Malangke. Malangke accounts for the great majority of the late pre-Islamic 'rest of Luwu' ceramic count (the number of identifiable tradewares, each represented either by a complete vessel, multiple sherds matched to the same vessel, or a single unmatched sherd) in Figure 7.

There is no oral tradition in support of Malangke as Luwuq's pre-Islamic capital, but the supporting archaeological evidence is overwhelming. Relying primarily on oral sources, Darmawan (2000) did accept a sixteenth century shift of the Luwuq capital to Malangke, but this was based on the hard evidence at Malangke of the commemorated grave of Datu Pattimang Matinroé ri Wareq, the Luwuq king who in 1603 became the first South Sulawesi ruler to embrace Islam. Survey by OXIS at the Malangke site of Tampung Jawa (= Javanese graves) revealed remains of a Javanese brick temple dated to the fourteenth or fifteenth century by the associated ceramics (Hakim 2000). Excavation by OXIS at 'old Pattimang' yielded slag and other debris from working iron – imported from the Luwu hinterland, no doubt including Lake Matano – and fragments of iron weapons. Comprehensive survey of Malangke's looted burial grounds pointed to an increase in the population of residents from around 3,000 in the fourteenth century to 10,000 in the fifteenth century and 15,000 in the sixteenth century, prior to a precipitous decline to fewer than 1,000 in the seventeenth century (when Luwuq's capital moved to Palopo, as is well documented historically). Thus, Luwuq would appear to have been based at Malangke by 1365, when its status as a significant polity was known to Majapahit Java (as mentioned in the *Desawarnana*—Robson 1995), but there is no credible evidence for pre-fourteenth century occupation at Malangke (Bulbeck and Caldwell 2000:73–77). If, as appears probable, Luwuq colonised the vacant Malangke estuary to exploit trading opportunities at the head of the Gulf of Bone, from where did the colonists come?

For the OXIS chief investigators, the answer to that question is Sengkang, on the upper Cenrana Valley. Early Luwuq's sister kingdom was Cina, whose capital (Allangkanangnge ri Latanete) contained a substantial population based on rice production by c. 1300 (Bulbeck and Caldwell 2008). As explained by Bulbeck and Caldwell (2000:103–05), locating early Luwuq in the Cenrana Valley explains a number of puzzles in Bugis historical texts. These include the claim by both Luwuq and Cina to Simpurusia as their founding progenitor, the toponymic match between the Lompoq settlement reportedly founded by Simpurusia and Lompo in Sengkang, the inclusion of three places near Sengkang amongst Luwuq's ancestral lands, and three reported royal marriages between Luwuq and Cina spanning the fourteenth century. From this perspective, Sawérigading's fabled trip to Cina to marry Wé Cudaiq would count as a jaunt downriver to the neighbouring kingdom (as it were) rather than the epic voyage as commonly understood.

Two other points in support of Luwuq's origin in the vicinity of Sengkang are worth noting. One is that there is considerable historical evidence for Luwuq's late pre-Islamic expulsion (primarily at the hands of Wajoq) from the Cenrana Valley, but no accounts whatsoever of Luwuq's intrusion into the valley (e.g. Pelras 1996). This suggests that no such intrusion ever occurred because Luwuq originated there. The second is that knowledge of the *La Galigo* stories is generally shallow in present-day Luwu, especially in Ussu and Cerekang, where the Tossuq stories reveal an idiosyncratic understanding of the acts and social relationships of the *La Galigo* characters. In contrast, Bugis groups along the Cenrana combine their reverence for the *La Galigo* material with a deep knowledge of its canonical accounts, and one group, the Tolotang Benteng in Amparita just northwest of Sengkang, consider *La Galigo* manuscripts sacred (Koolhof 2003). To the degree that *La Galigo* had an association with early Luwuq, this association was apparently with Luwuq in the Cenrana Valley, and was brought to Luwu when Luwuq colonised Malangke.

The blending of *La Galigo* mythology with Luwuq's early royal genealogy is evident in the *Lontaraqna Simpurusia* ('writing concerning Simpurusia') composed by at least the late seventeenth century (Caldwell 1988:26,34). This work presents three stories that greatly elaborate on the accounts of Luwuq's three archetypal rulers contained in the 'King List of Luwuq'. One of the stories' main themes is to emplace these rulers in the *La Galigo* cosmology. For instance, Simpurusia presented himself to Daeng Patotoé before returning to earth to commence the Luwuq royal line, and his descendants and their spouses made successive journeys between the upper-world, middle-world and underworld. In addition, the *Lontaraqna Simpurusia* claims that Anakaji's wife was descended from Majapahit royalty, and it mentions an array of characters whose names have coastal or riverine meanings, including *Wé Mattengngaémpong* ('in the middle of the waves') who was also queen of the crocodiles (Caldwell 1988:33–47). The reference to Majapahit Java and connotations of an estuarine environment evidently associate the *Lontaraqna Simpurusia* with Malangke.

The above interpretation of the *Lontaraqna Simpurusia* would agree with Macknight's proposal that the late pre-Islamic Bugis immigrants to Luwu lacked local origin mythology and so filled the lacuna through a local application of the *La Galigo* stories. The need to do this would have been particularly acute at Ussu and Cerekang, which are remote Bugis enclaves within a Mori-speaking area (Macknight 2004). Local adaptation of *La Galigo* lore is evident in such aspects as inserting the traditional Luwu staple of sago into the story on rice's origins, and identifying the *La Galigo* ancestral figures with the Cerekang River crocodiles. Despite the relatively subdued documentation that could be obtained for the late pre-Islamic archaeology of Ussu and Cerekang (Figure 7), there is good reason to believe that these communities were involved in the transport of smelted iron from Lake Matano to Malangke, and the main places involved in this transport fall within the bounds of the Ussu and Cerekang sacred places (Bulbeck and Caldwell 2000). Following Malangke's abandonment, the economic function of iron transport was lost, and memories of the trade merged with the *La Galigo*-based ancestor worship practised by the Tossuq. Nonetheless, the fourteenth to sixteenth century trade in iron probably formed the original core of the sacred places in the Cerekang landscape, the abodes of the ancestors to where adepts expect to return upon their own demise.

The Tossuq beliefs, which enliven the Tossuq landscape with mystic ancestral associations, are not unusual by the standards of the belief systems in many rural communities. For these communities, their surroundings resonate with cultural memories and the physical manifestation of social relationships (see contributions in Stewart and Strathern 2003). What may be remarkable is that other South Sulawesi Bugis communities do not try to place the founding events of the *La Galigo* within their own lands, even though they may match the Tossuq in honouring the *La Galigo* stories as the wisdom of the ancestors (Koolhof 2003). These points explain the temptation amongst scholars of South Sulawesi history to place the origins of Bugis 'high culture' in Ussu and Cerekang, and to believe this memory was retained with the spread of Bugis high culture, explaining why other Bugis groups do not claim specific *La Galigo* origins for their own community. The alternative interpretation of this state of affairs, consistent with the views of Caldwell, Macknight and the author, is that the 'distancing' of the *La Galigo* mythology – separated from historical times by an 'age of anarchy' (Macknight 1993) – removed it from the ownership of any single community and allowed it to flourish as an 'encyclopaedia' of Bugis cultural knowledge (Koolhof 1999). The concept of common origins would have assisted Bugis communities to unite into kingdoms and their kingdoms to forge alliances (of variable longevity), as amply documented historically (e.g. Andaya 1981; Caldwell 1988; Pelras 1996). From that perspective, the beliefs of the Tossuq in their specific *La Galigo* origins would reflect the isolation of the Tossuq from mainstream Bugis society rather than any primordial status of Ussu and Cerekang (Macknight 2004).

Scholars of South Sulawesi agree on accepting historical associations for the Ussu and Cerekang sacred places; the divergence in opinion is whether these associations date back to the early second millennium CE or a later time. Both perspectives would affront the Tossuq as reifying, in the past, what they view as the eternal present, although I suspect the Tossuq interpret the trickle of outsiders who enquire about their beliefs as confirmation of the veracity or at least the importance of their worldview. A more serious issue in terms of appropriation of Indigenous 'knowledge', a charge frequently laid against archaeology (e.g. Watkins 2000; Bruchac 2005; Nicholas 2005; Smith and Wobst 2005), may be the scholarly disagreement over the reality of the 'Age of *La Galigo*'. The Bugis and other Indonesian scholars associated with the OXIS project (e.g. Darmawan 2000; Sumantri 2000; see also Nayati 2003) have retained the traditional view, which is regarded as a sign of being knowledgeable on Bugis history, that the Age of *La Galigo* hosted the incubation of Bugis high culture. It is the Westerners associated with the OXIS project (including Liebner 2003) who have shifted toward demoting the Age of *La Galigo* to the status of myth. Overt conflict however is largely avoided by the Indonesian and Western scholars addressing their publications to different audiences.

Indonesian and Western scholars would agree that the Ussu and Cerekang sacred places enclose historical and perhaps even prehistoric sites, potentially covering much or all of the Common Era. Changes to Tossuq attitudes, or a revision of which particular locations are sacred and secret, may allow detailed investigation in the future of these sites and their exact historical significance. Fortunately for heritage interests, Tossuq practices in safeguarding their sacred sites amount to 'best practice' in terms of heritage protection, minimising the potential for conflict between heritage professionals and local community interests (see Miura 2010; Byrne 2011). For the moment we can feel secure that the Tossuq community is protecting a selection of its ancestral dwellings from the looter's spade and the developer's hoe, which have inflicted so much damage on archaeological sites across South Sulawesi.

Evidence for an early twentieth century date for the protected status of at least some of the Ussu and Cerekang sacred sites, following the construction of the Palopo-Malili highway, suggests that the Tossuq may allow external events to change the associations they ascribe to any particular place. As became clear to me on numerous occasions during fieldwork, Tossuq adepts view their dream experiences as sources of mystic knowledge, and so there is the potential for them to literally dream up a reconfiguration of their sacred landscape. None of the Ussu and Cerekang sacred places, with the possible exception of part of Bola Merajae, lie within officially gazetted forest reserve (Figure 2). Further, Cerekang residents are divided on the issue of protection of their forests, as revealed by a survey of attitudes towards a government plan to appropriate community land along the lower Cerekang for fishpond development. Forty-seven percent of respondents opted for exploitation of their forests in any way that had an economic benefit, 29 percent preferred controlled exploitation, and just 24 percent elected blanket protection (Gunawan 2005). As noted previously, around 80 percent of Cerekang families have a primary economic dependence on gardening, which revealed itself during fieldwork in terms of the expansion of cleared land near the Palopo-Malili highway. Prohibition of the exploitation of certain belts of secondary forest within this undulating terrain currently assists maintenance of ecological diversity of this zone most at risk to forest clearance. However, community desires for local economic development may find themselves reflected in the dreams of Tossuq leaders, permitting change in the sacred lore. Accordingly, the medium to long-term stability of the Tossuq sacred landscape in its currently documented form (Figure 2) would be a topic for future research.

Conclusion

The Tossuq sacred places of Ussu and Cerekang are an interesting case of local heritage beliefs that intersect with a celebrated history, that of Luwuq, reputedly the founder kingdom of South Sulawesi. The conventional scholarly approach has been to use these beliefs to write an ersatz protohistory of Luwuq's origins in Ussu, as done for instance by Pelras (1996) and Darmawan (2000). The OXIS project uncovered archaeological and early textual evidence that would place Luwuq's origins in the Cenrana Valley, well to the south of Luwu. This more critical perspective would imply that Ussu and Cerekang were second-hand recipients of *La Galigo* mythology, and so place the Tossuq beliefs in their correct historical perspective. This at least is the message taken away by Western scholars associated with the OXIS project, although to my knowledge none of the Indonesian scholars associated with OXIS have followed suit. Neither the Westerner nor the Indonesian perspective would deny the Tossuq the right to treat their sacred sites as they see fit, particularly when their conservation interests are aligned with those of heritage professionals. Tossuq prohibition of clearance and exploitation of their sacred forests also helps protect the ecological diversity of the undulating terrain near the Palopo-Malili highway in the Ussu and Cerekang area. However, the Tossuq beliefs that enhance cultural and ecological heritage preservation near Ussu and Cerekang may be susceptible to alteration or dilution, and their medium to long-term efficacy would be a matter for follow-up research.

Acknowledgements

I extend my gratitude to my fellow team members during the Ussu and Cerekang fieldwork, notably Ian Caldwell, Bagyo Prasetyo, Moh. Ali Fadillah, Iwan Sumantri, Tanwir Wolman and Budianto Hakim. I also thank Geoff Hope, Campbell Macknight and Doreen Bowdery for their advice on the Ussu and Cerekang sites during a short visit in 1999. Campbell Macknight and Peter Lape provided valuable comments on earlier versions of the text.

References

Andaya, L.Y. 1981. *The heritage of Arung Palakka*. Martinus Nijhoff, The Hague.

Badan Koordinasi Survey dan Pemataan Nasional (Bakosurtanal). 1992. Peta Rupabumi Indonesia 1:50.000, Lembar 2113-33, Malili, Edisi 1, Bogor.

Basri, H. and Siojang, B. 2003. Sawérigading versi Sulawesi Tengah. In: Rahman, N., Hukma, A. and Anwar, I. (eds), *La Galigo: Menelusuri jejak warisan sastra dunia*, pp. 98–131. Hasanuddin University, Divisi Ilmu Sosial dan Humaniora, Pusat Kegiatan Penelitian/Barru Kabupaten government, Makassar.

Braam Morris, D.F. van. 1889. Het landschap Loewoe. *Tijdschrift van het Bataviaasch Genootschap van Kunsten en Wetenschappen* 32:497–530.

Bruchac, M.M. 2005. Earthshapers and placemakers: Algonkian Indian stories and the landscape. In: Smith, C. and Wobst, H.M. (eds), *Indigenous archaeologies: Decolonizing theory and practice*, pp. 56–80. Routledge, London.

Bulbeck, D. 2003. The archaeology of the major sites in Ussu/Cerekang. In: Rahman, N., Hukma, A. and Anwar, I. (eds), *La Galigo: Menelusuri jejak warisan sastra dunia*, pp. 467–484. Hasanuddin University, Divisi Ilmu Sosial dan Humaniora, Pusat Kegiatan Penelitian/Barru Kabupaten government, Makassar.

Bulbeck, D. 2009. The application of Darwinian cultural evolutionary theory to ceramics: the case of 'soft pottery' from Luwu, South Sulawesi, Indonesia. In: Muscio, H. and Lopez, G. (eds), *Theoretical and methodological issues in evolutionary archaeology: Toward an unified Darwinian paradigm/questions*, pp. 3–11. BAR International Series 1915, Oxford.

Bulbeck, D. 2010. Uneven development in southwest Sulawesi, Indonesia during the Early Metal Phase. In: Bellina, B., Bacus, E.A., Pryce, T. and Christie, J. (eds), *50 years of archaeology in Southeast Asia: Essays in honour of Ian Glover*, pp. 152–169. River Books, Bangkok.

Bulbeck, D. and Caldwell, I. 2000. *Land of iron: The historical archaeology of Luwu and the Cenrana Valley. Results of the Origin of Complex Society in South Sulawesi Project (OXIS)*. University of Hull Centre for South-East Asian Studies, Hull.

Bulbeck, D. and Caldwell, I. 2008. *Oryza sativa* and the origins of kingdoms in south Sulawesi: evidence from rice husk phytoliths. *Indonesia and the Malay World* 36:1–20.

Bulbeck, D. and Prasetyo, B. 1998. Survey of pre-Islamic historical sites in Luwu, South Sulawesi. *Walennae* 1/I:29–42.

Byrne, D. 2011. Thinking about popular religion and heritage. In: Miksic, J.N., Goh, G.Y. and O'Connor, S. (eds), *Rethinking cultural resource management in Southeast Asia: Preservation, development, and neglect*, pp. 3–14. Anthem Press, London.

Caldwell, I. 1988. *South Sulawesi A.D. 1300–1600: Ten Bugis texts*. Unpublished PhD thesis. The Australian National University.

Caldwell, I. 1993. Report on fieldwork in South Sulawesi. *Baruga* 9:6–8.

Caldwell, I. 1994. The pre-Islamic capitals of Luwu'. Unpublished paper presented at the *Asian Studies Association of Australia Biennial Conference*, Perth, 13–16 July 1994.

Caldwell, I. 1998. The chronology of the king list of Luwu' to AD 1611. In: Robinson, K. and Paeni, M. (eds), *Living through histories: Culture, history and social life in south Sulawesi*, pp. 29–41. The Australian National University, Canberra.

Darmawan, M. 2000. Identitas budaya Luwu: tinjauan ringkas. In: Fadillah, M. and Sumantri, I. (eds), *Kedatuan Luwu: Perspektif arkeologi, sejarah dan antropologi*, pp. 47–59. Lembaga Penerbitan Universitas Hasanuddin, Makassar.

Darmawan, M., Pawiloy, S., Dirawan, G. and Andayarani, D. 1999. OXIS: The Origin of Complex Society in South Sulawesi (A Social Anthropological Perspective). Universitas Negeri Makassar, Ujung Pandang.

Darmawan M. and Dirawan, G. 2003. Nature and culture (studi awal tentang konsep lingkungan dalam La Galigo). In: Rahman, N., Hukma, A and Anwar, I. (eds), *La Galigo: Menelusuri jejak warisan sastra dunia*, pp. 280–287. Hasanuddin University, Divisi Ilmu Sosial dan Humaniora, Pusat Kegiatan Penelitian/Barru Kabupaten government, Makassar.

Druce, S., Bulbeck, D. and Mahmud, I. 2005. A transitional Islamic Bugis cremation in Bulubangi, south Sulawesi: its historical and archaeological context. *Review of Indonesian and Malaysian Affairs* 39:1–22.

Fachruddin, A. 1999. *Ritumpanna Wélenrénngé: Sebuah episode sastra Bugis klasik Galigo*. École française d'Extrème-Orient, Jakarta.

Fadillah, M.A. 2000. Arkeologi dan Sejarah kuna Wotu: catatan survei dan eksksavasi. In: Fadillah, M.A. and Sumantri, I. (eds), *Kedatuan Luwu: Perspektif arkeologi, sejarah dan antropologi*, pp.159–195. Lembaga Penerbitan Universitas Hasanuddin, Makassar.

Fadillah, M. and Sumantri, I. (eds). *Kedatuan Luwu: Perspektif arkeologi, Sejarah dan antropologi*. Lembaga Penerbitan Universitas Hasanuddin, Makassar.

Gunawan, H. 2005. *Desentralisasi: Ancaman dan harapan bagi Masyarakat Adat. Studi kasus masyarakat adat cerekang di kabupaten Luwu timur, propinsi Sulawesi selatan*. Centre for International Forestry Research, Jakarta.

Hakim, B. 2000. Fragmen bata di Tampung Jawa, Malangke: jejak gilde Majapahit di ibukota Luwu. In: Fadillah, M. and Sumantri, I. (eds), *Kedatuan Luwu: Perspektif arkeologi, sejarah dan antropologi*, pp.103–113. Lembaga Penerbitan Universitas Hasanuddin, Makassar.

Koolhof, S. 1999. The '*La Galigo*': A Bugis encyclopaedia and its growth. *Bijdragen tot de Taal-, Land- en Volkenkunde* 155:362–387.

Koolhof, S. 2003. The *La Galigo*: A Bugis encyclopedia and its growth. In: Rahman, N., Hukma, A. and Anwar, I. (eds), *La Galigo: Menelusuri jejak warisan sastra dunia*, pp. 4–33. Hasanuddin University, Divisi Ilmu Sosial dan Humaniora, Pusat Kegiatan Penelitian/Barru Kabupaten government, Makassar.

Liebner, H. 1998. Four oral versions of a story about the origin of the Bajo people of southern Selayar. In: Robinson, K. and Paeni, M. (eds), *Living through histories: Culture, history and social life in south Sulawesi*, pp.107–133. The Australian National University, Canberra.

Liebner, H. 2003. Berlayar ke tompoq tikkaq: sebuah episode *La Galigo*. In: Rahman, N., Hukma, A. and Anwar, I. (eds), *La Galigo: Menelusuri jejak warisan sastra dunia*, pp. 373–414. Hasanuddin University, Divisi Ilmu Sosial dan Humaniora, Pusat Kegiatan Penelitian/Barru Kabupaten government, Makassar.

Macknight, C.C. 1993. *The early history of south Sulawesi: Some recent advances*. Monash University Centre of Southeast Asian Studies Working Papers 81, Clayton.

Macknight, C.C. 2004. South Sulawesi before AD 1600. Unpublished paper presented at the Asian Studies Association of Australia conference, 29 June–3 July, 2004, Canberra.

Miura, K. 2010. World heritage sites in Southeast Asia: Angkor and beyond. In: Hitchcock, M., King, V.T. and Parnwell, M. (eds), *Heritage Tourism in Southeast Asia*, pp.134–150. NIAS Press, Copenhagen.

Nayati, W. 2003. Pemanfaatan lingkungan alam bagi pemenuhan kebutuhan hidup masa lalu di Sulawesi: refleksi mitos *La Galigo*. In: Rahman, N., Hukma, A. and Anwar, I. (eds), *La Galigo: Menelusuri jejak warisan sastra dunia*, pp. 289–302. Hasanuddin University, Divisi Ilmu Sosial dan Humaniora, Pusat Kegiatan Penelitian/Barru Kabupaten government, Makassar.

Nicholas, G.P. 2005. The persistence of memory; the politics of desire: archaeological impacts on Aboriginal peoples and their response. In: Smith, C. and Wobst, H.M. (eds), *Indigenous archaeologies: Decolonizing theory and practice*, pp. 81–103. Routledge, London.

Nourse, J.W. 1998. Sawerigading in strange places: the *La Galigo* myth in Central Sulawesi. In: Robinson, K. and Paeni, M. (eds), *Living through histories: Culture, history and social life in south Sulawesi*, pp. 107–133. The Australian National University, Canberra.

Pelras, C. 1996. *The Bugis*. Blackwell Publishers, London.

Pelras, C. 2006. *Manusia Bugis* (Translated into Indonesian by A. Rahman, A. Hasriadi and N. Sirimorok). École française d'Extrème-Orient, Jakarta.

Reid, A. 1990. Luwu: land where the gods descended. In: Volkman, T. and Caldwell, I. (eds), *Sulawesi: The Celebes*, pp.106–109. Periplus Editions, Singapore.

Robson, S. 1995. *Desawarnana (Nagarakrtagama) by Mpu Prapanca*. KILTLV Press, Leiden.

Smith, C. and Wobst, H.M. 2005. Decolonizing archaeological theory and practice. In: Smith, C. and Wobst, H.M. (eds), *Indigenous archaeologies: Decolonizing theory and practice*, pp. 5–16. Routledge, London.

Stewart, P.J. and Strathern, A. (eds). 2003. *Landscape, memory and history: Anthropological perspectives*. Pluto Press, London.

Sumantri, I. 2000. Persepsi masyarakat terhadap penelitian arkeologi di Luwu Utara. In: Moh, A. and. Sumantri, I. (eds), *Kedatuan Luwu: Perspektif arkeologi, sejarah dan antropologi*, pp.237–249. Lembaga Penerbitan Universitas Hasanuddin, Makassar.

Watkins, J. 2000. *Indigenous archaeology: American Indian values and scientific practice*. AltaMira Press, Walnut Creek .

13

Cultural heritage and its performative modalities: Imagining the Nino Konis Santana National Park in East Timor

Andrew McWilliam, The Australian National University, Canberra, Australia

Introduction

In East Timor the struggle for national independence was hard won and required a unity of shared purpose from the broad community of resistance. Part of the task of sustaining that sense of unity in the post-independence Democratic Republic of Timor Leste is the imaginative work of commemorative symbols that enjoin citizens within a common narrative of nation. My paper looks at one such commemorative symbol: the establishment in 2007[1] of the Nino Konis Santana National Park in the densely forested eastern portion of the island. The legislation creates the first National Park in an independent Timor-Leste and carries with it a complex range of associations, expectations and attributions. Within that complex, I am interested in the questions it raises around what I call the competing performative modalities of connection and significance. The tension in this case arises between state-making projects of institutional governance acting in the wider public interest and the more prosaic place-making projects of customary communities' resident within the National Park itself from which they derive household sustenance and material needs.

Towards an understanding of this mutual entanglement of state and local sensibilities, I wish to draw upon two inter-related analytical perspectives. One develops a useful distinction highlighted by anthropologist, Janet Hoskins, between history and heritage as ideal types of contrasting interpretations of the past. In her view: '[T]he historical past is a linear time line, with non-repeating events by individual actors discontinuous with the present.' This historical consciousness: 'creates a finished chapter that may be reopened and reread but not re-written.' In contrast the heritage of the past, rather than a line of unique occurrences, 'forms an array of established and shared sequences that may be instantiated in various new and transformed ways. People with a cultural heritage consciousness see themselves acting in the place of ancestors, reproducing their practices and continuing a pattern of timeless reciprocities' (Hoskins 1993). The distinction here between history and heritage is not mutually exclusive however, and people may sustain both forms of orientations towards the past, but it is equally evident that in doing so they induce very different kinds of understandings about the ontology of past events.[2]

---

1   Legislation establishing the creation of the NKSNP was formally enacted in the dying days of the Fretilin government (3 August 2007) under the signature of the Minister of Agriculture Fisheries and Forestry at the time.

2   The distinction between history and heritage has of course been subject to much debate and definition (e.g. see Harvey 2001 for a discussion of the history of heritage). Although narrower in its formulation than other representations of the debate, I find Hoskins' take on the matter to be particularly instructive in terms of the material at hand. There are also resonances with other views such as those of Lowenthal who has written extensively on the changing role and perceptions of the past in shaping contemporary lives (e.g. Lowenthal 1985). His observation that the 'creed of heritage answers needs for ritual devotion', rather than the facts of the matter, is consistent with the distinction Hoskins seeks to make (1998).

I would like to compare and complement this view with a second analytical perspective taken from another anthropologist, Charles Zerner who writes of the 'performative modalities of customary attachments to particular local environments' (2003:3). These modalities can include poetics and mythologies of connection, ecological knowledge of plants and animals and the various material and ideational attachments to place reproduced by local communities over time. As diverse forms of knowledge and practice they work to authorise and legitimate local attachments to place in customary terms. Characteristically however, from the perspective of state regulatory agencies, local knowledge in this form confronts what Zerner calls the 'challenge of translation'. This is the problem of translating these diverse cultural performative modalities into legible forms recognised by state regulatory institutions as 'legal rights', or otherwise 'legally valid representations of entitlement that still maintain their integrity and sensibility' (Zerner 2003). Unfortunately the history of these attempts, where they have been recognised as relevant issues by governments, has been a poor one and frequently resulted in the denial or erasure of customary entitlements in the interests of 'the wider public good' (see McWilliam 2007); in Zerner's terms, their distinctive logics, metaphors and modalities defined away (2003:17).

In the following discussion I seek to adapt these two analytical approaches to argue that the contemporary development of the Nino Konis Santana National Park is part of an imaginative and interpretive struggle over the commemorative shape of national history in the context of a living emplaced local heritage.

Inscribing natural values

Following its ratification, the public proclamation of the Nino Konis Santana National Park (NKSNP) in August 2008 was enacted with much ceremonial protocol and fanfare within the Park itself. The attendance of politicians, international dignitaries, media, armed security forces and government officials along with invited groups of local Fataluku villagers cloaked the event with the symbolic authority of the state, while the then president, Xanana Gusmao himself, arrived by helicopter and was greeted with traditional dancing and ritual speech.

The Park covers a terrestrial area of some 1236 km^2 in the most easterly district of East Timor known as Lautem. It includes a similarly extensive area of coastal waters and fringing coral reefs (556 km^2) that forms a designated protected marine zone. The Park incorporates Lake Ira lalaru, the largest freshwater lake on the island of Timor and the densely forested Pai cao^3 mountain range on its southern flanks. The extensive and largely unpopulated stretches of monsoon and evergreen canopy rainforest on its northern and southern reaches are integral features and form a unique zone of lowland tropical rainforest.

In recognition of the long history of settlement in the region by Fataluku-speaking local communities, the government has sought to follow the conventions of the International Union for the Conservation of Nature (IUCN) to establish the National Park in a way consistent with principles of IUCN Conservation Level 5. This category is one that intentionally recognises the role of human interaction in the reproduction of environmental values, and so allows for multi-purpose activities and the development needs of resident Fataluku villages within defined zones of activity. Specifically the intent of the government in establishing the initiative was to manage the Park 'as an internationally recognised protected area where the traditional interactions of people and nature are maintained in a way that protects the environment' (MAFP 2006). As the boundaries of the Park incorporate the resident populations and lands of six Timorese villages (suku),^{4} it is evident that much of the area will be zoned for settlements and agriculture, but there are also plans for protected and no-take areas.

---

3   In Fataluku orthography the letter c is pronounced 'ch'. Hence, Pai cao or 'pig's head'.

4   They comprise the villages of Bauro, Poros, Mehara, Tutuala, Muapitine and Lorehe 1, with a total resident population of around 15,000 (see census 2004).

The creation of the National Park has been many years in the making and represents the sustained efforts, aspirations and commitment of many Timorese championing the cause both within and outside government. There are two key aspects of the Park that arguably justified its creation from the collective perspective of the National Government. The first of these elements is the rich complexity of natural values and high biodiversity flora and faunal species found in the park area. This has been confirmed in numerous surveys of its forested and marine environments that have consistently identified important endemic and endangered species. In fact, regard for the natural environmental values has its origins in Portuguese colonial rule with the establishment of the Colonial Forestry Service in 1924 (Cardoso 1933). Admittedly this was driven more by interests in commercial grade timber supply than its unique biodiversity, but the seeds of a forestry reserve system was established at the time and included a silviculture research station in the dense lowland canopy forests of Lorehe and Paicao.[5]

Recognition of the need to conserve the remaining forest reserves in East Timor emerged during the post-war colonial period. Silva (1956:89) for example, called for the protection of primary forest across the region as a matter of urgent necessity, but it was during Indonesian rule after 1975 that the forested Pai cao mountains were designated as a conservation reserve (*Cagar alam*). In 1993 under a Biodiversity Action Plan for Indonesia, the region was recognised as one of 40 representative areas. Finally, during the UN Transitional Administration in East Timor (Untaet 1999–2002) the eastern part of Lautem was legislated as one of 17 so-called, Wild Protected Areas which sought to conserve its impressive environmental qualities and build a network of protected environmental zones across the country (UNTAET 19/2000) (see McWilliam 2003). The present formulation of the region as a high quality biodiverse land and seascape thus reinforces a lengthy history of state inscriptions and authorised recognition of core natural values.

Landscapes of resistance

The second important basis for declaring the National Park is its historical and symbolic significance in the heroic struggle for national independence and there are at least three influential factors that might be seen as central elements in this consideration. The first is the name itself. Nino Konis Santana (1958–1998) was a local Fataluku school teacher from Tutuala who joined the armed struggle and rose through the ranks to become the leader of Timorese armed resistance known as Falintil[6] between 1993 until his untimely death in 1998. The name of the Park commemorates his sacrifice and resonates with local aspirations for the recognition of Fataluku contributions to independence while speaking symbolically to the suffering of the nation as a whole.

The forested region with its tangled jungle and rugged karst topography also provided refuge and protection to another former guerrilla leader, Xanana Gusmao especially during the dark days of the late 1970s following the near destruction of the Falintil forces as a result of the Indonesian encirclement and bombing campaign in the Matebian mountains in neighbouring Baucau district (Budiarjo *et al.* 1984). Xanana and his supporters spent years living in the forests provisioned and protected by local Fataluku village sympathisers who risked reprisals and worse in their efforts to maintain the small groups of guerilla combatants. During the early 1980s while Xanana lived under secluded protection in the village of Mehara, he is credited with formulating the clandestine resistance strategy and the shift to a largely urban based civilian resistance struggle that ultimately proved unstoppable.

---

5   The lucrative trade in sandalwood that attracted Portuguese and other interests to Timor from the late sixteenth century by this time was much degraded and in 1926 export of the fragrant timber was officially banned. Sandalwood is a prominent endemic species in the park, particularly in the fallowed garden lands of the monsoon rainforest that covers much of the north east coast and hinterland. It is not a feature of the dense wet rainforests of the Paicao ranges.

6   From the Portuguese, Forças Amadas da Libertação Nacional de Timor-Leste.

The third and related factor is that the forested eastern region remained one of the key sites of continuing armed resistance against the Indonesian military throughout the 24 years of occupation (see also Pannell and O'Connor 2005). For this reason the area was known in the lingua franca, Tetun, as the *rai funu nafatin* ('the land of continuous war'). In this heroic historical perspective then the Park commemorates the forest as a landscape of warfare and struggle; one inscribed with markers and memories of conflict and suffering, hideouts and refuges, secret trails and ambush sites. As a 'landscape of memory' (Hviding and Bayliss Smith 2000) the Park offers a scenic living monument to national liberation and one which Fataluku residents recognise as their specific contribution to the 'historical record'.

Taken together these two influential guiding values, high biodiversity and resistance history, certainly find some support among the resident local Fataluku communities. Many have been active in the events it commemorates and make use of the bio-diversity it celebrates. But if recent resistance history represents an important episode in their lives, it is only one chapter in generations of connection and reproduction of the forested landscape. As a proposed conservation park with plans for ecotourism and restrictive protections over high value portions of the forest and reefs zones, many local people express a degree of ambivalence and anxiety over the direction of future government regulatory management.

Cultural landscapes and unsettling histories

One of the ironies of the values attributed to the National Park through its state authorised creation' is that while the region has long been recognised for its natural biodiversity, it is by no means a pristine wilderness or primary canopy rainforest. On the contrary, while pockets of primary rainforest survive in steep ravines or isolated peaks, the broader park reflects a highly enculturated mosaic of aged and long fallow regrowth, of former swidden gardens and settlement sites (McWilliam 2007). The forest floor is littered with the remnants of past occupation by former Fataluku households and communities. Old stone garden walls, aged coconut palms, crumbling evidence of house foundations as well as numerous massive stone walled fortifications on hilltops with their old stone graves and pottery shards providing rich evidence of extensive past occupation. In reality the now unsettled areas of the National Park and its forested interior owe more to long term processes of population displacement than any active processes of nature conservation.

The process of displacement has its origins in early twentieth century policies of the Portuguese Colonial Government that sought administrative enclosure by drawing the scattered forest Fataluku into more concentrated residential clusters.[7] The movement was a gradual one, disrupted by the brief Japanese occupation of the island during World War II, but eventually it saw dozens of small residential communities in the forested northern and southern zones of the park permanently relocated to the main roads or to main towns where they were closer to development services and state regulatory surveillance and where, for the most part, they still reside. Places like Loequeiro, Poros, Somoco, Bauro, Asalaino and Sepaerara now cluster along the Tutuala road, but previously occupied residential sites in the northern forested plateau. Similarly, the whole of the settled highlands of Nari (Planalto de Nari in the central north) were emptied out and its residents relocated to contemporary roadside settlements near the northern coast. They include: Pairara, Punu, Moro, Ipairira and Soekili. In the southern forested zone, relocated settlements include, Ira Ara, Lere Loho, Pehe Fitu, Lupuloho, Muapitine, Veterr, and Vero among others, and in many cases are the result

---

7   This heightened period of Portuguese military effort was part of a general campaign to extend their authority and effective control across the whole of the territory. It mirrored the kind of military-backed pacification wars promulgated by the Colonial Dutch government across the eastern islands at much the same time and directed to similar ends. We might view the displacement process of these and later Indonesian government programs, as part of a characteristic strategy of states to impose structures of surveillance and economic development as part of a territorial impulse for encompassment and control (see for example Scott 1998; Li 1999).

of later Indonesian resettlement policies that followed in the wake of military occupation as part of a wider campaign to distance the community from the subversive work of the guerilla resistance (Falintil).

The significance of this protracted history of displacement is not simply a distancing from food gardens, economic resources and some putative 'archive of past habitation and sociality' (Fairhead and Leach 1996:113), but very much a sense of separation from sites of economic livelihood, spiritual agency and moral authority (McWilliam 2007:173). Many Fataluku are keenly conscious of these processes of separation and attempt to overcome their loosening hold through active practices of re-engagement with ancestral sites and the manifold resources they contain. For the most part however, this has not led to a resettlement of ancestral places and arable lands, as the distance from markets, schools and other services as well as existing built investments, now make such a move for the majority, improbable and undesirable. Nevertheless, the significance of these lineage-based, ancestral land attachments remain a vital source of household and community identity and well being; an association that is sustained and reproduced through a range of strategic, practical and ritualised actions (McWilliam 2011).

Fataluku land attachments

There are multiple ways of appreciating Fataluku attachments to the deep forests and coasts of the Park area. For the purposes of illustration I would highlight two general domains of activity that provide some of the lived enactments of these Fataluku connections to their natural resources. The first and very prominent activity in which most people engage, is the gathering and harvesting of various wild growing foods that help sustain people's livelihoods and provide important and regular supplements to the seasonal harvesting of maize and other food staples cropped in nearby food gardens. In this regard there are multiple food resources available, from hunted game (especially deer, pigs, possums, monkeys, bats and snakes) and harvested marine foods (fish, shellfish, sea snakes, sea worms (*meci*), river shrimp, sea turtles,[8] turtle eggs), as well as seaweeds and the diverse marine species collected through evening practices of reef gleaning. The hunting of game also forms part of a ritualised corpus of practices associated with harvesting the natural bounty of the land, including the gathering in of staple grain crops, the production of distilled palm liquor or seasonal harvesting of fish aggregations (*api lere*) and other living marine food (e.g. *meci, Eunice virides*) (see Palmer and Carvahlo 2008).

The forest and its rich tropical plant growth also provide a whole range of tubers, fruits, leafy vegetables, wild honey, gums, resins, timber, rope, thatch and rattan. Use is made of the multiple resources derived from the wild growing arenga palm which produces the distinctive black roof thatch of Fataluku traditional houses (*fia*), as well as prodigious amounts of the milky syrup which is distilled into a strong clear liquor (*tua haraki*) and consumed with enthusiasm as a pastime and during ceremonial gatherings. The natural environment also provides a diverse pharmacology of plants, gums, fruits and barks that are widely used for their medicinal properties, and which are sought out as poultices for wounds and herbal remedies for all manner of identified illnesses (see McWilliam 2008). Medicinal plants represent a store of closely kept local knowledge with a deep experiential genealogy developed over generations of self reliance and self medication among Fataluku forest communities (see also Collins *et al.* 2007 for discussion of forest-based plant medicines used in Lautem to alleviate wounds and fevers among armed resistance fighters during the Indonesian military occupation when allopathic remedies were scarce). As a storehouse of food, the living environment of the national park ensures that Fataluku communities never suffer the draining privations of drought and food shortages experienced in other areas of Timor.

---

8   The coastal fringe of the NKSNP is a rich nesting ground for a range of sea turtle species that have been regularly captured and consumed. Turtle meat is smoked on racks over smouldering coals and savoured as a delicacy (see Edyvane *et al.* 2009).

A second important realm of Fataluku attachment to the natural domain is the enduring significance of spiritual connections to ancestral and non-human spirit agencies. Fataluku ideas of spirit and potency are an esoteric and complex arena to which the following comments offer only brief reflections. Suffice to say that for all Fataluku communities residing within the boundary of the National Park, the living environment is always one where the agency of spirits and their often unobserved influence has a potentially powerful impact on everyday life. For these reasons the process of accessing and utilising the natural domain and drawing on its resources and protection, is one that needs to be approached with a degree of respect, caution and clear purpose. Fataluku describe the power of the spirit world with the generic term, *téi* which expresses a complex range of qualities associated with the sacred but also ideas of power, danger and dread. Places and forms that are marked as *téi*, (*téinu* pl.) are generally placated through invocatory prayers, sacrificial offerings and sometimes simple avoidance and in so doing people seek to reproduce a range of mutually beneficial ritual relationships (see also Pannell 2006; McWilliam 2011).

For the majority of resident Fataluku, key associations with the landscape are shaped by their membership to specific named clans or more specifically, paternally ordered origin groups and lineages known as *Ratu*. There are dozens of named *Ratu* in the region and their dispersed membership convenes periodically in commensal gatherings around marriages, funerals and other clan ceremonies to celebrate their ties to one another. It is through the mythologies and narrative traditions of ancestral settlements and their trajectory of place-making across the land that the customary claims of *Ratu* and their constituent households are founded. In a recent study, Raimundo Mau, who works within the National Directorate of Forestry responsible for coordinating the development of the National Park, has noted that all of the Park's land and resources are traditionally divided into these clan-centric claims of ownership and connection (2010:53–54,62). Group names such as Cailoro Ratu, Renu Ratu, Puitical Ratu, Aca Cao Ratu, Lavera Ratu, Pair Ratu, Konu Ratu, Kati Ratu, Marapaki, Latu Loho, Vacumura, Masipan, Serelao and Naza Ratu and many others, are linked through complex mythologies and oral histories of residence and other forms of tangible and intangible connections to bounded areas of land and resources within the Park. They represent the culmination of generations of settlement, warfare, alliance and mobility across the region.

An important element that sustains Fataluku Ratu agnatic groupings is the close attention to, and ritual communication with the ancestors of the house and clan – known generically as the *calu ho papu* ('grandfathers' and 'great-grandfathers').[9] All ceremonial and life cycle events for example are simultaneously an opportunity to 'feed' (*fané*) the ancestors and invoke their blessings, notwithstanding a general Fataluku orientation to Catholicism. Ancestral protections and blessings form part of an intrinsic set of approaches among most Fataluku to healing and warding off illness, seeking protective blessings against misfortune, and giving thanks for successful harvests and good fortune.

In these cases people make use of purpose-built household shrines, known as *aca kaka* or *lafuru téi* (sacred hearth).[10] Most households maintain these separate structures which are derived from the paternal ancestors of the husband and are understood as an extension ultimately of the earliest origin hearth of the named clan. In this respect, house shrines represent the domestic expression of an extensive network of spatialised spiritual connections that link living members of the group with each other and to their ancestral origins (McWilliam 2011:12). Through sacrifice, prayer invocations and commensality at venerated places across the landscape, members of the group reproduce the memorialised record of ancestral settlement histories and their emplaced connections and entitlements.

---

9   Fataluku recognise at least 5 levels of upper generational agnates, Palu, Calu, Papu, Cuci and Macua [F, FF, FFF, FFFF, FFFFF]. These names may be publically known but beyond that they are said to be *téi* and may not be voiced. I have heard it said that the collective ancestral body may be referred to as the *arafura*.

10   The sacred hearth forms a separate boxed structure with sand and ash base on which are placed three hearth stones and several small forked sticks that are used in the cooking of sacrificial offerings and accompanying prayers.

According to conventional Fataluku ideas, the efficacy and power of spirit agency increases as one moves away from the contemporary domestic space to more spatially, and hence, temporally distant sites of emplaced spirit. Here at various material markers of the ancestral presence – first landing sites, ancestral footsteps (*ia mari tuliya*), old graves (*calu luturu*) and abandoned settlements (*lata paru*), altar posts commemorating mythic events (*ete uru ha'a*) or emplaced land spirits (*têi/muacawa*) – the enactment of ritual sacrifice engages a wider ancestral field and for this reason requires a higher level of ritual expertise to perform the necessary invocations. Within the boundaries of the park there are regular annual invocations undertaken at communal sites of origin to invoke the blessings of clan ancestors and participate in the commensal rituals of sociality that reconfirm group identity and solidarity. Thus, through the multiple, overlapping and criss-crossing network of connections to ritual sites across the memorialised Fataluku landscapes of the National Park, contemporary households and communities reproduce their cultural heritage through active performative engagements with their past. Collectively they provide exemplary illustrations of the kinds of perspectives highlighted in the complementary representations cultural heritage offered by Hoskins and Zerner.

Park futures

The cultural and economic significance of the National Park to contemporary Fataluku residents demonstrates in part some of the influential reasons why the decision was made to establish a multi-purpose conservation park. It is certainly the case that the formation of the park had its beginnings in part via the progressive environmental policies promoted and implemented during the Transitional Administration of the United Nations in East Timor [Untaet 1999–2002]. But the question remains to what extent the ideas and practices reproduced by local Fataluku in their everyday livelihoods and smallholder projects of economic engagement, can be accommodated in the management arrangements of the Park. Here things become rather less clear.

Table 1 shows the IUCN vision of what Conservation Parks might become. It demonstrates the shift that has occurred in the management of public reserves and conservation zones: one that moves from a top down people-averse system to more a participatory, people-inclusive model that acknowledges the multiple interests of stakeholder communities. The legislative framework for the Nino Konis Santana Park is very much consistent with this evolving appreciation of new ways to manage conservation.

Table 1. IUCN contrasting approaches to park organisation.

Command Approach	Consultative Approach
Planned and managed against people	Run with, and in some cases by local people
Run by Central Government	Run by many partners
Set aside for conservation	Run also with social and economic objectives
Managed without regard to local community	Managed to help meet needs of local people
Developed separately	Planned as part of national, regional and International systems
Managed as islands	Developed as networks – strictly protected areas buffered and linked by green corridors
Established mainly for scientific protection	Often set up for scientific, economic and cultural reasons
Managed mainly for visitors and tourists	Managed with local people more in mind
Managed reactively within short time frames	Managed adaptively with a long term perspective.
About protection	Also about restoration and rehabilitation
Viewed primarily as a national asset	Viewed as a community asset
Viewed exclusively as a national concern	Viewed also as an international concern.

Source: Management Guidelines for IUCN Category V Protected Areas. Series 9, 2002:14.

But while the decision of the Timor-Leste Government to promote this kind of park is certainly applauded, especially the inclusive participatory philosophy it expresses, the reality is that since its formal public declaration in 2008, very little has happened in terms of the organisational development of the Park. The Park remains to a significant degree a work in progress, and little more than a 'paper park.'[11] Plans for the establishment of an advisory board and management committees, inter-governmental coordination and agreements on the official conservation zoning of the various multi-purpose uses of the Park, or indeed concerted and continuing research efforts and consultation with local communities remain at a very preliminary stage.[12]

The reasons for this inertia are not entirely clear but are likely to be related to competing development priorities in the fledgling nation which struggles to overcome widespread poverty. There is also a general lack of institutional capacity and insufficient funding allocated to the National Directorate of Forestry, the agency charged with overall supervision and coordination of the National Park. One might also add that there is a marked absence of any community demand for action on this front reflecting the general point that the momentum to establish the conservation zone derives from interests external to the Park environment itself. In this sense one can speak of the Park as doubly imagined, a concept that still really only exists on paper as a symbolic commemoration of history, and perhaps a well meaning gesture towards environmental conservation and the economic possibilities that might develop in its wake.

A second issue that remains largely unresolved in terms of park management, is the question of formal rights and entitlements of communities with claims and connections to areas within the conservation zone. This is especially relevant to land ownership and the division of tenure rights between the State and local Fataluku groups who assert forms of customary attachment. The question is complicated by chronic delays in the formal acceptance of a national land law which will provide the necessary legal clarity and definitive regulatory framework for establishing entitlements to land. The decision to delay the current iteration of the new draft law until after the 2012 National Elections only extends that uncertainty into the future.[13] But it means that, in this context, opportunities for local Fataluku households and communities to assert claims over customarily held lands are very much constrained (see also McWilliam 2007). While in most cases, cultivated land around existing settlements and residential land is likely to be recognised and converted into clear legal entitlements, this is by no means clear for the unsettled former ancestral lands in the distant canopy forest areas, much of which is still designated as a 'Wild Protected Area' under former UNTAET legislation with a range of restrictive access and use provisions (McWilliam 2004; UNTAET Regulation19/2000).[14]

Government Land and Property (DTP) staff often draw a distinction between *Lei* ('law and legal entitlements') and *Kultura* ('cultural interests'). The former provides secure entitlements subject to compensation for loss and the various protections accorded legally constituted rights. The latter represents a weaker form of negotiable customary interests subject to public consultation that may or may not attract binding obligations. If history is any guide, 'cultural interests' (*Kultura*) and related customary claims are often unsuccessful in the face of state regulatory simplifications,

---

11   The IUCN popularised the phrase 'paper park', which is generally held to be a largely unmanaged protected area or one where the threat of degradation continues unabated (IUCN 1999). http://www.worldwildlife.org/what/globalmarkets/forests/WWFBinaryitem7370.pdf [accessed: 29/4/12].

12   An exception to this general statement is the thesis of Raimundo Mau (2010) who has developed a detailed zonation model for the Nino Konis Santana Park that seeks to address the different and overlapping environmental and community values of the region.

13   In February 2012, The President of Timor-Leste, Jose Ramos Horta, vetoed the passage of the parliamentary land law legislation on equity grounds.

14   My understanding is that the process of zoning within the broader boundaries of the Park will eventually supercede the need for retaining the Wild Protected Area legislation.

which, as Zerner has observed, often work to deny or subsume any formal relation between what he calls the performative modalities of custom and the tightly scripted letter of the law. The ascription of *Kultura* in this context can also be read as a devaluation of the authority of local heritage in the face of the rigorous historical determinants of law authorised by the state.

That being the case, one sense of the prospective directions that might be taken by the government in this regard can be read in one of the articles (Chapter V) of the 2012 iteration of the draft National Land Law.[15] Article 24 makes provision for two categories of recognition for community based land ownership; namely, *Community Protection Zones* and *Community Property*. The expressed purpose of these provisions is to safeguard the common interests of local communities and the lands and natural resources on which they depend. Drawing on elements of Mozambique land law, the category of community protection is specifically designed to avoid land speculation or unscrupulous economic activity by third parties, and is meant to ensure that significant development is always preceded by consultations that work to the benefit of the local community. According to the wording of the draft, the provision is intended to respect gender equity and 'socio-cultural sustainability' in use of natural resources. Importantly the provision also allows privatised titles of land within the boundary of the community protection zone to remain unaffected.

The second possibility of a Community Property provision may be even more significant. Under the draft law the definition of this category of land title refers to any immovable property which is of common and shared use. Potentially, the customary community in this case could secure an ownership title of the collective ancestral property, issued in its name, which thenceforth would be inalienable and unseizable (Draft Law footnote 7).

These progressive provisions within the draft land law indicate that lawmakers and politicians in Timor-Leste may be willing to go some way in recognizing customary claims and connections to ancestral lands that persist in varying degrees among local communities. The provisions would certainly accommodate and provide a much stronger legal protection to the kinds of cultural heritage practices and attachments that are sustained among Fataluku ritual communities within the Nino Konis Santana National Park. What remains unclear in these draft provisions is how they might work in practice and whether in fact they will ever be implemented in law. In a recent World Bank options paper, Fitzpatrick (2010) has provided a detailed analysis of the implications and unspoken complications inferred by the provisions of article 24, highlighting a range of risks and pitfalls of regulation, not the least of which include the potential for communal disputes and disenfranchisement, definitional problems and governmental capacity to oversee the process (see also Fitzpatrick *et al.* 2012). As Dove has observed in a commentary on the vulnerability of forest communities: '[t]he problem is not that (they) are poor in resources … but that they are politically weak; and the problem is not that the forest is environmentally fragile, but that it is politically marginal' (2012:191). The comparatively low and declining population numbers within the NKSNP, make this relation even more problematic.

For the majority of Fataluku households and communities resident within the National Park broader issues of park management remain secondary to the task of securing everyday livelihoods and the maintenance of social obligations. Reproduction of these relationships including the religious anchoring of ancestral blessings will continue to inform Fataluku decision making and orientation towards the region now encompassed within the National Park. The challenge, for the Timorese government is to find practical and supportive ways to accommodate the living cultural heritage values of locally resident Fataluku communities within a governance framework

---

15   See Draft Land Law Version 3, http://timorlestelandlaw.blogspot.com/2010/02/timor-leste-draft-law-for-special.html If the law is ever ratified it is likely that there would need to be an implementing regulation covering its application (see Fitzpatrick 2010:1).

for the conservation of environmental values and its significance as a commemorative symbol of nationalism and the resistance struggle. Five years after the legislative creation of the Park, this remains the pressing challenge.

Acknowledgements

Fieldwork in East Timor has been made possible over a number of visits under the auspices of The Australian National University and an Australian Research Council Grant entitled "Waiting for law, land, custom and legal regulation in East Timor". I extend my appreciation to Timorese colleagues and numerous local Fataluku advisors for their assistance and generosity. I also wish to thank two anonymous reviewers for critical comment and suggestions on the paper.

References

Budiarjo, C. and Liem S.L. 1984. *The war against East Timor.*Zed Books, London.

Cardoso, J. 1933. Notas Florestais de Colónia de Timor. *Boletim Geral das Colónias* 9.

Collins, S. X., Martins, A., Mitchell, A., Teshome and Arnason, J. 2007. Fataluku medicinal ethnobotany and the East Timorese military resistance. *Journal of Ethnobiology and Ethnomedicine* 3(5).

De Silva, H.L.E. 1956. *Timor e o cultura do café.* Memórias Série de Agronomia Tropical I, Ministerios do Ultramar.

Dove, M. 2012. A political-ecological heritage of resource contest and conflict. In: Daly, P. and Winter, E. (eds), *Routledge Handbook of Heritage in Asia*, pp. 182–198. Routledge, London.

Edyvane, K., McWilliam, A., Quintas, J., Turner, A., Penny, S., Teixeira, I., Pereira, C., Tibirica, Y. and Birtles, A. 2009. *Coastal and marine ecotourism values, issues and opportunities on the north coast of Timor Leste – final report.* Ministry of Agriculture and Fisheries, National Directorate of Tourism, Government of Timor Leste.

Fairhead, J. and Leach, M. 1996. *Misreading the African landscape: Society and ecology in a forest-savanna mosaic.* Cambridge University Press, Cambridge.

Fitzpatrick, D. 2010. Policy options for regulating community property and community protection zones in Timor-Leste, Draft Report, World Bank.

Fitzpatrick, D., McWilliam, A.R. and Barnes, S. 2012. *Property and Social Resilience in Times of Conflict: Land, custom and law in East Timor.* Ashgate, Farnham UK.

Harvey, D.C. 2001. Heritage pasts and heritage presents: temporality, meaning and the scope of heritage studies. *International Journal of Heritage Studies* 7(4):319–338.

Hoskins, J. 1993. *The play of time: Kodi perspectives on calendars, history and exchange.* California University Press, California.

Hviding, E. and Bayliss-Smith, T. 2000. *Islands of rainforest: Agro-forestry, logging and eco-tourism in Solomon Islands.* Aldershot, Burlington.

IUCN. 1999. *Threats to forest protected areas: Summary of a survey of 10 countries.* A research report from IUCN The World Conservation Union for the World Bank/WWF Alliance for Forest Conservation and Sustainable Use. http://www.worldwildlife.org/what/globalmarkets/forests/WWFBinaryitem7370.pdf [accessed: 29/4/12].

Li, T.M. (ed.) 1999. *Transforming the Indonesian Uplands:Marginality, power and production.* Harwood Academic Publishers, Singapore.

Lowenthal, D. 1985. *The past is a foreign country.* Cambridge University Press, Cambridge.

Lowenthal, D. 1986. Fabricating heritage. *History and Memory* 10(1):5–24.

McWilliam, A. R. 2003. New beginnings in East Timor forestry management. *Journal of Southeast Asian Studies* 34(2)307–327.

McWilliam, A. R. 2006. Fataluku forest tenures and the Conis Santana National Park (East Timor). In: Reuter, T. (ed.), *Sharing the earth, dividing the land: Territorial categories and institutions in the Austronesian world,* pp.253–75. ANU E-Press, Canberra.

McWilliam, A. R. 2007. Customary claims and the public interest: On Fataluku resource entitlements in Lautem. In: Kingsbury, D. and Leach, M. (eds), *East Timor: Beyond independence,* pp.168–75. Monash Asia Institute Press, Melbourne.

McWilliam, A. R. 2008. Fataluku healing and cultural resilience in East Timor. *Ethnos* 73(2): 217–240.

McWilliam, A. R. 2011. Fataluku living landscapes. In: McWilliam A. R. and Traube, E.G. (eds), *Land and life in Timor Leste: Ethnographic Essays*, pp.61–86. ANU E-Press, Canberra.

MAFF. 2006. *Draft proposal for Declaration Nino Konis Santana National Park.* Ministry of Agriculture Forestry and Fisheries, Timor-Leste.

Mau, R. 2010. *Ecosystem and community based model for Zonation in Nino Konis Santana National Park, Timor-Leste.* Unpublished thesis, Institute Pertanian Bogor.

O'Connor, S. 2003. Nine new painted rock art sites from East Timor in the context of the Western Pacific region. *Asian Perspectives* 42:96–128.

O'Connor, S. and Pannell, S. 2006. Cultural heritage and the Nino Conis Santana National Park, Timor Leste: A preliminary survey. Unpublished manuscript.

Pannell, S. 2006. Welcome to the Hotel Tutuala: Fataluku accounts of going places in an immobile world. *The Asia Pacific Journal of Anthropology* 7(3)203–219.

Pannell, S. and O'Connor, S. 2005. Toward a cultural topography of cave use in East Timor: A preliminary study. *Asian Perspectives* 44(1):193–206.

Palmer, L. and de Carvalho, D. 2008. National building and resource management: The politics of nature in Timor Leste. *Geoforum* 39:1321–1332.

Scott, J. C. 1998. *Seeing like a State: How certain schemes to improve the human condition have failed.* Yale University Press, New Haven.

Zerner, C. A. 2003. Introduction - moving translations: poetics, performance and property in Indonesian and Malaysia. In: Zerner, C. (ed.), *Culture and the question of rights: forests, coasts and seas in Southeast Asia,* pp. 1–23. Duke University Press, Durham.

14

The dynamics of culture and nature in a 'protected' Fataluku landscape

Sue O'Connor, The Australian National University, Canberra, Australia

Sandra Pannell, James Cook University, Cairns, Australia

Sally Brockwell, The Australian National University, Canberra, Australia

Introduction

In his work on 'Landscape and Memory', the historian Simon Schama reminds us that 'landscape is the work of the mind. Its scenery is built up as much from strata of memory as from layers of rock' (1995:7). Focusing upon those elemental physical features evident in the Western landscape tradition, namely wood, water, and rock, Schama examines the layers of social memory and visual representation, to reveal the many historical associations and varied cultural meanings of 'natural' objects and places. In the history of protected area management, most notably exemplified by the 1972 World Heritage Convention, the idea of 'heritage' often mediates the expression of the nature-culture relationship that Schama refers to here (see Pannell 2006a).

Indeed, in recent years, we have seen an increasing number of articles exploring 'the intersections of culture, politics, performance and nature' in customary relationships to land and sea in Indigenous communities in Southeast Asia (Zerner 2003a:xi; Pannell 2006a; also see papers in Zerner 2003b). At the heart of these papers lies the question of rights and the issue of how Indigenous and other communities can most effectively engage with the processes and institutions affecting their access and use of areas they regard as their ancestrally given homelands. In the discourse of protected area management, one of the more pressing issues is also about the 'imaginative and material simplifications imposed on the landscape of customary tenure' (Zerner 2003a:12) and upon local people and their practices. As the work of Anna Lowenhaupt Tsing among Meratus Dayaks demonstrated, their disenfranchisement from their forest lands and livelihoods, in part, rested upon the characterisation of these lands as 'wild', and thus free of any prior property claims, making them 'fair game for commercial exploitation' (2003:30). As Pannell has argued, similar simplifications and dichotomous classifications came into play in the fashioning and invocation of '*sasi*' as a traditional, Moluccan-based 'sustainable resource management strategy' (Pannell 1997). In the 1990s, '*sasi*' was vaunted by environmentalists, government agents, and academics alike not only as a magic cure to the ills ailing eastern Indonesia's declining fisheries, but also as means to address local equity and social justice issues (see Bailey and Zerner 1992; Zerner 1994). As Charles Zerner observed, it didn't seem to matter that the representation of *sasi* as an 'intentionally conservationist institution' was inaccurate (1994:104). What really mattered, however, was the impact of this fiction on the lives of local people, when they were perceived to fail the 'ecologically sustainable' parameters imposed by this reworking of *sasi*.

With the lessons of *sasi* in mind, in this paper we adopt a similar perspective to Schama and Zerner (as well as others) in examining the layering and porosity of 'nature' and 'culture' in the emplaced cultural practices of Fataluku people in Timor-Leste. Informed by Tsing's comments about the need to "take very seriously other ways of understanding and making 'rights' and 'forests'" (2003:27), we also look at the problems of accommodating this world view, and the customary rights and interests associated with it, within the structural frameworks of the nation-state and within newly introduced heritage legislation, where regulations governing land use are predicated not only on the dichotomization of nature and culture, but also upon the separation of these concepts, and often times, upon the disavowal of one at the expense of the other (O'Connor *et al.* 2011a). The tensions between the new structures of nationhood and Fataluku understandings of the world also reproduce a higher level duality that exists in Timor-Leste, namely, the division between customary rights and traditional lore and that of parliamentary governance and civil law.

Since the formal achievement of independence in 2002, the government of Timor-Leste has embraced the democratic ideal of statehood for the wider public good and in some areas of the country the government has moved towards the gazettal of land for this purpose (see McWilliam 2011). In 2007, the new nation's first National Park (Figure 1) was declared over an area of some 68,000 ha of land and 55,600 ha of seascape, situated at the eastern end of the island, in the Lautem District. Named after a former Falintil commander, who was born within its boundaries, the Nino Konis Santana National Park commemorates his service and that of the many East Timorese resistance fighters who lived in the forests during the Indonesian occupation, in the period from 1975 until 1999. While the Nino Konis Santana National Park was gazetted primarily on the basis of its 'natural' values (see Whistler 2001; Trainor *et al.* 2003; McWilliam 2007a; O'Connor *et al.* 2011a), the Park is home to more than fifteen thousand Fataluku-speaking people, who regard many of the identified 'natural' features of the Park as socially significant places within a wider Fataluku cultural landscape. These sites and spaces range from ancestral settlements, graves and shrines, through to geographic features, such as rock outcrops and freshwater springs, which are mythologically significant or empowered by supernatural or non-human beings; and not least, the forest as whole, which is not only used to extract an extensive variety of resources, but it is also used as the location for the performance of rituals and ceremonies.

The Park also contains a rich and diverse array of evidence of human occupation, dating from ancient prehistoric to historic times. This includes the earliest evidence of modern human colonisation in the Wallacean Archipelago (O'Connor *et al.* 2002; O'Connor 2007; O'Connor *et al.* 2010a), the world's earliest evidence for systematic pelagic fishing and the manufacture of fish hooks (O'Connor *et al.* 2011b), world-class rock art sites, including the oldest known engraving site in Southeast Asia (O'Connor 2003; Lape *et al.* 2007; O'Connor *et al.* 2010b), traditional fortified settlements (Lape 2006), excellent examples of colonial architecture, dating from the Portuguese period of occupation, and sites associated with the Japanese occupation of the island during the later part of World War II. Some of these sites have national and even international heritage significance due to their antiquity, rarity, research and/or aesthetic values, but many also hold intense local significance for Fataluku people due to their mythological status, ancestral connections and/or contemporary cultural associations. As the former home to many Falantil freedom fighters, including President Xanana Gusmão, during the period 1975–1999, the Park can also be identified as having considerable local and national value as a landscape of political resistance and social liberation (Pannell and O'Connor 2005).

Contrary to the ideas underpinning environmental preservation and the creation of nature-based protected areas (see Luke 1995; Katz 1998), to Fataluku people, land which is unoccupied and cultural sites which are not regularly attended are regarded as dangerous places, as the relationship between the human and the non-human world requires ongoing physical interaction and regular ritual communication. In common with so many Indigenous groups around the world, Fataluku people also believe that their cultural practices, rituals and familiarity with the non-human beings which co-occupy their lands are essential to the reproduction of the 'natural' world, now supposedly protected by the declaration of the Nino Konis Santana National Park. This has led to a desire on the part of some Fataluku clan groups now living in the roadside settlement of Tutuala to return to their ancestral homelands within the Park (Figure 1). These are the families, or descendants of families, who were compulsorily resettled by the administration during the Portuguese and Indonesian occupation of Timor-Leste. Their desire to reoccupy ancestral homelands, like so many other Fataluku cultural practices which were previously focused upon the forested region now within the Park, is not regarded by the Ministry of Agriculture and Fisheries (MAF) (formerly Ministry of Agriculture, Forestry and Fisheries, MAFF) as compatible with the Park's existence. In this regard, the MAF aims to curtail residential and agricultural expansion into forested areas, while it currently limits the hunting of forest animals. We argue here that the very policies which exclude local people from occupation and use of the resources within the forested areas of the Park are putting at risk both the cultural sites and the natural resources that they seek to protect.

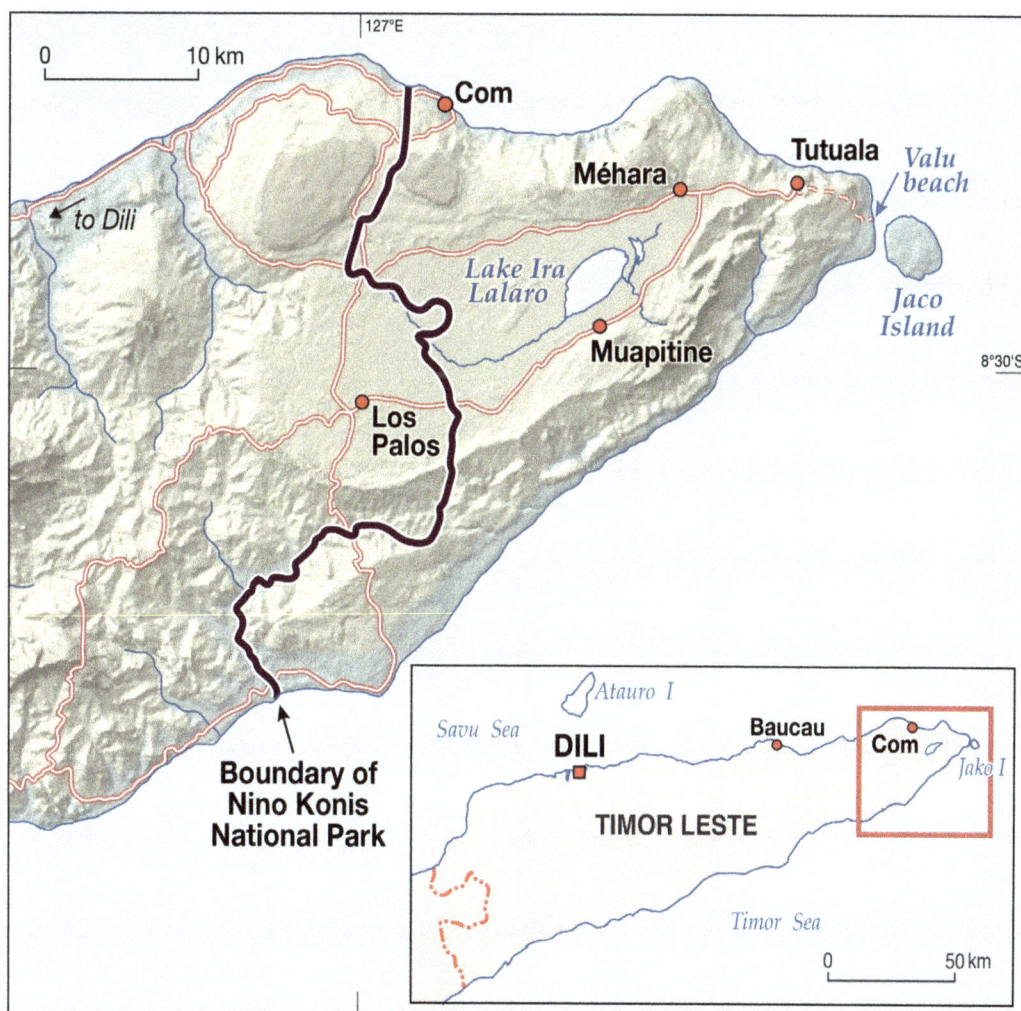

Figure 1. Timor-Leste, Nino Konis Santana National Park and locations mentioned in the text.

Source: CartoGIS, College of Asia and the Pacific, ANU.

Fataluku people: Living in the Park

With very few exceptions, the people living within the Park are speakers of Fataluku, a Papuan language. Although the antiquity of Papuan languages in Timor and the relationship of Fataluku to the other Papuan languages spoken in Timor-Leste, being Bunaq and Makasai, is still unclear, it is widely agreed that the Papuan languages are unrelated to the Austronesian languages spoken throughout West Timor and most of the western half of Timor-Leste (Figure 1) (McWilliam 2007b; Schapper 2012).

For the Fataluku residents who live in and around the six villages (*suco*) now encompassed within the Park boundaries, many of the natural resources of the forest constitute important elements in local subsistence strategies. These communities have few opportunities to participate in the cash economy. For example, 87.7% of household economic activity centres on food crops and livestock farming, for the most part, based upon shifting mixed crop swidden gardens and the raising of domestic livestock, such as chickens, cattle, water buffalo, pigs and goats (Mau 2010:1) (Figure 2). The daily diet is largely vegetarian, relying on staple crops obtained from swidden gardens, such as maize, cassava, and various kinds of yams and beans (Pannell and O'Connor 2010).

Figure 2. Garden clearing within the forest in the Nino Konis Santana National Park.

Source: Sue O'Connor.

Hunting is an integral part of the local economy. People rely on the forests for *cuscus* (a species of possum), monkeys and feral animals, such as deer and wild pigs, to supplement their protein supplies, and these 'wild' resources are particularly important in times of seasonal and politically triggered food shortages (Pannell and O'Connor 2010). Apart from some species of birds (Trainor *et al.* 2003), bats and small murids, none of the animals that inhabit the forests are endemic to the island of Timor; they are all human introductions brought from areas to the east or west during late prehistoric or historic times (O'Connor and Aplin 2007). Many other kinds of resources, aside from wild game, are acquired from within the Park including plant foods such as nuts, fruits and tubers, medicinal species (Collins *et al.* 2007), bamboo, canoe trees, and wood and fibrous plant materials used for building and for making the diversity of items of material culture used by households every day. In fact, up until the gazettal of the Park in 2007, resources used by Fataluku people on a daily basis were unofficially, but widely, acknowledged by the government's representatives at the regional and district level as being the traditional property of local Fataluku people, in much the same way in which the area of the Park is said by Fataluku people to include their customary lands and waters.

In view of the distance between most of the permanent settlements in the Park and the coastline, few Fataluku communities regularly engage in coastal subsistence activities or rely wholly on marine resources as food staples, although a diverse range of coastal resources from the beach and the sea are exploited on a seasonal or episodic basis (Figure 3). The antiquity of fishing in this region of Timor-Leste, however, is demonstrated by the fish bone assemblage from *Jerimalai* shelter, located within the Park boundaries, which dates back to 42,000 years ago and which contains a broad spectrum of species including pelagic and reef fish (O'Connor *et al.* 2011b).

The few groups of Fataluku people who specialise in fishing have boats and temporary shelters located in the Park at Valu Beach, at the eastern point of Timor island (Figure 4). Today this activity seems to be primarily undertaken by a local fishing co-operative, whose members catch and cook fish for the tourists who drive from Dili to spend the weekend camped on the beach, or stay in the nearby eco-resort, which was built shortly before the Park was gazetted.

Figure 3. Children collecting and processing sea urchins at a beach near Com within the Park.

Source: Sue O'Connor.

A number of other marine resources are collected intensively on a seasonal schedule from the coastal zone and the seas contained with the Park. At times of extreme low tide, whole families camp overnight on the beach and comb the reef using kerosene lanterns or torches to collect the multitude of edible marine creatures. In February and March, the sea worms (*mechi*) spawn. This event lasts only two days in each month but is a time when the local communities come together (see also Palmer and Carvalho 2008).

Figure 4. Valu Beach. The boats of the fisherman are kept at the back of the beach and storage racks and caverns for their possessions are made in the limestone caves at the back of the beach.

Source: Sue O'Connor.

Green sea turtles come onshore at Valu and the adjacent beaches to lay their eggs seasonally. Prior to the establishment of the Park, some families would camp on the beaches when this event was imminent and keep watch throughout the night in order to collect both the eggs and capture the nesting turtles. Hunting of turtles and the collection of turtle eggs is now prohibited under the rules governing use of the Park.

This all too brief description of Fataluku livelihood needs and ecological knowledge regarding the forests and seas now encompassed by the National Park, in part, gives some idea of the social and economic significance of this landscape to local communities, located in and around the protected area. However, as we discuss in the following sections, the significance of the Park area to Fataluku people extends beyond their material use of forest and sea-based resources.

The socio-political dynamics of documenting heritage values in the Nino Konis Santana National Park

The Nino Konis Santana National Park is administered by the 'Protected Areas and National Parks, National Directorate of Forestry', within the MAF. It includes approximately 300 km2 of lowland tropical and monsoon forest surrounding the Paichao mountain range – one of the largest continuous tracts of forest on the island of Timor. The Park also includes coral reef areas surrounding Jaco Island, which have been identified as possessing high biodiversity values (Edyvane *et al.* 2009a). It was largely for the values of the reef, the forest and its birdlife that the Park was declared. However, as noted above, the Park is a far cry from a 'pristine block of natural heritage values and primary forest' (McWilliam 2007a:168). As previously mentioned, the Fataluku communities resident within the Park depend on its land and resources for their ongoing livelihood. The Park also encloses an existing cultural landscape of Fataluku named places and ancestral spaces (McWilliam 2011; O'Connor *et al.* 2011a). Indeed, as McWilliam (2007a:168) notes, 'ironically much of the forested area is composed of former swidden gardens and settlement sites. Its existence as a contemporary canopy forest is, to a significant degree, the result of the reluctant disengagement of local Fataluku farming communities in response to external pressures applied by successive colonial governments, especially in the form of re-settlement policies and restrictive security arrangements'.

The Nino Konis Santana National Park has been identified as a 'Category V Protected Landscape/ Seascape'. According to IUCN's (World Conservation Union) Guidelines for Protected Area Management Categories, Category V Protected Landscape/Seascapes are defined as an 'area of land, with coast and sea as appropriate, where the interaction of people and nature over time has produced an area of distinct character with significant aesthetic, ecological and/or cultural value ... ' (IUCN 1994:22). Reflecting new ways of imagining forests and seas, beyond the historical classification of them as 'natural' or 'wild' spaces (see Tsing 2003), and trying to accommodate human uses, key management objectives for 'Category V Protected Landscapes/Seascapes' include the maintenance of the 'harmonious interaction of nature and culture', providing support for 'lifestyles and economic activities which are in harmony with nature' and the 'preservation of the social and cultural fabric of the communities concerned' (IUCN 1994:22). In this respect the IUCN 'Protected Landscape/Seascape' category has much in common with the World Heritage category of 'Cultural Landscapes' (UNESCO 2005:14) and thus the Park should be able to accommodate the full spectrum of cultural activities undertaken by local communities.

Although there have been a number of surveys within the Park, conducted under the auspices of the MAF (e.g. Edyvane *et al.* 2009a, 2009b), they have primarily aimed at assessing the natural values and biodiversity of the Park, and they have only fleetingly dealt with cultural sites; with a focus mainly on churches, built during the Portuguese era (Edyvane *et al.* 2009b:41–49). Archaeological sites are not mentioned at all, although 'Archaeology' is incorporated as a heading under 'History and Development', in an example of a protected area management plan for the Park (Edyvane *et al.* 2009a:95). To date, no systematic cultural mapping has been undertaken with local communities within the Park. However, as noted above, archaeological and anthropological research by O'Connor, Lape, McWilliam, Pannell, Spriggs, and Veth, undertaken between 2000 and 2005, has identified the area as having a range of outstanding cultural heritage values at the local, national and international level (cf. O'Connor 2003; Spriggs *et al.* 2003; O'Connor and Veth 2005; Pannell and O'Connor 2005; Veth *et al.* 2005; Lape 2006; McWilliam 2006;, O'Connor 2006; O'Connor and Pannell 2006; Pannell 2006b; O'Connor and Oliveira 2007;O'Connor *et al.* 2010a and 2010b; O'Connor *et al.* 2011a and b).

In 2006, two of the authors of this paper, O'Connor (an archaeologist) and Pannell (an anthropologist), were commissioned by the NSW Department of Environment and Conservation (DEC NSW), in partnership with the then Ministry of Agriculture, Forestry and Fisheries (MAFF), to undertake a desktop study of cultural heritage places within the area under consideration for inclusion within the Park.[1] This study was to be provided by DEC to the MAFF

---

1   The boundaries on the map provided to O'Connor and Pannell by DEC NSW officers prior to the desktop were provisional, however, the boundaries of the Park as gazetted were not greatly different from the boundaries as provisionally mapped and the sites we discuss here are within the NCSNP boundaries as gazetted.

in 'East Timor' as a preliminary background document on the cultural heritage values of this area. The DEC consultancy brief to the authors stated that the desktop study was to be only the first stage of a cultural heritage survey, and would be followed by a field-based study. Although the desktop report produced by O'Connor and Pannell (2006) was based on archaeological and anthropological field work and research conducted over many months and years in this region, and while informed by 'the knowledge, memory or interest of the local informants involved' (O'Connor and Pannell 2006:5), this research was, largely driven by the context of specific academic research agendas. As we made clear in the report to DEC, the archaeological surveys undertaken were not 'designed on the basis of 'sampling' of landscape units or total area' and they were not 'in any way systematic' (O'Connor and Pannell 2006:5). In a similar fashion, the anthropological research undertaken with Fataluku people in this area by no means exhausted the tangible and intangible cultural heritage values of the landscape within the Park.

Basically, archaeological research involved community consultation in areas likely to contain the types of sites of interest to archaeologists. In this respect, we would describe caves, rockshelters and fortified sites to those members of the local Fataluku community who had been nominated to work with us. These people usually included representatives of the state, nominated by the *chefe de suco* ('village head'), and representatives of the landowning *ratu*,[2] nominated by senior *ratu* customary law and ritual leaders (also referred to here as the 'lord of the land'). Our Fataluku counterparts then 'identified sites [being caves, rockshelters and fortified sites] with which they were familiar and which they thought might match our description, and which were regarded by them as appropriate for us to record' (O'Connor and Pannell 2006:5). Having established relationships with particular individuals/landowning *ratu* members in areas with identified archaeological research potential, the physical *loci* of our research was largely confined to these areas. Our report made clear that the research had been focused on the sub-district (*posto*) of Tutuala, the village (*suco*) of Méhara and its associated hamlets (*aldeia*), and upon the landscape surrounding the harbour-side village of Com (*suco*). Thus, while some areas were extremely well surveyed and areas of high cultural values known to *ratu* members were documented, other areas of the Park were not even visited and, as such, community consultation was not undertaken in respect of these particular areas. The report and an electronic site register, containing site locations, descriptions and informant contacts for over 200 sites was given to DEC NSW in 2006. The 2006 report advocates that, 'further community-based research on the cultural values of the Nino Conis Santana National Park is required and strongly recommended' (2006:4). It also recommended that the Ministry (MAFF) undertake consultation and engage in a partnership with the Office of the Secretary of State of Culture within the Ministry of Education (then 'Education, Culture, Youth and Sport'), whose officers had independently recorded a number of cultural heritage sites within the area identified for the Park, and who held a site register dating from the period of Indonesian governance, when their Ministry was established.

Correspondence with DEC officers informed us that the field survey which was to follow the desktop study would be delayed until late in 2006 due to the pressures and work commitments of the Timor-Leste MAFF staff but that it would follow later in the year. Fieldwork was then delayed by political turmoil in 2006 and 2007. To the best of our knowledge, no cultural heritage surveys or cultural landscape mapping exercise has yet been undertaken by the Ministry, or under their remit since this time, apart from a five day survey by Edyvane and McWilliam in 2008, which did not extend east of Com (Edyvane *et al.* 2009b).

---

2   McWilliam (2011:65) defines *ratu* as 'a dispersed, exogamous, paternal 'house of origin' and as the 'fundamental social institution that informs the organization of Fataluku society'.

In 2009 O'Connor, Aplin, Dos Santos and St Pierre carried out archaeological surveys aimed at locating palaeofaunal deposits and speleothems which might provide information on biological extinctions and past rainfall regimes in this part of Timor-Leste. In the course of this reconnaissance, a number of ancient petroglyphs were identified and documented within *Lené Hara* cave (O'Connor *et al.* 2010b). O'Connor and McWilliam also undertook a brief forest survey to locate stone walled fortified settlements in the inland, densely forested areas of the Park. The 2009 survey was carried out under a research permit from the MAF (previously MAFF), but it was not instigated by that Ministry.

The layering and porosity of nature and culture in Fataluku landscapes

Of the 200 plus sites recorded to date within the Park, most have contemporary socio-cultural significance to Fataluku people (O'Connor and Pannell 2006). They include: a wide range of ritual and mythological sites, specialist subsistence/resource sites, rock art sites, occupation sites including caves and shelters, fortified walled village sites, surface scatters of stone artefacts and/ or pottery and shell; burial sites and water sites. Almost none of these site types are mutually exclusive and in almost all cases they are integral to contemporary cultural practice. For example so called 'ritual sites' include settlement and forest-based locality guardian/sacrificial sites, ancestral landing or first footprint sites (*ia mari tulia*), the 'stone boats' of immigrant *ratu* (*loiasu mataru*), ancestral graves (*calu lutur* or *narunu*) located within and outside of former settlements, 'increase' sites associated with warfare and past headhunting ventures, constructed signs or markers, which are said to be empowered, and which warn strangers that particular forest trees or other natural products are 'owned', and ancestral mythological sites associated with the first autochthonous people to emerge with the land and their sacred animals. In addition, there are a number of individually named stones or 'natural' landscape features, which feature in local *ratu* origin myths, and which serve to mark land boundaries or commemorate significant past events (O'Connor and Pannell 2006:35). Ritual sites are found throughout the Park – in old villages, caves, near water sources, and throughout the forest. These sites are often physically marked and are identified by Fataluku people as sacred or *téi*, being places where non-human beings are said to live and influence people's action in the immediate locale.

Sites are not regarded as 'cultural archives'; rather many are seen as 'alive' or as animated beings in the landscape and in peoples' lives, past and present. In this regard, they are said to have agency and efficacy. They connect the present to the past, the creation of the land and the ancestors to today, and they may assist or act against humans. According to local people, these kinds of sites do not need heritage protection. Indeed, local beliefs posit an inverse relationship to that proffered by environmental conservation measures, whereby these *téi* ('hot' or 'sacred') sites are acknowledged as empowered to provide protection for humans and the environment in general. While, the MAF may not regard these sites as having 'national' significance, being the cultural heritage of just one of the many ethnolinguistic groups in Timor-Leste, from the point of view of Fataluku people, some of these sites, such as *Ili Kérékéré*, a large rock overhang containing painted art and ritual sites, are considered to be the source of all the peoples and lands comprising the island of Timor and they are thus regarded by Fataluku people as of manifest national significance (see Pannell and O'Connor 2013).

As well as showing that there are large numbers of socially and culturally important places within the Park, the research summarised in the report to DEC NSW demonstrated that the division between nature and culture, which so often underpins the creation and management of protected areas, is not meaningful in terms of Fataluku people's beliefs and their understanding of the local environment. As previously indicated, the environment holds multiple levels of meaning for Fataluku people. Many of the so-called 'natural' features found within the Park are ascribed culturally specific names. Names for an identified area or geographic feature often relate to the names of ancestral settlements situated within the area. Conversely, cultural sites or former settlements may take their names from topographic features in the landscape, which in turn have mythological significance. For example, the former settlement of *Mua Mimiraka*, which also gives its name to the surrounding area, takes its name from the 'red' (*mimiraka*) 'earth' (*mua*) found in this locality. Along the northeast coast, from Tutuala to Tanjung *Tëi* ('Cape Hero'), locales are also known by the names ascribed to the uplifted limestone terraces, identified as seven distinct levels or *téni* (O'Connor and Pannell 2006:8). In fact, the whole island of Timor is anthropomorphised. Jaco Island (*Totina*) is identified by Fataluku people in Tutuala as the "head of the land", while the rest of the island constitutes the "body" (O'Connor and Pannell 2006:16). In the following section, we provide some further examples of significant Fataluku sites and landscapes within the Park and discuss the ways in which nature and culture exchange and interact in the agency of these places.

Figure 5. *Ili Kérékéré* shelter overlooks the sea. One of the new *téi* posts can be seen in the centre of the cave opening.

Source: Sue O'Connor.

In the beginning: The creation of Fataluku landscapes

Fataluku mythology holds that after the initial creation of the earth, and the coming into being of the first man and women, who are said to have given rise to the original Tutuala-based *ratu*, other settlers came from across the water by boat. They were attracted by large candles made out of bee's wax gathered from the honeycombs hanging from the entrance of the rockshelter *Ili Kérékéré* (Figure 5). Out of this wax the ritual and ceremonial head of the original clan group or the 'lord of the land' (*mua ocawa*) fashioned candles, which he placed in the entrance of the shelter facing out to sea. The candles were lit as a sign that this far-eastern part of the island was now occupied. Following the lighting of the candles, and their sighting by people situated further east, new settlers came from across the sea and, according to local traditions, they were given the right to farm certain tracts of land and form new clan or *ratu* groups (Pannell 2006b). In the village of Tutuala, more than 24 *ratu* groups have been recorded. All but one *ratu* is acknowledged as originating from elsewhere, notably from one of the many islands situated to the east and northeast of the island of Timor, which now form the Province of Maluku, Indonesia. *Tutuala Ratu* is identified as the autochthonous *ratu* in the village, and senior people in the *ratu* are acknowledged as the 'lord of the land' (*mua ocawa*) of the Tutuala region. This position carries with it many ceremonial duties and obligations (O'Connor and Pannell 2006:12), including the maintenance of ritual sites in the rockshelter known as *Ili Kérékéré*.

Ancestral landing sites

The ancestral, 'first landing sites' of the immigrant groups are often marked by 'stone boats'. These limestone outcrops, located along the coast, are identified by Fataluku people as actual boats that turned to stone after the arrival of their ancestors. With one exception, all of the stone boats we recorded within the Park are naturally ocurring rocky outcrops, albeit sometimes augmented with ritual platforms and placed objects or offerings. The one exception is the stone boat belonging to members of *Ratu Koawatca*. Reminiscent of the stone boats found in the Tanimbar Islands in Maluku (McKinnon 1991) this is a built structure made from massive blocks of coraline limestone, located at *Oirata Latamoko*. It is identified as a representation of the ancestral landing site and original boat of *Koawatca* ancestors at *Manuméri Hoiku*. The boat-shaped stone arrangement is also identified as the 'grave' of the first ancestors of this *ratu*.

Téi

Throughout the forests and former settlements of the Tutuala region, locales may also be known by the name of an individual *téi* site. *Téi* sites, which mark the presence of non-human entities, are considered to be inherently dangerous, and they are associated with a range of protocols and prohibitions (see Pannell and O'Connor 2013). *Téi* beings are acknowledged by Fataluku people as being able to transmogrify into a number of zoomorphic forms, or even human forms. In the district of Lautem, they are marked by a specific configuration of physical structures, most commonly by platforms built of layered stones or coral blocks, and a central carved wooden post, a plain, upturned tree root, or even a stone plinth (Figure 6). Of all of the Fataluku sites we recorded, *téi* occupied sites best embody the conceptual ambiguity or porosity of 'nature' and 'culture', 'wild' and 'domestic' in Fataluku beliefs. These unpredictable, shape-shifting entities are said to "guard" areas and their associated *ratu* members from strangers and sickness, however, if transgressed or neglected they may cause illness, madness or death. According to Fataluku beliefs, they must be regularly visited by *ratu* members, who are required to 'feed' them with rice, eggs, palm spirit and the meat of animals, such as pigs sacrificed for this purpose. These ceremonies are credited with promoting fertility and well-being, not just for humans, but for the environment as a whole. Entering settlement sites and all other places identified as, and marked by *téi*, without the custodial *ratu* member(s), who can speak to the *téi* being, is regarded by local people as dangerous, and invites all manner of catastrophes (Pannell 2006b; McWilliam 2011; Pannell and O'Connor 2013). Areas of the landscape without people, and where such rituals have been neglected, are similarly dangerous. As such, the cultural beliefs and social practices associated with *téi* have wider ecological effects. As this suggests, for Fataluku people, nature and culture do not exist as separate realms of meaning or practice. For them, the landscape of the Tutuala region bears testimony to the indivisible relationship that exists between the physical environment and local traditions, between nature and culture (O'Connor and Pannell 2006:10,12).

Figure 6. *Téi* at Moa Mimi raka.

Source: Sue O'Connor.

Figure 7. The president *téi Titiru*.

Source: Sue O'Connor.

This is well illustrated by our informants' description of the efficacy of the 'president' *téi* at *Titiru*, situated near *Ili Kérékéré* rockshelter (Figure 7). The *téi* at *Titiru* is said by members of Tutuala *ratu* to be the 'biggest head and bones' (*téi ilafai chau hafa*) of all 'the *téi* in Timor'. At the centre of the *Titiru* stone platform stand two stones, the larger of the stones is said to be the husband of the smaller one. While identified as 'human', this *téi* is said to be able to shape-shift and it may appear in many forms. The language of statehood is further extended to a number of other nearby *téi*, which are said to act as the 'deputy' and the 'commander' of *Titiru*. While the identification of this *téi* as the 'president' captures its paramount rank in the local order of *téi*, it also links it to the political struggle by the people of Timor-Leste for independence.

For Fataluku people in Tutuala, the *téi* at *Titiru* is ultimately responsible for the defeat of the Indonesian military and various militia groups in the later part of 1999. A desperate Xanana Gusmão is said to have personally requested assistance from the present 'lord of the land' to dispel the foreigners. Upon receiving this request, and with a photograph of the Falantil leader in hand, the 'lord of the land' sacrificed a pig at *Titiru* in September 1999 and requested the *téi* to emerge from the earth and take action against the Indonesians and others who were supporting them. Out of its hole, the *téi* started to 'eat' the enemy, while Xanana is said to have been imbued with the thoughts and power of the *téi*, the two creating an unbeatable front to the Indonesian forces. When the Indonesians left, the 'lord of the land' returned to *Titiru* and sacrificed another pig to 'calm' the *téi* down and entice it to enter its hole once again, satiated with food and drink, and the 'blood and flesh' of the enemy. As a postscript to this event, Xanana Gusmão visited Tutuala in 2005. The then 'lord of the land' had recently replaced the carved *téi* posts, which had been stolen from *Ili Kérékéré* during the Indonesian occupation. A ceremony was held at *Ili Kérékéré* to mark the occasion and the role of Fataluku people and the empowerment of their traditions in the national victory (Figure 8).

Figure 8. Ceremony following the replacement of a new *téi* post at *Ili Kérékéré*.

Source: Sue O'Connor.

Lupurasa

Throughout the Fataluku area, important subsistence resources such as lontar palms (*kakala oco*), bamboo stands, coconut palms, former garden sites (*pala*) and even forest trees with large honey combs or those designated as future building material, may be marked by ritual signs called *lupurasa*. According to Fataluku informants, *ratu* groups possess their own distinctive *lupurasa* signs. These may include wooden structures used to hang young coconuts, monkey skulls or fragments of woven cloth known to be associated with a particular *ratu*. In their simplest form they may be merely tied to the trunk of the tree. *Lupurasa* warn others against using or taking the resources they mark. *Lupurasa* are ritually empowered by senior men of the *ratu* group. Ignoring their warning is said to result in injury, sickness, and sometimes death (O'Connor and Pannell 2006:36). *Lupurasa* provide a clear case where the distinction between nature and culture blurs. Trees so marked may be naturally occurring forest species, for example, a tree designated for future use as a canoe. However, as it will take many generations to mature, the tree may be passed down as inherited property. This provides a cultural dimension to the concept of a 'biodiversity bank', as resources thus identified are regarded as cultural repositories for future social uses.

Rock art

More than 25 painted rock art sites and one ancient engraving site have now been recorded within the Park boundaries (O'Connor 2003; Lape *et al.* 2007; O'Connor *et al.* 2010b). No doubt many more await discovery. The rock art is predominantly found on the walls of the uplifted limestone terraces which overlook the sea, but some is also located in the deeper darker areas of caves. Although they can identify and they have names for many of the painted motifs, our Fataluku informants do not regard the rock art as resulting from human agency. They believe that it pre-existed the advent of humans, spontaneously appearing 'in the beginning,' at the same time as the first land emerged from the sea.

The exception to this are the images of boats which are said to depict the craft of arriving foreigners who were welcomed by the first 'lord of the land'. These were reproduced on the walls of the cave to show that people had arrived and they had been allowed to settle (Lape *et al.* 2007:4). Exactly who reproduced the boats on the walls is somewhat vague and usually meets with the response, 'the ancestors'. The status of the dominant motifs, small human anthropomorphs shown in a variety of active poses often holding weapons, is also somewhat vague (Figure 9).

The rock art motifs readily identified by Fataluku people include boats, horses, cockerels and recurrent geometric motifs referred to as 'eagles wings', '*poko*' and '*faria*'. *Poko* are one of the most commonly occurring geometric motifs. They are made up of a symmetrical arrangement of joined sets of concentric semi-circles or chambers with a feather or rayed line emanating from the upper part of the central form. They are sometimes elaborate and occur as alternating bands of black, red and yellow and sometimes simply in red pigment. *Poko* are rather obscure. They are said to represent a type of sacred receptacle which is no longer used. Rayed circles are another common motif and are often identified as *faria*, the Fataluku word for eye blinkers on horse bridles. Elaborate forms again use alternating bands of colour (O'Connor 2003). The reproduction of some of these motifs by the original autochthonous *ratu* is permitted 'in accordance with the ancestors' and they are woven into sacred clothes to be exchanged at marriage ceremonies and worn on ceremonial occasions. Some motifs are also carved into the wooden facing boards of the *ratu* houses and reproduced on modern concrete graves (Figure 10) (O'Connor and Pannell 2006; O'Connor and Pannell 2013).

Figure 9. Rock art at *Suntaleo*. These small human figures are shown holding a variety of weapons including spears and a bow and arrow. They are painted in red pigment.

Source: Sue O'Connor.

Figure 10. Detail of carved geometric designs on a derelict ratu house in the Nino Konis Santana National Park. The maintenance and use of these sacred houses was actively discouraged during the Indonesian occupation of Timor-Leste.

Source: Sue O'Connor.

In 2009, a number of extremely weathered petroglyphs, in the form of human faces, were discovered in *Lené Hara* cave within the Park. Although archaeological excavations and rock art recording at this cave has a long history, beginning with Portuguese exploration (Almeida and Zbyszewski

1967), the petroglyphs had escaped notice by researchers and they were also apparently unknown to the traditional owners. Their discovery was interpreted by Fataluku people as the outcome of a 'supernatural' event in which the ancestors revealed themselves (O'Connor *et al.* 2010b). As they were carved into speleothems, the engravings were able to be dated by the Uranium Thorium dating method and proved to be over 10,000 years old (O'Connor *et al.* 2010b).

Occupation sites

This category includes rockshelters and caves (*veraka*) used in prehistory and during the Japanese and Indonesian occupation of the island, as well as caves occupied by ancestral figures, *téi* , and other spirit figures, former walled settlements (*lata irinu*) and other old villages (*lata*), and surface concentrations of shell, stone and pottery.

Caves and rockshelters

The Park includes extensive areas of uplifted limestone and caves and rock shelters occur throughout it. These include some of the most scientifically significant sites in Island Southeast Asia. The cave known to local people as *Lené Hara* (also *Lené Ara*) is an internationally significant site, based on its scientific values. It is not only one of the oldest occupation sites in Island Southeast Asia, but it also houses a considerable display of rock art, including painted art and the recently discovered petroglyphs described above (O'Connor *et al.* 2010b). *Lené Hara* is highly significant to Fataluku people, being identified as the next 'settlement' occupied after the ancestors of *Tutuala Ratu* left the site known as *Ili Kérékéré*. It is also a resource zone where young men come to collect bird's nest which they sell to Chinese traders in the nearby town of Los Palos.

Lené Hara cave also has historical significance as a narrative space on changing colonial and post-colonial research perspectives. In the early 1960s, the Portuguese archaeologist, Almeida, excavated two trenches and described some of the art. The resulting brief publication on the excavation focuses on the stone artefacts and draws comparisons with European artefact 'types' (Almeida and Zbyszewski 1967). In the late 1960s, *Lené Hara* was visited by Ian Glover and John Mulvaney from The Australian National University. Glover's assessment of the pottery and stone artefacts drew comparisons with the archaeology of other sites in Timor-Leste and the broader Southeast Asian region, rather than with Europe (Glover 1972 (1):40-1, (2): plate 3.2; Glover 1986:7). Glover also described and photographed 'an interesting structure, a forked wooden pole set in a low semi-circle of stones, reminiscent of the spirit shrines of central Timor' (Glover 1986:17). The structure he described is the *Lené Hara téi* (Figure 11). This large stone platform is built against a massive speleothem column. It has a prominent upright stone marking the *téi* centre. As well as the *téi*, *Lené Hara* is divided by an extensive length of stone walling, which includes a gateway with standing stones. None of these features were noted by Almeida and yet they must have been just as prominent at the time of his excavation in the early 1960s as they were when Glover visited a few years later, and as they are today. Almeida's excavation was carried out as part of an official Portuguese government 'mission'. It would appear that, as a result, no permissions were sought from the landowning *ratu*, no ceremonies were undertaken by the 'lord of the land' prior to or following the excavations, and all vestiges of Fataluku customary ownership were regarded as unworthy of documentation.

O'Connor, Spriggs and Veth undertook excavations at *Lené Hara* in 2000 and, in association with Pannell, again in 2002 (O'Connor *et al.* 2002; 2010a). In 2000, the excavations were undertaken under the direction of the state appointed head of Tutuala, the *kepala desa* (Indonesian, *chefe de suco*, Tetum).[3] At the time, the 'lord of the land' for Tutuala *ratu* was consulted about the

---

3   The excavation took place prior to the establishment of Portuguese and Tetum as the official languages of Timor-Leste.

excavation and he gave his permission and a local field team was subsequently employed. No ceremony was performed at the *Lené Hara téi* as we were unaware that any was required. Towards the end of the excavation we were approached by a group of angry young men. We asked our Fataluku-speaking translator what the problem was and she informed us that this group of men were in fact the close relatives of the 'lord of the land', who wanted to know why they had not been given work with us. Unbeknown to us, the field crew we had employed was entirely selected from the family and associates of the *kepala desa*. Thus, our team had represented the state, as much as the *kepala desa* and his associates represented this entity, but it did not necessarily represent the community of people with customary rights to the site.

Figure 11. *Lené Hara* cave showing the speleothem with the *téi* platform during the excavations in 2002.

Source: Sue O'Connor.

The 2002 field season involved the excavation of three test pits in different parts of the *Lené Hara* cave. In this year, the team, consisting of two archaeologists and an anthropologist (respectively, O'Connor, Veth and Pannell), was careful to ensure that our community-based consultations were more inclusive. The local field crew was chosen by the 'lord or the land' and by senior members of Tutuala *ratu*, in consultation with the *chefe de suco*. The appropriate rituals were carried out at the *téi* prior to and following the excavations. The rituals were performed to accompany the 'opening' and the 'closing' of the earth and they involved the sacrifice of a pig, and offerings of lontar spirit and rice to the *Lené Hara téi*. This was considered necessary to ensure the safety of all those involved in the excavation, as well as the safety of their respective families. In one of the excavation pits, Pit D, a human cranium was uncovered (O'Connor *et al.* 2010b). The cranium was not exhumed. After consultation with the 'lord of the land' and other landowners, the pit was backfilled with the cranium *in situ* (O'Connor and Pannell 2006:40–41; O'Connor *et al.* 2010a).

Despite our efforts, when we returned in 2003 we were told that some Fataluku people blamed the excavations at *Lené Hara*, specifically the exposure of human skeletal remains, for the death of the 'lord of the land' earlier in the year.

What transpired at *Lené Hara* over the course of two field seasons aptly demonstrates the complexity of carrying out research or heritage assessment, and the porosity of the concepts of nature and culture, in this part of Timor-Leste. It is clearly a cave of natural origin, yet it is conceptualised as a settlement or village by Fataluku people (and by archaeologists as well, who regard it as a site of previous human occupation), it is also seen as part of a wider resource zone, along with other nearby caves, where young men come to collect birds' nests, and it is a *téi* place, and thus acknowledged as a powerful and potentially dangerous locale.

According to our research findings, another rockshelter in the area known as *Jerimalai* was first occupied more than 42,000 years ago (O'Connor 2007). It is the oldest modern human settlement site in Island Southeast Asia, east of the Sunda Shelf, and it is unique on a world scale due to its abundant evidence of maritime resource exploitation. *Jerimalai* contains the world's earliest evidence for systematic pelagic fishing and the manufacture of fish hooks (O'Connor *et al.* 2011b). The assemblage was also very rich in shell decorative artefacts, including shell beads and pendants made of several different species. This shelter does not have the same level of significance to Fataluku people as *Lené Hara*, as it is not central to local origin myths and nor is it empowered by a non-human presence. But like the many other shelters in the area, it is used by men while on hunting expeditions in the secondary forest surrounding it. *Jerimalai* contains abundant evidence of contemporary use, such as fragments of metal plates, coconut bowls, recent hearths, and wooden racks for cooking game (O'Connor and Pannell 2006:21; O'Connor 2007).

The caves identified as *Matja Kuru* 1 and 2 similarly have abundant evidence of human use from ancient times through to today. The finds in the excavated assemblages indicate that periodically the caves served as base-camps for people exploiting a wide variety of resources from the surrounding area. These include long neck turtle from the freshwater lake to the south, monkeys and cuscus from the forests between the lake and the coast, as well as fish and shellfish from the coast over eight km to the north (Spriggs *et al.* 2003; Veth *et al.* 2005; O'Connor 2006, 2010). Like *Jerimalai*, the *Matja Kuru* caves also contained abundant evidence of contemporary Fataluku use, including a pottery still for making palm spirit, and the remains of hearths and racks, used by hunters pursuing game in the forest (Pannell and O'Connor 2005).

Walled settlements

A number of old stone fortified settlements have been located within the Park. Fataluku oral traditions, as well as the physical features of the settlements, indicate that they were primarily built for defence purposes. These features include the massive nature of the walls themselves, the fact that the walls encircle and enclose the settlements and ritual areas inside, the very narrow entrances, and their location on bluffs and/or cliff top edges with clear views to the coast and often sheer precipitous drops off one or more sides (Figure 12). Inside the fortified walls are raised stone platforms, which are often over a metre high and rectangular in form. They are said to be the graves of ancestors (*calu lutur* or *narunu*) (Figure 13). Other areas inside the fortified sites are identified as circumscribed spaces called *sépu*, where ceremonies and dances were performed (Lape 2006; O'Connor *et al.* 2012). Site maps produced for two of these settlements, *Locami* (*Lo Chami*) and *Lori Lata*, show some of these features (Lape 2006).

Figure 12. Massive walls surrounding a fortified settlement in within the forested region of Nino Konis National Park.

Source: Sue O'Connor.

Several such sites in the Tutuala and Com areas have been test-pitted and their occupation dated. These include *Tutuala, Lori Lata, Lopomalai, Ili Mimiraka, Mua Mimiraka*, and *Tutun* (*Tutunca'u*) near Tutuala and *Macapainara* near Com. Results of radiocarbon dating on marine shell and charcoal and OSL dating of pottery from the sites suggest that the fortified structures began to be constructed after 1300 AD (Lape 2006; O'Connor *et al.* 2012). They appear to have continued to be made and used into the historic period, and some were still in use up until the middle of the 20th century (O'Connor *et al.* 2012; O'Connor and Pannell 2006:25; Pannell 2006b). As previously indicated, while these ancestral settlements are no longer permanently occupied, *ratu* members regularly return to them to perform rituals and ceremonies at the *téi* and grave sites located within them (O'Connor and Pannell 2006:37). Even Jaco Island (*Totina*), which is said to have no permanent sources of freshwater, has at least two large fortified village sites.

Figure 13. Monumental ancestral grave at the fortified settlement of *Macapainara* near Com.

Source: Sue O'Connor.

The forest survey undertaken by McWilliam and O'Connor in 2009 located many more stone walled fortified settlements in the remote, densely forested areas of the Park, including *Mapulo*, the ancestral village of Nino Konis Santana himself (Figure 14). The forest floors throughout this part of the Park are criss-crossed with stone walling bearing witness to the density of past human occupation of this region.

Scatters of stone artefacts, pottery sherds and shell

Scatters of stone artefacts (*nelukala*), pottery sherds and shell are commonly found surrounding resource areas, such as those adjacent to freshwater springs located in the limestone terraces and at the back of coastal embayments and beaches. It is difficult to ascertain whether these scatters represent the remains of community activities at a particular point in time or whether they are palimpsests representing thousands of years of episodic human visitation. While these sites are recognised by local people as resulting from human activities, they are not of themselves considered to hold 'cultural' significance. However, they often co-occur with important cultural sites and ceremonial aggregation areas, such as at water sources, ancestral settlement sites, and at *téi* sites.

Figure 14. Ancesestral graves and internal walling at Mapulo the ancestral village of the Falantil commander Nino Konis Santana. Vardemar Cabral, a relative of Nino Konis Santana, is clearing weeds from around the graves.

Source: Sue O'Connor.

The open area at the back of Valu Beach provides a good example of this kind of tangible and intangible symbolic nexus. It contains a sparse scatter of pottery, stone artefacts and shell, as well as broken and abandoned contemporary artefacts, such as cigarette lighters, thongs and broken floats. Today this back beach zone is used by the fisherman servicing the tourists who come to camp, swim and snorkel on the weekends. We were told that during the historic period families would bring their goods to trade here during specific times of the year when people from the island of Leti and other islands to the east came to exchange pottery, goats and other goods, such as corn. The shallow overhangs and back beach area on Jaco Island, which face Valu Beach, are said to have been delegated by the 'lord of the land' to the incoming traders from Leti, and thus this area is known as *Leti Mo*. The Valu Beach site was excavated in 2004 with the assistance of Abilio da Silva (from the Ministry of Education) (Figure 15). Consistent with Fataluku oral traditions, it has pottery sherds dated between 1800 BP and the present (O'Connor and Pannell 2006:22–23).

Burial sites

Prior to the resettlement of the local population at roadside locations, the deceased were buried within the walled compound of former settlements. In these settlements, graves (*narunu* or *calu lutur*) are marked by a rectangular platform of layered stones (O'Connor *et al.* 2012). Former settlements contain the graves of unknown distant ancestors, as well as more recently deceased, known individuals (O'Connor and Pannell 2006:40). At the ancestral settlement complex of *Nari*, located between Com and Los Palos, current senior *ratu* members have recorded a long genealogy of named individuals for the ancestral graves However, the more distant of these are considered *lulik*[4] or 'taboo' and their names are thus not to be spoken.

---

4   *Lulik* (Tetum but also used as a generic term in other languages in Timor-Leste. *Lulik* has usually been translated as 'sacred' or 'taboo' (Hull 2002:227, Hicks 2004:25). Bovensiepen (2011:47–8) states that amongst Idaté speakers, *lulik* is used as an adjective, noun and

Figure 15. Senior *ratu* members conduct closing ceremony following the excavation at Valu Beach site in 2004.

Source: Sue O'Connor.

Most burial sites consist of large raised rectangular stone structures, sometimes marked with status goods, such as buffalo horns or ceramic bowls or plates of Chinese or European origin (Figure 13), but some burial caves are also known. A burial cave recorded in an area of forest

---

verb and 'can designate an avoidance relationship, but also refers to a potent source of power, prosperity and danger, usually associated with the land and the ancestors'.

within the Park, south of the settlement of Méhara, contained remains of secondary burials and an eggshell blue celadon ceramic bowl. In more recent (but no longer permanently occupied) open settlements, graves are marked by concrete structures, which are sometime decorated with traditional motifs (known as *lakulili*), and they are generally located on the edge of the residential area. Here again we see the porosity of nature and culture. As noted above, one of the most commonly occurring motifs in the rock art of the region is known as *poko*. These motifs are not regarded as a product of human agency but rather they are said to have been come into being following the creation of the land itself and thus have strong ritual connections for *ratu* members and they are sometimes reproduced on the graves of *ratu* members.

In Méhara, the graves of a number of people killed during the Indonesian occupation are also located within the household compound. In the Tutuala region, the graves of a number of prominent, named Fataluku antecedents and ancestral figures featured in local origin narratives are located outside of settlement complexes. On Jaco Island, we also recorded a massacre site dating from 1984. According to Fataluku informants, Indonesian forces killed six local people, and threw their bodies into a sinkhole in the karst limestone. While some of the skeletal remains of these murdered individuals were still present in October 2002, the remains of four of the murdered individuals had been removed by family members just prior to this date.

Water sources

In the Tutuala region, we recorded current and previously used, but now dry, water (*ira*) sources known to informants. In this area, water sources include; permanent springs (*ira ina*), emerging at around an altitude of 300 m in the uplifted limestone terraces; a number of spring-fed sources along the coast only accessible at low tide; seasonal rockholes (*piar ira*) (often protected with a rock lid); ephemeral creeks and rivers (*wéru*); temporary rain-filled depressions (*luri*), and drip seepage (*cupucupu*) and drip-fed pools in caves (O'Connor and Pannell 2006:38).

In Com and in areas surrounding Tutuala drinking water is obtained from several springs located on the raised limestone terraces behind the village. Using an elaborate system of channels, water from these springs is used to irrigate nearby crops and supply water to households now settled along the coast road leading to the harbour at Com. In Méhara, drinking water is obtained from a number of localised springs and, when lake levels permit, from nearby Lake Iralaloro. The hamlet of Malahara obtains its drinking water from the lake outlet before it goes underground and resurfaces in the sea (*tahi calu*) near Loré on the south coast. In the Com, Méhara, and Tutuala regions, permanent water sources have mythological and ritual significance. For example, the discovery of permanent spring-fed water is often attributed to a dog (*iparu*) in local origin narratives. The water sources associated with these accounts are often marked with *téi* structures (O'Connor and Pannell 2006:39). Reflecting their *téi* status, as places associated with a range of ritual and ceremonies restrictions and responsibilities, permanent springs are often fenced or walled off, not only protecting the *téi* related values of the sites, but also the water itself from contamination by animals. At the ancestral settlement complex of *Nari* we recorded an elaborate walled well. This water feature is marked by a *téi* and it is part of a stone arrangement which is said to be a symbolic representation of the ancestral *ratu* structure for this area (Figure 16). In a landscape and a country in which water availability is naturally restricted and many people regularly experience water shortages in the dry season months, one could argue that the identification of water sources as *téi* places, associated with mythological events, affords a form of protection which is far more intelligible to local people than an IUCN protected area category or national legislation. In a similar manner, as McWilliam has documented, the local designation of forested groves as '*lulik*' or 'sacred' in other parts of the island has certainly contributed to their preservation over the years (McWilliam 2011).

Figure 16. *Lulira* at the settlement of *Nari* is both a water site and an ancestral ritual site. It includes a deep stone walled well and elaborate stone arrangements which include a *téi, téi Mauresi* and a stone arrangement that symbolises the structure of the ancestral ratu. Kati ratu is said to be shown at the centre of the arrangement.

Source: Sue O'Connor.

Discussion

From a conservation perspective, the Nino Konis Santana National Park should be regarded as an anthropogenic landscape. As we have discussed at length, much, if not all, of the forested areas of the Park have been altered and shaped by thousands of years of human occupation and use. With very few exceptions, this is the situation found throughout Timor-Leste (see for example, Metzner 1977 for the Baucau and Viqueque areas). The identification of the Park as a 'Category V' 'Protected Landscape/Seascape' theoretically makes it possible to recognise this human-environment interaction and its long history. Under UNESCO provisions, Category V protected 'landscapes/seascapes' provide for the preservation of lifestyles and support for customary economic activities. That said, many Fataluku traditional subsistence activities have been banned or curtailed by the regulations governing Park use. The current restrictions on hunting will certainly adversely affect the kind of subsistence flexibility that has enabled local people to deal with chronic environmental uncertainty for thousands of years and more recent acute political instability (Pannell 2011). The limits on hunting will also reduce the regularity with which Fataluku people are able to visit the more remote areas of the Park in order to maintain their ancestral connections and ritual responsibilities.

Since Independence, Timor-Leste has witnessed a resurgence in customary law, beliefs and ritual. For example, despite the impoverished circumstances that most communities currently live in, scarce resources are being directed towards the construction of new 'ancestral' houses, which were destroyed during the period of Indonesian governance, and 'pagan' rituals that were once banned, are now being revitalised. In many respects, this resurgence is at odds with the attempts made by the Timor-Leste national governments to impose 'cultural uniformity' on the social fabric of life and their efforts to impress public good models on the land and resources of the Park. In order for the Park to be a 'success' it will need to be viewed by Fataluku people as their Park. It is hoped that local *ratu* involvement will become a central part of future negotiations, zoning decisions and management models.

In this regard, there is some evidence that the national government is serious about acknowledging local people and local beliefs in the management of natural resources and environment. For example, as part of the 'Regional Fisheries Livelihood Program for South and Southeast Asia' (RFLP), sponsored by the Spanish Government and the United Nations Food and Agriculture Organization, fishing communities in Timor-Leste are being encouraged to document 'traditional laws', known as '*tara bandu*'. According to a recent RFLP newsletter, the laws and customs identified as '*tara bandu*' function to 'restrict access to and use of natural resources and spaces and as such constitute a traditional resource protection and management mechanism' (RFLP 2012:1). Communities are being encouraged to submit written versions of their 'traditional laws' to the national authorities, which are apparently 'seeking to recognise the usage and enforcement of traditional laws' in the management of coastal resources (RFLP 2012:1). Local people's familiarity with their own laws and customs is identified by RFLP officers as one of the ways to overcome the problems associated with local compliance with national fisheries legislation (RFLP 2012). While this may be the case, government acknowledgement of local communities as the traditional owners of natural resources and environments, is also one way to counter the un-regulated use of resources and areas. Ironically, protected area zoning which 'locks up' 'wilderness' and 'locks out' the traditional custodians, opens the floodgates to illegal timber cutters and other resource raiders. In the case of the Nino Konis Santana National Park, the restriction of Fataluku people's traditional occupation and use of the Park, coupled with the near absence of enforcement agents, in the form of Park rangers, has all the potential to create yet another 'tragedy of the commons' (Hardin 1968).

While the RFLP initiative can be regarded as a significant step towards community-based resource management, the written codification of local belief systems and related social behavior, and reliance upon national or local government systems to acknowledge and foster these traditional practices, raises a number of issues in its own right. One is reminded here of the discussion which took place in the 1990s regarding the concept of '*sasi*', identified in the literature as a Moluccan-based 'marine resource management institution' (see Bailey and Zerner 1992). In this regard, Pannell's comments regarding '*sasi*' equally apply to the RFLP initiative involving '*tara bandu*':

> Our understanding of *sasi* is further obstructed if we impute to it a set of values, a range of functions or an efficacy which are neither apparent or intended … There are inherent problems of translation when we attempt to read Indigenous practices and knowledges in terms of the discourses of non-Indigenous environmental management approaches. (Pannell 1997:292)

As Pannell points out, and following Aboriginal examples from northeast Arnhem Land and Cape York Peninsula, perhaps what is needed is a reformation of the fundamental tenets and orientation of current environmental and resource management strategies, what Donald Brenneis calls a 'new conception of rights' (2003:219). We suggest here that a different approach may have as its focus the recognition of cultural values and social practices as the primary means of ensuring some forms of biological diversity and ecotopic integrity. In this approach, Indigenous expressions of what their culture is and how it should be maintained, their rights relating to the ownership and use of lands and seas, natural resources, cultural property and local knowledge, and their interests in economic development and political recognition would constitute the basis for the criteria and intentions of management (Pannell 1997:302–303). In the case of the Nino Konis Santana National Park, with careful planning, the cultural resources of the Park hold enormous potential to provide sustainable livelihoods for the Fataluku communities who live within its boundaries. In the meantime, however, in the far eastern reaches of Timor-Leste, government inaction has provided local people with some respite from the discourse and dictates of nature-based protected area management.

Postscript

As of March 2012 there has still been no formal zoning for the Park and no initiatives have been implemented for managing, protecting or interpreting the rock art and other cultural sites within the Park. Valu Beach continues to attract regular tourists from Dili, who stay at the eco resort. These tourists come predominantly to swim and snorkel on the coral fringed beaches of Valu and Jaco Island. Some also go to explore the caves and rock art sites of *Lené Hara* and *Ili Kérékéré*. Currently, there is no Park-based system to manage tourist visits to these cultural sites or to ensure that visitors are guided by the appropriate land owners. No information on the rock art or the cultural values of the Park is available as signage or literature or on the MAF's website. The few Park Guards employed by Protected Areas within MAF have received some training on the natural heritage values of the Park and the MAF regulations and measures designed to protect it, but none on the cultural heritage values of the Park. No wide-scale community consultation has been undertaken and no attempt has been made at mapping the contemporary cultural values and ancestral ties of the different *ratu* groups living within the Park.

Acknowledgements

The research for this paper was supported by an Australian Research Council Discovery Grant to O'Connor (DP0556210), to O'Connor and McWilliam (DP0878543) and a Linkage Grant to O'Connor and Byrne (LP0776789). The Australian National University and the Rainforest

Cooperative Research Centre, James Cook University are also thanked for research support to O'Connor, Brockwell and Pannell. Peter Lape is thanked for his involvement in the field during the rock art recording. In Timor-Leste we extend our thanks and appreciation to staff of the Ministry of Education and the Ministry of Agriculture and Fisheries, who assisted us during our surveys. Permissions for the archaeological excavations, rock art recording and surveys were obtained from the Ministry of Education and Culture. We would particularly like to thank Sr Virgílio Simith, Sra Cecília Assis and Sr Abilio da Silva for facilitating this process. We also wish to acknowledge the support of the Director of Protected Areas and National Parks, National Directorate of Forestry in the Ministry of Agriculture and Fisheries, Manuel Mendes, who gave permission for O'Connor, McWilliam and Aplin to work within the Nino Konis Santana National Park following its formal declaration in 2008. We would particularly like to acknowledge the support of the people of Tutuala, Com and Méhara, without whom this research would not have been possible.

References

Almeida, A. De and Zbyszewski, G. 1967. A contribution to the study of the prehistory of Portuguese Timor-lithic industries. In: Solheim II, W.G. (ed.), *Archaeology at the Eleventh Pacific Science Congress: Papers presented at the XI Pacific Science Congress, Tokyo, August–September 1966*, pp. 55–67. Asian and Pacific Archaeology Series, Honolulu.

Bailey, C. and Zerner, C. 1992. Community-based fisheries management institutions in Indonesia. *Maritime Anthropological Studies* 5(1):1–17.

Bovensiepen, J. 2011. Opening and closing the land: Land and power in the Idaté highlands. In: McWilliam, A. and Traube, E. (eds), *Land and life in Timor-Leste: Ethnographic essays*, pp. 47–60. ANU E-Press, Canberra.

Brenneis, D. 2003. Toward new conceptions of rights. In: Zerner, C. (ed.), Culture and the question of rights: Forests, coasts, and seas in Southeast Asia, pp. 212–235. Duke University Press, Durham.

Collins, S., Martins, X., Mitchell, A., Teshome, A. and Arnason, T. 2007. Fataluku medicinal ethnobotony and the East Timorese military resistance. *Journal of Ethnobotany and Ethnomedicine* 3(5) Published online Jan 22. 2007.

Edyvane, K., Cavalho, N., Penny, S., Fernandes, A., de Cunha, C.B., Amaral, A.L., Mendes, M and Pinto, P. 2009a. *Conservation values, issues and planning in the Nino Konis Santana Marine Park, Timor Leste – final report*. Ministry of Agriculture and Fisheries, Government of Timor Leste, Dili.

Edyvane, K., McWilliam, A., Quintas, J., Turner, A., Penny, S., Texeira, I., Pereira, C., Tibirica, Y. and Birtles, A. 2009b. *Coastal and marine ecotourism values, issues and opportunities on the north coast of Timor Leste – Final Report*. Ministry of Agriculture and Fisheries, National Directorate of Tourism, Government of Timor Leste, Dili.

Glover, I. 1972. *Excavations in Timor: A study of economic change and cultural continuity in Prehistory*. Unpublished PhD thesis. Australian National University.

Glover, I. 1986. *Archaeology in Eastern Timor, 1966–67*. Australian National University, Canberra.

Hardin, G. 1968. The tragedy of the commons. *Sciences* 162:1243–1250.

Hicks, D. 2004. Tetum ghosts and kin: Fertility and gender in East Timor (2nd edition), Waveland, Long Grove.

Hull, G. 2002. Standard Tetum-English dictionary, (3rd edition), Sebastiao Aparicio da Silva Project in Association with Insituto Nacional de Linguistica, Timor Leste. Winston Hills.

IUCN. 1994. *Guidelines for protected area management categories*. IUCN, Gland.

Katz, C. 1998. Whose nature, whose culture? Private productions of space and the 'preservation' of nature. In: Braun, B. and Castree, N. (eds), *Remaking reality: nature at the millenium*, pp. 46–64. Routledge, London.

Lape, P. 2006. Chronology of fortified settlements in East Timor. *Journal of Island and Coastal Archaeology* 1(2):285–97.

Lape, P.V., O'Connor, S. and Burningham, N. 2007. Rock art: A potential source of information about past maritime technology in the Southeast Asia-Pacific region. *The International Journal of Nautical Archaeology* 36 (2):238–53.

Luke, T.W. 1995. The nature conservancy or the nature cemetery: Buying and selling 'perpetual care' as environmental resistance. *Capitalism Nature Socialism* 6:1–20.

McKinnon, S. 1991. *From a shattered sun: Hierarchy, alliance and exchange in the Tanimbar Islands*. University of Wisconsin Press, Wisconsin.

McWilliam, A. 2006. Fataluku forest tenures in the Conis Santana National Park in East Timor. In Reuter, T. (ed.), *Sharing the earth; Dividing the land: Land and territory in the Austronesian World, land tenure in the Austronesian World*, pp. 253–76. ANU E-Press, Canberra.

McWilliam, A. 2007a. Customary claims and the public interest: Fataluku resource entitlements in Lautem. In: Kingsbury, D. and Leach, M. (eds), *East Timor: Beyond independence*, pp. 165–78. Monash Asia Institute Press, Melbourne.

McWilliam, A. 2007b. Austronesians in linguistic disguise: Fataluku cultural fusion in East Timor. *Journal ofSoutheast Asian Studies* 38:355–75.

McWilliam, A. 2011. Fataluku living landscapes. In: McWilliam, A. and Traube, E. (eds), *Land and life in Timor-Leste, ethnographic essays*, pp. 61–87. ANU E Press, Canberra.

Mau, R. 2010. *Ecosystem and community based model for zonation in Nino Konis Santana National Park, Timor-Leste*. Unpublished MS thesis. Bogor Agricultural University.

Metzner, J. 1977. *Man and environment in Eastern Timor: A geo-ecological analysis of the Baucau-Viqueque area as a possible basis for regional planning*. Monograph No. 8, Development Studies Centre, Australian National University, Canberra.

O'Connor, S. 2003. Nine new painted rock art sites from East Timor in the context of the Western Pacific region. *Asian Perspectives* 42:96–128.

O'Connor, S. 2006. Unpacking the island Southeast Asian Neolithic cultural package, and finding local complexity. In: Bacus, E.A., Glover, I.C. and Pigott, V.C. (eds), *Uncovering Southeast Asia's Past*, pp. 74–87. Singapore: NUS Press.

O'Connor, S. 2007. New evidence from East Timor contributes to our understanding of earliest modern human colonization east of the Sunda Shelf. *Antiquity* 81:523–35.

O'Connor, S. 2010. Continuity in shell artefact production in Holocene East Timor. In Bellina, B., Bacus, E.A., Pryce, T.O. and Wisseman Christie, J. (eds), *50 years of archaeology in Southeast Asia: Essays in honour of Ian Glover*, pp. 218–33. Rivers Books, Bangkok.

O'Connor, S. and Aplin, K.P. 2007. A Matter of balance: An overview of Pleistocene occupation history and the impact of the Last Glacial Phase in East Timor and the Aru Islands, eastern Indonesia. *Archaeology in Oceania* 42 (3):82–90.

O'Connor, S. and Oliveira, N.V. 2007. Inter and intra regional variation in the Austronesian painting tradition: A view from East Timor. *Asian Perspectives* 46 (2):389–403.

O'Connor, S. and Pannell, S. 2006. *Cultural heritage in the Nino Conis Santana National Park: A preliminary survey.* Unpublished report prepared for the Department of Environment and Conservation, New South Wales.

O'Connor, S., and Veth, P. 2005. Early Holocene *Trochus* shell fish-hooks from East Timor. *Antiquity* 79:1–8.

O'Connor, S., Spriggs, M. and Veth, P. 2002. Excavation at *Lené Hara* establishes occupation in East Timor at least 30,000–35,000 years on: Results of recent fieldwork. *Antiquity* 76:45–50.

O'Connor, S., Barham, A., Spriggs, M., Veth, P., Aplin, K. and St. Pierre, E. 2010a. Cave archaeology and sampling issues in the tropics: A case study from *Lené Hara* Cave, a 42,000 year old occupation site in East Timor, island Southeast Asia. *Australian Archaeology* 71:29–40.

O'Connor, S., Aplin, K., St. Pierre, E. and Feng, X. 2010b. Faces of the ancestors revealed: Discovery and dating of Pleistocene-aged petroglyphs in *Lené Hara* Cave, East Timor. *Antiquity* 84 (325):649–65.

O'Connor, S., Pannell, S. and Brockwell, S. 2011a. Whose culture and heritage for whom? Exploring the limitations of National Public Good Protected Area Models in East Timor. In: Miksic, J., Goh, G. and O'Connor, S. (eds.), *Rethinking cultural resource management in Southeast Asia: Preservation, development and neglect*, pp.39–66. Anthem Press, London.

O'Connor, S., Ono, R. and Clarkson, C. 2011b. Pelagic fishing at 42,000 BP and the maritime skills of modern humans. *Science* 334:1117–21.

O'Connor, S., McWilliam, A., Fenner, J.N. and Brockwell, S. 2012. Examining the origin of fortifications in East Timor: Social and environmental factors. *Journal of Island and Coastal Archaeology* 7:200-18.

Palmer, L. and do Amaral de Carvalho, D. 2008. Nation building and resource management: The politics of 'nature' in Timor Leste. *Geoforum* 39:1321–32.

Pannell, S. 1997. Managing the discourse of resource management. The case of *sasi* from 'southeast' Maluku, Indonesia. *Oceania* 67:289–307.

Pannell, S. 2006a. *Reconciling nature and culture in a global context: Lessons from the World Heritage List.* Rainforest CRC, Cairns.

Pannell, S. 2006b. Welcome to the Hotel Tutuala: Fataluku accounts of going places in an immobile world. Special Edition, *The Asia Pacific Journal of Anthropology* 7(3):203–21.

Pannell, S. 2011. Struggling geographies: Rethinking livelihood and locality in Timor-Leste. In: McWilliam, A. and Traube, E. (eds), *Land and life in Timor-Leste, ethnographic essays*, pp.217–39. ANU E Press, Canberra.

Pannell, S. and O'Connor, S. 2005. Toward a cultural topography of cave use in East Timor: A preliminary study. *Asian Perspectives* 44(1):193–206.

Pannell, S. and O'Connor, S. 2010. Strategy blurring: Flexible approaches to subsistence in East Timor. In: Hardy, K. (ed.), *Archaeological invisibility and forgotten knowledge*, pp. 115–30. British Archaeological Reports BAR International Series, Archaeopress, Oxford.

Pannell, S. and O'Connor, S. 2013. Where the wild things are: An exploration of sacrality, danger and violence in confined spaces. In: Moyes, H. (ed.), *Journeys into the dark zone: A cross cultural perspective on caves as sacred spaces,* pp. 317-30 University of Colorado Press, Boulder.

RFLP. 2012. Regional fisheries livelihood program in South and Southeast Asia E-newsletter, February 2012.

Schama, S. 1995. *Landscape and memory,* Knopf, New York.

Schapper, A. 2012. Finding Bunaq: The homeland and expansion of the Bunaq in central Timor. In: McWilliam, A. and Traube, E. (eds), *Land and life in Timor-Leste, ethnographic essays*, pp. 163–86. ANU E Press, Canberra.

Spriggs, M., O'Connor, S. and Veth, P. 2003. Vestiges of early pre-agricultural economy in the landscape of East Timor: Recent research. In: Kallen, A. and Karlström, A. (eds), *Fishbones and glittering emblems: Proceedings from the EurASEEA Sigtuna Conference*, pp. 49–58. Museum of Far Eastern Antiquities, Stockholm.

Trainor, C., Santana, F., Xavier, A., Xavier, F. and A. da Silva. 2003. Status of globally threatened birds and internationally significant sites in Timor-Leste (East Timor) Based on rapid participatory biodiversity assessments. Report to birdlife International – Asia Programme.

Tsing, L. 2003. Cultivating the wild: Honey-hunting and forest management in southeast Kalimantan. In Zerner, C. (ed.), *Culture and the question of rights: Forest, coasts, and seas in Southeast Asia*, pp. 24–56. Duke University Press, Durham.

UNESCO (United Nations Educational, Scientific and Cultural Organization). 2005. *Operational Guidelines for the Implementation of the World Heritage Convention*. UNESCO World Heritage Centre, Paris.

Veth, P., Spriggs, M. and O'Connor, S. 2005. The continuity of cave use in the tropics: Examples from East Timor and the Aru Islands, Maluku. *Asian Perspectives* 44 (1):180–92.

Whistler, A. 2001. Ecological survey and preliminary botanical inventory of the Tutuala Beach and Jaco Island protected natural areas, East Timor. Report for UNTAET.

Zerner, C. 1994. Transforming customary law and coastal management practices in the Maluku Islands, Indonesia, 1870–1992. In: Western, D. and Wright, R. (eds), *Natural connections: Perspectives in community-based conservation*, pp. 80–112. Island Press, Washington D.C.

Zerner, C. 2003. Moving translations: Poetics, performance, and property in Indonesia and Malaysia. In: Zerner, C.(ed.), *Culture and the question of rights: Forests, coasts, and seas in Southeast Asia*, pp.1–24. Duke University Press, Durham.

Zerner, C. (ed.) 2003. *Culture and the question of rights: Forests, coasts and seas in Southeast Asia*. Duke University Press, Durham.

Index

www.ingramcontent.com/pod-product-compliance
Lightning Source LLC
Chambersburg PA
CBHW051309270326
41929CB00029B/3460